T0341183

ABIOTIC STRESS TOLERANCE MECHANISMS IN PLANTS

Abiotic stresses in plants caused by salt, drought, temperature, active oxygen, and toxic compounds are the principal reasons behind reduced crop yield. Abiotic stresses lead to the generation of reactive oxygen species, which in turn damage the plant cells. The threat of global environment change and increasing population makes it necessary to generate crop plants that could withstand such harsh conditions. Much progress has been made in the identification and characterization of the mechanisms that allow plants to tolerate abiotic stresses. Better understanding of metabolic pathways and other factors involved in the production of compatible solutes and the identification of many transporters, collectively pave the ways towards genetic engineering in crop plants for improved stress tolerance. Abiotic Stress Tolerance Mechanisms in Plants is a new book focused on how plants adapt to abiotic stress and various biotechnological and molecular approaches that could improve the stress tolerance in crop plants for adequate food supply across the globe.

Gyanendra Kumar Rai obtained Ph.D. from University of Allahabad, Allahabad, UP, India. He is serving as Assistant Professor (Sr. Scale) Biochemistry at School of Biotechnology, Sher-e- Kashmir University of Agricultural Sciences and Technology of Jammu. His research areas include stress physiology, proteomics and molecular biology and nutritional physiology. He has published more than 40 research papers and has contributed in several books.

Ranjeet Ranjan Kumar obtained Ph.D. from Division of Biochemistry, Indian Agricultural Research Institute, New Delhi, India. He is currently Scientist (Sr. Scale) at Division of Biochemistry, Indian Agricultural Research Institute, New Delhi His research interests include abiotic stress tolerance mechanisms at molecular and biochemical level. He has published more than 25 research papers and has contributed in several books.

Sreshti Bagati obtained Ph.D. from Sher-e-Kashmir University of Agricultural Sciences and Technology, Jammu (J&K), India. Her research areas of interest are molecular breeding approaches to enhance abiotic stress tolerance and improve the quality of crops.

ABIOTIC STRESS TOLERANCE MECHANISMS IN PLANTS

Edited by

Gyanendra Kumar Rai

School of Biotechnology,
Sher-e- Kashmir University of Agricultural Sciences
and Technology of Jammu
JAMMU-180009 (J&K), India

Ranjeet Ranjan Kumar

Division of Biochemistry,
Indian Agricultural Research Institute
NEW DELHI-110012, India

and

Sreshti Bagati

School of Biotechnology,
Sher-e- Kashmir University of Agricultural Sciences
and Technology of Jammu
JAMMU-180009 (J&K), India

CRC Press is an imprint of the
Taylor & Francis Group, an **Informa** business

NARENDRA PUBLISHING HOUSE
DELHI (INDIA)

First published 2021
by CRC Press
2 Park Square, Milton Park, Abingdon, Oxon, OX14 4RN

and by CRC Press
6000 Broken Sound Parkway NW, Suite 300, Boca Raton, FL 33487-2742

© 2021 selection and editorial matter, Narendra Publishing House; individual chapters, the contributors

CRC Press is an imprint of Informa UK Limited

The rights of Rai et. al. to be identified as the authors of the editorial material, and of the authors for their individual chapters, has been asserted in accordance with sections 77 and 78 of the Copyright, Designs and Patents Act 1988.

Print edition not for sale in South Asia (India, Sri Lanka, Nepal, Bangladesh, Pakistan or Bhutan).

British Library Cataloguing-in-Publication Data
A catalogue record for this book is available from the British Library

Library of Congress Cataloging-in-Publication Data
A catalog record has been requested

ISBN: 978-0-367-75745-8 (hbk)
ISBN: 978-1-003-16383-1 (ebk)

CONTENTS

PREFACE

Since recent years, the population across the globe is increasing expeditiously; hence increasing the agricultural productivity to meet the food demands of the thriving population becomes a challenging task.Abiotic stresses pose as a major threat to agricultural productivity. Having an adequate knowledge and apprehension of the physiology and molecular biology of stress tolerance in plants is a prerequisite for counteracting the adverse effect of such stresses to a wider range. This book deals with the responses and tolerance mechanisms of plants towards various abiotic stresses. Until recently, various pervasive studies have been carried out on field and off field in the field of physiology and biochemistry to develop a better understanding of the abiotic stress responses of plants. However, the advent of molecular biology and biotechnology has shifted the interest of researchers towards unraveling the genes involved in stress tolerance. More effort is being made to understand and pave ways for developing stress tolerance mechanisms in crop plants. Several technologies including Microarray technology, functional genomics, on gel and off gel proteomic approaches have proved to be of utmost importance by helping the physiologists, molecular biologists and biotechnologists in identifying and exploiting various stress tolerance genes and factors for enhancing stress tolerancein plants.

This book would serve as an exemplary source of scientific information pertaining to abiotic stress responses and tolerance mechanismstowards various abiotic stresses for the post- graduate students, research workers, faculty and scientists involved inagriculture, plant sciences, molecular biology, biochemistry, biotechnology andrelated areas.

We would like to thank the authors for their interest and cooperation in this intriguing endeavor. We are grateful to Narendra Publishing House, Delhi for their support and technical advice in bringing out this book.

Gyanendra Kumar Rai
Ranjeet Ranjan Kumar
Sreshti Bagati

Abiotic Stress Tolerance Mechanisms in Plants, Pages 1–84
Edited by: Gyanendra K. Rai, Ranjeet Ranjan Kumar and Sreshti Bagati
Copyright © 2018, Narendra Publishing House, Delhi, India

1

GENETIC ENGINEERING OF CROP PLANTS FOR SALINITY AND DROUGHT STRESS TOLERANCE: BEING CLOSER TO THE FIELD

Pravin V. Jadhav[1], Prashant B. Kale[1], Mangesh P. Moharil[1],
Deepti C. Gawai[1], Mahendra S. Dudhare[1], Shyam S. Munje[1],
Ravindra S. Nandanwar[1], Shyamsundar S. Mane[1], Philips Varghese[2],
Joy G. Manjaya[3] and Raviprakash G. Dani[1]

[1] *Biotechnology Centre, Department of Agricultural Botany,*
Dr. Panjabrao Deshmukh Krishi Vidyapeeth, Akola-444 104, Maharashtra state, India
[2] *Agharkar Research Institute, Pune-411004, Maharashtra, India*
[3] *Nuclear Agriculture and Biotechnology Division, Bhabha Atomic Research Centre (BARC),*
Trombay, Mumbai-400085, Maharashtra, India
E-mail: jpraveen26@yahoo.co.in

Abstract

Abiotic stresses such as drought, salinity, heat and cold, are greatly affecting the plant growth and agricultural productivity and causes more than 50% of worldwide yield loss of major crops every year. Many efforts have been made by researchers to mitigate these stresses and to increase crop productivity under unfavourable environments. Traditional measures are niche-specific and exhibit tremendous vulnerability to the climactic conditions due to varied intensity of edaphic factors. Genetic engineering is one of the tools being used to develop crop plants which are tolerant to these stresses. It is imperative to understand the adaptive mechanism of plants to drought especially the type and expression levels of drought responsive genes. Several genes upregulated during abiotic stress have been exploited to develop drought tolerant plants. Genetic engineering of crop plants with stress responsive genes has been an effective method of generating drought tolerant plants. Keeping this in purview, the present chapter makes a serious attempt to accumulate the available information and finally come out with a sound strategy by deploying cutting edge science and modern biotechnological approaches paying special attention towards development of large scale cultivation of transgenic crops, which is deemed to be conclusive by halting menaces caused by biotic and abiotic stresses to enhance productivity especially at the interface of global climatic changes, which are amplifying alarmingly, which would lead to acute food and nutritional insecurity which would fail to sustain an livelihood.

Keywords: Abiotic stress, Genetic engineering, Nutritional insecurity.

1. Introduction

Abiotic stresses are serious threats to agriculture and the environment which have been exacerbated in the current century by global warming and industrialization. Worldwide approximately 70% of yield reduction is the direct result of abiotic stresses (Acquaah, 2007). According to FAO, globaly more than 800 million hectares of land are currently salt-affected, including both saline and sodic soils equating to more than 6% of the world's total cultivated area. Continuing salinization of arable land is expected to have overwhelming global impact, resulting in a 30% loss of cultivated land over the next 25 years it will be double by 2050. Currently, the world is losing three hector of arable land every minute due to soil salinity. Some of the most serious effects of abiotic stresses occur in the arid and semiarid regions where rainfall, high evaporation, low native rocks, saline irrigation water and poor water management etc., contribute in agricultural areas. The increasing pressure put on agricultural land by burgeoning human populations has resulted in land degradation, a cultivation shift to more marginal areas and soil types and heavier requirements for agricultural productivity per unit area. Additionally, climate change has worsen the frequency and severity of various abiotic stresses, particularly drought, salinity and high temperatures with significant yield reductions reported in major cereal species (Lobell and Field, 2007). Diseases, pests and weed competition losses account for 4.1% and 2.6% yield reductions, respectively, with the remainder of the yield reduction of 69.1% due to unfavorable abiotic stresses induced by problematic soils and erratic climate patterns. Certainly, some of these losses are caused by inherently unfavorable environments as well as by suboptimal management practices by farmers, often due to economic constraints or lack of training (Godfray *et al.*, 2010; Peleg *et al.*, 2011). Agricultural practices for improving crop productivity per unit area have, in many cases, accelerated the rate of land degradation, with particularly marked effects in irrigated areas. Irrigation has led to salinity across large tracts of agricultural land, with cases, such as in India, where it has reportedly led to the loss of seven million hectares from cultivation (Martinez-Beltran and Manzur, 2005). Higher yields are also only sustainable with higher nutrient use, and the heavy demand for fertilizers has caused rising cultivation costs for farmers worldwide. The environmental and economic consequences of increased nutrient use have been widely reported. For sustainability of crop production, there is a need to reduce the environmental footprint of food and fiber production, and nutrient use efficient crops are highly sought after.

2. How Crop Plants Sense the Stress: Signaling and Pathway

2.1. Abiotic Stress and its Recognition by Plant

As plants are sessile, it is tough to measure the exact force exerted by stresses and therefore in biological terms it is difficult to define stress. A biological condition, which may be stress for one plant may be optimum for another plant. The most practical definition of a biological stress is an adverse force or a condition, which inhibits the normal functioning and well being of a biological system such as plants (Jones and Jones, 1989).

A cell is separated from its surrounding environment by a physical barrier i.e. the plasma membrane. This membrane is permeable to specific lipid molecules such as steroid hormones, which can diffuse through the membrane into the cytoplasm and is impermeable to the water-soluble material including ions, proteins and other macromolecules. Primarily, the cellular responses are initiated by interaction of the extracellular material with a plasma membrane protein. This extracellular molecule is known as ligand or an elicitor and the plasma membrane protein, which binds and interacts with this molecule, is called a receptor. Various stress signals both abiotic as well as biotic serve as elicitors for the plant cell.

The stress is first perceived by the receptors present on the cell membrane of the plant (Figure 1), the signal is then transduced downstream and lead to the generation of second messengers like calcium, reactive oxygen species (ROS) and inositol phosphates.

These messengers promote the intracellular calcium level. This perturbation in cytosolic Ca^{2+} level is sensed by Ca^{2+} sensors. These sensors apparently lack any enzymatic activity and change their conformation in a calcium dependent manner. These sensory proteins interact with their respective interacting partners initiating a phosphorylation cascade and target the major stress responsive genes or the transcription factors regulating these genes. Eventually the products of these stress genes lead to plant adaptation and help the plant to survive even under stressed conditions. Thus, plant responds to stresses as individual cells and synergistically as a whole organism. Stress induced changes in gene expression in turn may participate in the generation of hormones like ABA, salicylic acid and ethylene. These molecules may amplify the initial signal and initiate a second level of signaling that may follow the same or different components of signaling pathway. Certain molecules also known as accessory molecules may not directly participate in signaling but participate in the modification of signaling components. These proteins include the protein modifiers, which may be added cotranslationally to the signaling proteins like enzymes for myristoylation, glycosylation, methylation and ubiquitination.

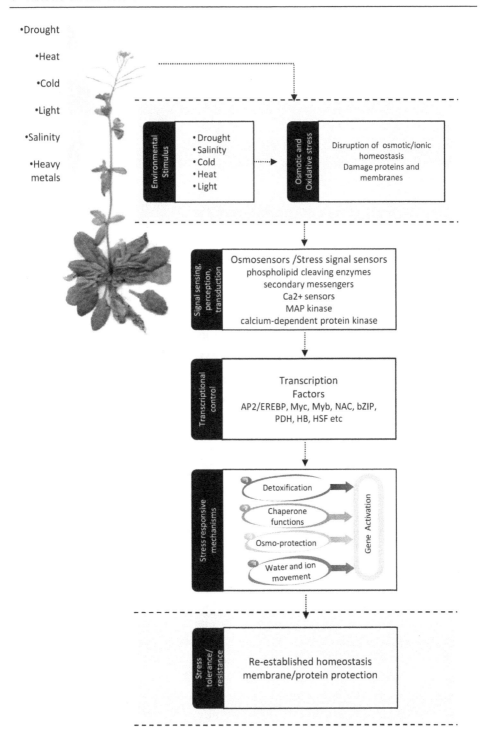

Fig. 1. Plant responses to abiotic stresses

3. Plant Response to Drought and Salinity Stress

3.1. Drought Stress and Agriculture

Drought is one of the most significant environmental stress affecting global agricultural production and massive efforts are being made by plant scientists to improve crop yields under limiting water availability (Cattivelli *et al.,* 2008). During the twentieth century, the world's population will be tripled from approximately 1.65 to 5.98 billion and population projections of 8.91 and 9.75 billion are expected to occur by 2050 and 2150, respectively. Developing countries like Africa and Asia account for approximately 80% of this growth and, with an estimated 800 million people in these countries already undernourished. The FAO predicts that a 60% increased world food production is required in the next two decades to sustain these populations.

Currently, agriculture accounts for approximately >70% of global water use and irrigation for up to 90% of total water withdrawals in arid nations (FAO, 2009a). Approximately, 40% of all crops produced in developing countries are grown on irrigated arable land, which accounts for only 20% of the total arable land in these nations (FAO, 2009c). The water withdrawal requirement for irrigation is expected to increase by 14% in developing countries by 2030 and strategies to reduce this demand by developing crops that require less irrigation will, therefore, play a vital role in maintaining world food supply. While within a few decades, the expanding world population will require more water for domestic, municipal, industrial and environmental needs (Hamdy *et al.,* 2003). This trend is expected to emphasize due to global climatic change and increased aridity (Vorosmarty *et al.,* 2000). Thus, to meet the projected food demands, more crops per drop are required (Condon *et al.,* 2004).

3.2. Nature of Drought and Plant Response

A plant requires water to complete its life cycle consisting of at least 70% water on a fresh weight basis. When water in the plant environment becomes deficient, plant transpiration cannot fully meet the atmospheric demand and plant water deficit evolves. Water deficit is a strain on the plant that causes damage and drives a network of gene responses. These are proportional to the rate of deficit. Exacerbate action of abiotic stress conditions can led to great losses in productivity due to crop stress. When subjected to water deficit plants go through a cascade of metabolic alterations started with reduction in photosynthetic pigments concentration. Physiological mechanisms of plant response to water stress are summarized in Figure 2. Deficient water level leads to removal of water from the membrane disrupts normal bilayer structure and results in the membrane becoming

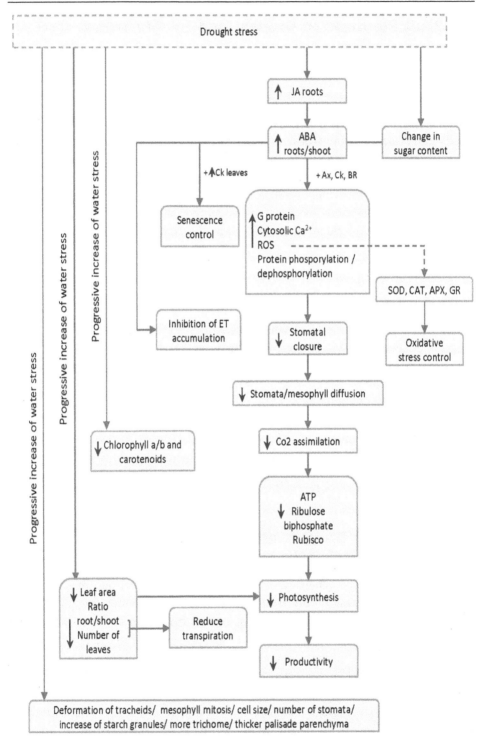

Fig. 2: Physiological mechanisms induced by water stress

exceptionally porous when desiccated. It induces imbalances in osmotic and ionic homeostasis, loss of cell turgidity, damage to structural and functional cellular proteins and membranes as well. Consequently, water-stressed plants wilt, lose photosynthetic capacity, and are unable to sequester assimilates into the targeted plant organs. Severe drought conditions result in significant yield loss and plant death. Revelation of plant drought tolerance and response mechanisms has been compounded at variable levels and forms of drought. Drought can be spatially and temporally variable; terminal, short-term, or sporadic; severe, moderate, or minor; and can occur at rates ranging from very sudden to gradual. The effects of drought and water deficit on crop productivity vary for different crops, macro and microenvironments across a single field, plant life stages, and the plant material to be harvested. Additionally, the effects of drought on crop productivity are often compounded by associated stresses such as salt, heat and other stress. The components of drought and salt stress cross talk with each other as both the stresses ultimately result in dehydration of the cell and osmotic imbalance. Drought and salt signaling encompasses three important parameters (Liu and Zhu, 1998) viz., 1) Reinstating osmotic as well as ionic equilibrium of the cell to maintain cellular homeostasis under stress; 2) control as well as repair of stress damage by detoxification signaling and 3) signaling to coordinate cell division to meet the requirements of the plant proteins under stress.

This stress also leads to activation of enzymes involved in the production and removal of ROS (Zhu, 2002; Cushman and Bohnert, 2000). Reduced cyclin-dependent kinase activity results in slower cell division as well as inhibition of growth under water deficit condition (Schuppler *et al.,* 1998). The physiological effects of drought on plants are the reduction in vegetative growth, in particular shoot growth. Leaf growth is generally more sensitive than the root growth. Reduced leaf expansion is beneficial to plants under water deficit condition, as less leaf area is exposed resulting in reduced transpiration. In accordance, many mature plants respond to drought stress by a process is known as leaf area adjustment by accelerating senescence and abscission of older leaves. Regarding root, the relative root growth may undergo enhancement, which facilitates the capacity of the root system to extract more water from deeper soil layers so as to cope with water stress.

4. Genetic Basis of Drought Tolerance

Drought tolerance is an inherent ability of the plant to sustain growth through altering cellular and metabolic levels. It involves co-ordination of physiological and biochemical alterations such as turgor maintenance, protoplasmic resistance and dormancy (Beard and Sifers, 1997). Plants respond to drought conditions by shifting the expression of a complex array of genes and synthesis of molecular chaperons

(Figure 3). Drought is interconnected with salinity, extreme temperatures, and oxidative stress and may induce similar cellular damage. For example, drought and/or salinization are manifested primarily as osmotic stress, resulting in the disruption of homeostasis and ion distribution in the cell (Serrano *et al.*, 1999; Zhu, 2001). Oxidative stress along with high temperature, salinity, or drought stress, causes the denaturation of functional and structural proteins (Smirnoff, 1998) activating similar cell signaling pathways (Zhu 2001, 2002) and cellular responses, such as the production of stress proteins, up-regulation of antioxidants and accumulation of compatible solutes (Wang *et al.*, 2003b). As discussed earlier, plants adapt to drought conditions by regulating specific sets of genes which vary depending on factors such as the severity of drought conditions and other environmental factors, and the plant species (Wang *et al.*, 2003b). These genes can be grouped into three major categories on the basis of their expression: 1) Genes involved in signal transduction pathways (STPs) and transcriptional control; 2) Genes with membrane and protein protection functions; and 3) Genes supporting water and ion uptake and transport (Vierling, 1991; Ingram and Bartels, 1996; Smirnoff, 1998; Shinozaki and Yamaguchi-Shinozaki, 2000). To date, success in genetic improvement of drought resistance have involved genetic manipulation using single or a few genes involved in signaling /regulatory pathways or that encode enzymes involved in the pathways (Wang *et al.*, 2003b). The disadvantage of this is that there are numerous interacting genes involved, and efforts to improve drought tolerance through manipulation of one or a few of them is often associated with other, often undesirable, pleiotropic and phenotypic alterations (Wang *et al.*, 2003b). These complex considerations, when coupled with the complexity of drought and the plant-environment interactions occurring at all levels of plant response to water deficit, demonstrate that the task plant researchers are faced with in engineering drought tolerant crop is dauntingly multi-faceted and extremely difficult. Genetic engineering using candidate genes for abiotic stress was found to be successful in model plants growing under controlled conditions and provided insights on the role of these genes in key physiological and biochemical processes (reviewed by Pardo, 2010; Vinocur and Altman, 2005). An exception was demonstrated by Rivero *et al.* (2007) who manipulated a leaf senescence gene. Leaf senescence is an avoidance strategy and is accelerated in drought-sensitive plants to decrease canopy size. In crop plants, accelerated senescence is often associated with reduced yield and is thought to be the result of an inappropriately activated cell death program. Therefore, suppression of drought-induced leaf senescence in tobacco plants was investigated as a tool to enhance drought resistance. Other drought avoidance trait has been investigated include stay-green and cuticular biosynthesis. Stay-green is a variable and quantitative trait, which generally refers to delayed senescence which has not yet been used to successfully produce transgenic plants with increased drought tolerance in the field.

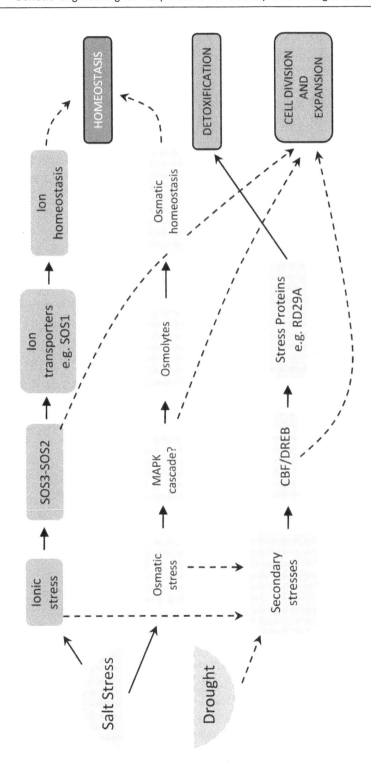

Fig. 3: The three aspects of drought and salt tolerance in plants (homeostasis, detoxification and growth control) and the pathways that interconnect them.

4.1. Salinity Stress: Complexity and its Impact on Agricultural Production

The negative effects of salt accumulation in agricultural soils have severely affected agricultural productivity in large swathes of arable land throughout the world and are affecting 45 million hectares of irrigated land which is expected to increase due to climate change and various irrigation practices as well (Roy *et al.,* 2014). In the majority, salinity comes from natural causes due to salt accumulation over long periods of time. High salinity causes both hyperionic and hyperosmotic stress and can lead to plant demise. Sea water contains approximately 3% of NaCl and in terms of molarity of different ions, Na^+ is about 460 mM, Mg^{2+} is 50 mM and Cl^- around 540 mM along with smaller quantities of other ions. Salinity in a given land area depends upon various factors like amount of evaporation (leading to increase in salt concentration), or the amount of precipitation (leading to decrease in salt concentration). Weathering of rocks also affects salt concentration. Inland deserts are marked by high salinity as the rate of evaporation far exceeds the rate of precipitation.

The significant portion of the cultivated land is fetching saline due to deforestation or excess irrigation and fertilization (Shannon, 1997). As drier areas in particular need intense irrigation, there is extensive water loss through a combination of both evaporation as well as transpiration. This process is known as evapotranspiration and as a result, the salt delivered along with the irrigation water gets concentrated, year-by-year in the soil. This leads to huge losses in terms of arable land and productivity as most of the economically important crop species are very sensitive to soil salinity. These salt sensitive plants, also known as glycophytes include rice (*Oryza sativa*), maize (*Zea mays*), soybean (*Glycine max*) and beans (*Phaseolus vulgaris*) which greatly affects the food supply.

More than 20% of the world's irrigated land producing one third of world's food supply is presently salt affected (Ghassemi *et al.,* 2006). With the expected increase in world population, the ultimate aim of the salt tolerance research is to enhance the ability of plants to maintain growth and productivity under saline condition relative to non-saline soils to minimize the effect of salinity on yield. Growth reduction under salinity is due to distinct processes related either to the accumulation of salts in the shoot or independent of shoots salt accumulation. Salinity reduces the ability of plants to absorb water, causing rapid reductions in growth rate, along with as array a suite of metabolic changes identical to those caused by water stress (Muuns, 2002). High salt concentration (Na^+) in particular which deposit in the soil can alter the basic texture of the soil resulting in decreased soil porosity and consequently reducing the soil aeration and water conductance. The basic physiology of high salt stress and drought stress overlaps with each other. High salt depositions in the soil generate a low water potential zone in the soil making it increasingly difficult for the plant to acquire both water as well as nutrients.

4.1.1. Effect of Salinity Stress on Plant Cell

High salt concentration induces several deleterious consequences on plant cell. First, salt stress causes an ionic imbalance (Zhu *et al.,* 1997). When salinity results from an excess of NaCl, the most common type of salt stress, the increased intracellular concentration of Na^+ and Cl^{2+} ions is deleterious to cellular systems (Serrano *et al.,* 1999). In addition, the homeostasis of not only Na^+ and Cl^2, but also K^+ and Ca^{2+} ions is disturbed (Hasegawa *et al.,* 2000a, b; Rodriguez-Navarro, 2000). As a result, plant survival and growth depend on adaptations that re-establish ionic homeostasis, there-by reducing the duration of cellular exposure to ionic imbalance. Second, high concentrations of salt impose hyperosmotic shock by decreasing the chemical activity of water and causing loss of cell turgor. This negative effect in the plant cell is thought to be similar to the effects caused by drought.

Third, salt-induced water stress reduction of chloroplast stromal volume and generation of reactive oxygen species (ROS) are also thought to play important roles in inhibiting photosynthesis (Price and Hendry, 1991). These can be summarized as,

1) Disruption of ionic equilibrium: Influx of Na^+ dissipates the membrane potential and facilitates the uptake of Cl- down the chemical gradient.

2) Na^+ is toxic to cell metabolism and has deleterious effect on the functioning of some of the enzymes (Niu *et al.,* 1995).

3) High concentrations of Na^+ causes osmotic imbalance, membrane disorganization, reduction in growth, inhibition of cell division and expansion.

4) High Na^+ levels also lead to reduction in photosynthesis and production of reactive oxygen species (Yeo, 1998).

The generic functions of K^+, role of Ca^{2+} and SOS pathway in relation to imparting salt stress tolerance, loss of water due to salinity stress, role of osmolytes and DNA unwinding enzymes imparting stress tolerance are the various aspects of salinity stress which needs to be understood. Where sodium (Na^+) is deleterious for plant growth, K^+ is one of the essential elements and is required by the plant in large quantities.

4.1.2. *Generic Role of K⁺*

The K^+ imparts three major functions i.e. 1) it is required for maintaining the osmotic balance; 2) K^+ has a role in opening and closing of stomata, and 3) it acts as an essential co-factor for various enzymes like the pyruvate kinase, whereas Na^+ is not.

4.1.3. *Transpirational Flux*

Movement of salt into roots and shoots is a product of the transpirational flux required to maintain the water status of the plant (Flowers and Yeo 1992). As common proteins transport Na^+ and K^+, Na^+ competes with K^+ for intracellular influx (Amtmann and Sanders, 1999; Blumwald *et al.*, 2000). Many K^+ transport systems have some affinity for Na^+, i.e., Na^+/K^+ symporters. Thus external Na^+ negatively impacts intracellular K^+ influx. Most cells maintain relatively high K^+ and low concentrations of Na^+ in the cytosol. This is achieved through a coordinated regulation of transporters for H^+, K^+, Ca^{2+} and Na^+.

The plasma membrane H^+-ATPases serves as the primary pump that generates a proton motive force driving the transport of other solutes including Na^+ and K^+. Increased ATPase-mediated H^+ translocation across the plasma membrane is a component of the plant cell response to salt imposition (Watad *et al.*, 1991). K^+ and Na^+ influx can be differentiated physiologically into two categories, one with high affinity for K^+ over Na^+ and the other for which there is lower K^+/Na^+ selectivity. The Na^+/K^+ transporter and K^+ transporters with dual high and low affinity may contribute substantially to Na^+ influx.

4.1.4. *Role of Ca²⁺ in Relation to Salt Stress*

For decades it has been shown that another ion, Ca^{2+} has role in providing salt tolerance to plant. Externally supplied Ca^{2+} reduces the toxic effects of NaCl, presumably by facilitating higher K^+/Na^+ selectivity (Liu and Zhu, 1998). High salinity results in increased cytosolic Ca^{2+} that is transported from the apoplast as well as the intracellular compartments (Knight *et al.*, 1997). This transient increase in cytosolic Ca^{2+} initiates the stress signal transduction leading to salt adaptation.

4.2. Water Loss due to Salinity Stress

A major consequence of NaCl stress is the loss of intra-cellular water. To prevent this water loss from the cell and protect the cellular proteins, plants accumulate many metabolites that are also known as "compatible solutes". These solutes do not inhibit the normal metabolic reactions (Breesan *et al.*, 1998). The metabolites observed with an osmolyte function are sugars, mainly fructose and sucrose, sugar alcohols and complex sugars like trehalose and fructans. In addition charged metabolites like glycine betaine proline and ectoine are also accumulated. The accumulation of these osmolytes, facilitate the osmotic adjustment (Delauney and Verma, 1993). Water moves from high water potential to low water potential and accumulation of these osmolytes make the water potential low inside the cell and prevent the intracellular water loss.

4.3. Mechanisms of Salinity Tolerance

As salinity induces different effects on a plant, similarly plant tolerate stress through diverse mechanisms. Plants respond to salt stress at three different levels, i.e., cellular, tissue and whole plant level. Cell-based mechanisms of ion homeostasis and the synthesis of osmoprotectants are essential determinants for salt tolerance. Osmotic tolerance, which is regulated by long distance signals that reduce shoot growth and is triggered before shoot Na^+ accumulation; secondly, ion exclusion, where Na^+ and Cl^- transport processes in roots reduce the accumulation of toxic concentrations of Na^+ and Cl^- within leaves; and finally, tissue tolerance, where high salt concentrations are observed in leaves but are compartmentalized at the cellular and intracellular level (Figure 4). Inadequate information is available about tolerance to the 'osmatic phase'. This process may involve rapid and long-distance signaling, perhaps via ROS waves (Mittler *et al.*, 2011), Ca^{2+} waves, or even long distance electrical signaling (Maischak *et al.*, 2010). ROS are generated in the chloroplast by direct transfer of excitation energy from chlorophyll to produce singlet oxygen, or by univalent oxygen reduction at Photo-system I, in the Mehler reaction (Allen, 1995). Differences in osmotic tolerance may take place due to differences in long-distance signaling, or it may vary in the initial discrimination between salt or the response to the signals. This is still an area of salinity research with many unknowns, and further research is needed to acquire a better understanding of osmotic tolerance.

Relatively, the mechanism of ionic phase is well explored which is mediated through the accumulation of Na^+ and Cl^- in the leaf blade. Tissue tolerance to Na^+ and Cl^- requires compartmentalization of Na^+ and Cl^- at cellular level and intera cellular level to avoid accumulation of toxic concentration within cytoplasm. Plants can be reduced toxicity by limiting accumulation of toxic ions in the leaf blades, and/or by increasing their ability to tolerate the salts that they have failed to exclude from the shoot, such as by compartmentation into vacuoles during the ionic phase. Both the mechanisms involve range of transporters and their controllers at plasma level and tonoplast (Plett and Moller, 2010). Osmoprotection and osmotic adjustment

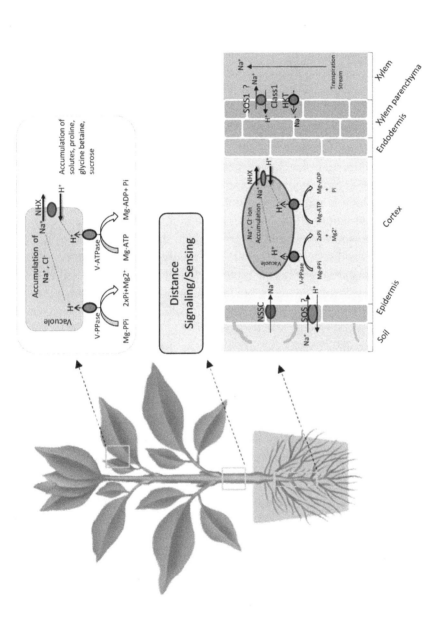

Fig. 4: The three main mechanisms of Salinity tolerance in a crop plants. Tissue tolerance where high salt concentrations are found in leaves but are compartmentalized at the cellular and intracellular level (especially in the vacuole), a process involving ion transporters, proton pumps and synthesis of compatible solutes. Osmotic tolerance, which is related to minimizing the effects on the reduction of shoot growth and may be related to as yet un known sensing and signaling mechanisms. Ion exclusion, where Na+ and Cl- transport processes, predominantly in roots, prevent the accumulation of toxic concentrations of Na+ and Cl within leaves. Mechanisms may include retrieval of Na+ from the xylem, compartmentation of ions in vacuoles of cortical cells and/or efflux of ions back to the soil

requires the synthesis of compatible solutes and higher level controls to co-ordinate transport and biochemical processes having role in both. To date, there is neither any evidence that these mechanisms are mutually exclusive nor that a particular plant is committed to only one strategy (e.g. a plant may have ion exclusion as its primary tolerance mechanism at moderate salinity).Moreover, salt tolerance is regulated throughout the plant development and control at tissue level, plant tolerance responses at one stage of development are not necessarily the same at other stages (Johnson *et al.,* 1992).Therefore, the mechanisms of tolerance at specific stages of plant development must be studied in order to understand the biochemical events that play important roles in the responses to salt stress (Borsani *et al.,* 2001a). Similarly, it is apparent that salt tolerance can be complex and involve many genes, as it has been critical for several decades, for example, programs intended to introgress salt tolerance by means of traditional breeding methods appear to have frequently failed because of the multigenic nature of tolerance (Flowers and Yeo, 1995; Roy *et al.,* 2014). It is therefore necessary not to study the molecular basis of salt tolerance as a particular trait in itself, but to study the behavior of traits that are hypothesized to contribute to salinity tolerance.The most intensively studied trait is the exclusion of Na^+ from leaf blades, mainly because it is relatively straight forward to phenotype. Focusing on this has led to significant increases in salinity tolerance, as measured by yield in the field, at least in durum wheat having poor Na^+ exclusion when grown in highly saline sites where yield has already been greatly reduced (James *et al.,* 2012).

4.4. Determinants for Salt Tolerance

In spite of considerable effort through breeding programmes, progress to enhance salt tolerance has been slow. Classic genetic studies have demonstrated that the ability of plants to tolerate salt stress is a quantitative trait involving the action of many genes.As a result, it has been difficult to obtain salt tolerance in crop plants by traditional methods (Foolad and Lin, 1997). Considering limitations of conventional approaches, various transgenictechnologies have been used to improve stress tolerance in plants (Allen, 1995).This situation has been complicated by the fact that the main character selected in crop plants has been productivity, which is also a complex trait. Therefore, the integration of genes required to increase salt tolerance in a specific genotype is difficult without affecting other important multigenic traits like flowering, fruit quality and dry matter production (Flowers *et al.,* 2000).

Various attempts were made to ascertain salt tolerance determinants in plants, the first approach being recognition the metabolic processes to tolerate NaCl. Establishment of ion homeostasis is an essential requirement for plants to survive under salt stress conditions. Plant cells respond to salt stress by increasing Na^+ efflux at the plasma membrane and its accumulation in vacuole.

Therefore, proteins and targeted genes involved in these processes can be considered as salt tolerance determinants. Salt tolerance requires not only adaptation to Na^+ toxicity but also the acquisition of K^+ whose uptake is affected by high external Na^+ concentration by the plant cell. The uptake of K^+ is affected by Na^+ due to the chemical similarities between both ions. Potassium is an essential nutrient being the major cationic inorganic nutrient in most terrestrial plants. Therefore, K^+ transport systems involving good selectivity of K^+ over Na^+ can also be considered as an important salt tolerant determinant (Rodriguez-Navarro, 2000).

Also, genes regulating the accumulation of compatible osmolytes can be considered as salt tolerant determinants. The synthesis of compatible osmolytes, the organic compounds are thought to mediate osmotic adjustment, protecting sub-cellular structures and oxidative damage by their free radical scavenging capacity (Smirnoff, 1993; Hare et al., 1998). Thus, salt, drought, and to some extent cold stress cause, an increased biosynthesis and accumulation of absicisic acid (ABA) (Koornneef et al., 1998; Taylor et al., 2000). ABA role in osmotic stress tolerance is well known and has been exhaustively reviewed (McCourt, 1999; Rock, 2000; Zhu, 2002). However, earlier evidences revealed that ABA could be involved in the control of ion homeostasis. This increased level of ABA content was accompanied by an improved K^+/Na^+ ratio (Bhora et al., 1995). Also, the transport and accumulation of K^+ in higher plant roots has been shown to be regulated by ABA (Roberts, 1998). Recent reports indicate that ABA regulates K^+ channel activity in maize and *Arabidopsis* roots, indicating ABA regulation of K^+ transport in roots is at least in part, ion channel-mediated (Roberts and Snowman, 2000).

With the advent of molecular biology, a more recent approach used to determine mechanisms of salt tolerance has been the identifications of cellular processes and genes whose expression is altered by salt stress (Zhu et al., 1997; Hasegawa et al., 2000a, b). Identification of such salt-regulated genes allowed a better understanding of the complexity of salt tolerance in higher plants (Bray, 1993; Hasegawa et al., 2000a, b). The generalized assumption was that a gene regulated by salt stress would probably be imparting the tolerance. The various stress responsive genes can be broadly categorized as early and late induced genes (Figure 4). Early genes are induced within minutes of stress signal perception and express transiently. Various transcription factors are grouped under early genes as the induction of these genes does not require synthesis of new proteins and signaling components are already primed. In contrast, most of the other genes, which are activated by stress more slowly, i.e. after hours of stress perception are included in the late induced category. The expression of these genes is often sustained.

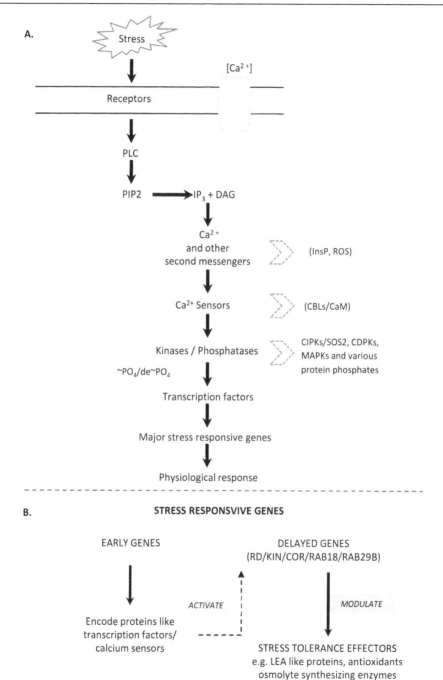

Fig. 5: Generic signal transduction pathway as well as the expression of early and late genes in response to abiotic stress signaling. (A) Represents the overview of signaling pathway under stress condition; (B) Early and delayed gene expression in response to abiotic stress signaling. Various genes are triggered in response to stress and can be grouped under early and late responsive genes.

These genes include the major stress responsive genes such as RD (responsive to dehydration) / KIN (cold induced) /COR (cold responsive), which encodes and modulate the proteins needed for synthesis, for example LEA-like proteins (late embryogenesis abundant), antioxidants, membrane stabilizing proteins and synthesis of osmolytes.

Stress-induced gene expression can be broadly categorized into three groups: (1) genes encoding proteins with known enzymatic or structural functions, (2) proteins with as yet unknown functions, and (3) regulatory proteins. Initial attempts to develop transgenics (mainly tobacco) for abiotic stress tolerance involved "single action genes" that would confer increased tolerance to salt or drought stress Stress-induced proteins with known functions such as water channel proteins, key enzymes for osmolyte (proline, betaine, sugars such as trehalose, and polyamines) biosynthesis, detoxification enzymes, and transport proteins were the initial targets of plant transformation. However, that approach has overlooked the fact that abiotic stress tolerance is likely to involve many genes at a time, and that single-gene tolerance is unlikely to be sustainable.

Many of these regulated genes have been reported but the most direct procedure to demonstrate their importance in salt tolerance is through functional genome analysis. This approach is not easily accomplished in crop plants. *SOS* is one such gene whose molecular nature has been recently determined. The *SOS1* gene encodes a protein showing significant sequence similarity to plasma membrane Na^+ /H^+ antiporters from bacteria and fungi (Shi *et al.*, 2000). The *SOS2* gene encodes a serine/threonine protein kinase with an N-terminal catalytic domain similar to that of the yeast SNF1 kinase (Liu *et al.*, 2000). The *SOS3* gene encodes a Ca^{+2} binding protein and has itsgreatest sequence homology with the yeast calcineurin B subunit and a neuronal calcium sensor, both of which are activated by Ca^{+2} (Liu and Zhu, 1998). The *SOS4* encodes a piridoxal (PL) kinase that is involved in the biosynthesis of PL-5-phosphate which is an active form of vitamin B6 (Shi *et al.*, 2002b).

5.   Rational Strategies for Developing Drought and Salt Stress Tolerance in Crop Plants

The response to stress depends on the duration and severity of the event, as well as the age and developmental stage of the plant, which varies with the species and genotype level (Bray, 1997). For crop plants, tolerance to abiotic stresses is measured by yield loss rather than survival. A key to progress towards breeding better crops under stress has been to understand the changes in cellular, biochemical and molecular machinery that occur in response to stress. Modern molecular techniques involve the identification and use of molecular markers that can enhance breeding programs. However, the introgression of genomic portions (QTLs) involved in stress tolerance often brings along undesirable agronomic characteristics from the donor parents. This is because of the lack of a precise knowledge of the key genes underlying the QTLs. Therefore, the development of genetically engineered plants by the introduction and/or over expression of selected genes seems to be a viable option to hasten the breeding of "improved" plants. Intuitively, genetic engineering would be a faster way to insert beneficial genes than through conventional or molecular breeding. Also, it would be the only option when genes of interest originate from cross barrier species, distant relatives, or from non-plant sources. Indeed, there are several traits whose correlative association with resistance has been tested in transgenic plants.

Therefore, a second "wave" of transformation attempts to transform plants with the third category of stress-induced genes, namely, regulatory proteins has emerged. Through these proteins, many genes involved in stress response can be simultaneously regulated by a single gene encoding stress inducible transcription factor (Kasuga *et al.,* 1999), thus offering possibility of enhancing tolerance towards multiple stresses including drought, salinity, and freezing.

Interestingly, genetic engineering allows controlling the timing, tissue-specificity, and expression level of the introduced genes for their optimal function. This is an important consideration if the function of a given gene is preferred only at a specific time, in a specific organ, or under specific conditions of stress (Katiyar *et al.,* 1999). The most widely used promoters in generating transgenic plants are constitutively expressed; this may have serious deleterious effects on the plant. However, in cases where the gene expression needs to be tailored to a specific organ or a specific time, such constitutive promoters may not be a suitable choice, especially for the stress-induced genes. Accordingly, the efforts to generate transgenic plants make use of gene cassettes drivenby stress-induced promoters. With an increasing number of stress genes becoming available and genetic transformation becoming a routine procedure, characterization ofstress-induced promoters has taken a firm footing (Bhatnagar *et al.,* 2008). Unfortunately, limited amount of published reports involving the assessment of transgenic plants under abiotic stresses has shown effect of the transgene under growth conditions that are unlikely to occur in the natural environmental conditions. Therefore, it is needed to set down basic guidelines on the protocols to be followed to conduct a rigorous evaluation of the transformed plants to abiotic stress. Since, most of the reports published so far has emphasized on model plants, this has hindered our ability to readily translate the discoveries into improved yield in crop plants. This can lead to use of transgenic plants as a sources of new cultivars (or their germplasm as new sources of variation in breeding programmes) and they could also exceedingly useful as proof-of-concept tool to dissect the function and interplay of gene networks for abiotic stress tolerance.

6. Resource Species Used for the Identification of Abiotic Stress Tolerant Genes

Various organisms have been employed to identify processes or genes associated with salt tolerance. These organisms vary from prokaryotic such as *Escherichia coli*, unicellular eukaryotic organisms such as *Saccharomycescerevisiae*, to halophytic land plants such as *Mesem-bryanthemum crystallinum*, and glycophytic plantssuch as rice, tomato, *Puccinellia tenuiflora,A.thaliana* and many more.

6.1. Saccharomycescerevisiae

Beside its genetic amenability, *S.cerevisiae* shares basic ion transport mechanisms with plants. Therefore, it can be used as an excellent model system for the study of salt tolerance at a cellular level. Genetic analysis has been very successful in elucidating salt stress tolerance determinants in yeast (Toone and Jones, 1998; Serrano *et al.,* 1999). A number of salt-sensitive yeast mutants have been identified and the cloning of the corresponding genes has shed light on the nature of many genes that are essential for salt tolerance (Brewster *et al.,* 1993; Mendoza *et al.,* 1994; Toone and Jones, 1998; Serrano *et al.,* 1999).

Yeast responds to NaCl by activating several signal transduction pathways. Two of these pathways are needed to sense the osmotic stress induced by NaCl by different osmosensors like SLN1, a transmembrane histidine kinase, and SSK1, the sensor and response regulator respectively, of a two-component system (Maeda *et al.,* 1994).The second osmosensor, SHO1, is an independent trans-membrane protein containing a SH3domain (Maeda *et al.,* 1995). Both osmosensors connect to a Mitogen Activated Protein Kinase (MAPK) that modulates the pathways that converges at the PBS2 kinase (a MAPKK) that phosphorilates the HOG1 kinase leading eventually to the induction of several defence genes including the glycerol biosynthetic genes and the *ENA1* (Na^+efflux pump) (Serrano, 1996; Toone and Jones, 1998). Plant proteins homologous to those of the HOG pathway in yeast have been identified. A putative MAPK from *Pisum sativum*(PsMAPK), which is 47% identical to Hog1p, functionally complements the salt growth defect of the *hog1* yeast mutant (Popping *et al.,* 1996). The *Arabidopsis* gene *ATHK1* was identified by its sequence homology to the yeast osmosensor SLN1 (Urao *et al.,* 1999). Over-expression of *ATHK1* suppressed the lethality of the temperature-sensitive osmosensing-defective yeast mutant *sln1-ts*. The third signal transduction pathway for tolerance to NaCl in yeast is specific for the ionic component of

NaCl stress which regulates ion homeostasis and involves the protein phosphatase calcineurin (Mendoza *et al.,* 1994). However, it is speculated that it could be a vacuolar cation exchanger that releases Ca^{+2} in ex-change with Na^+ (Serrano, 1996). Calcineurin is a Ca^{+2} and calmodulin-dependent protein phosphatase consisting of a catalytic A subunit (CnA) and a regulatory B subunit (CnB) possessing four high affinity EF-hand calcium-binding sites and full activation of CnA requires calcium-CnB and calcium-calmodulin dependent complexes (Klee *et al.,* 1988). Calcineurin regulates Na^+, K^+ and Ca^{+2} homeostasis (Nakamura *et al.,* 1993; Mendoza *et al.,* 1994).

Despite biochemical evidence for a calcineurin-like activity in plants (Luan *et al.,* 1993), the identification of the specific plant oriented gene has been unsuccessful so far. Notwithstanding the identification of CnB-like proteins in plants, none can interact with the yeast CnA, suggesting that they are not functional homologues (Kudla *et al.,* 1999). However, functional complementation of the calcineurin yeast mutant with *Arabidopsis* identified *AtGSK1*, which encodes a GSK3/shaggy like protein kinase (Piao *et al.,* 1999).

6.2 Halophytes

A halophyte is a plant that grows in waters of high salinity, coming into contact with saline water through its roots or by salt spray, such as in saline semi-deserts, mangrove swamps, marshes and sloughs, and seashores. It is now clear that the halophytes's tolerance to NaCl is not the result of unique adaptive mechanisms or metabolic processes that are unique to these plants (Yeo, 1998; Glenn *et al.,* 1999). It seems that the biochemical mechanisms leading to salt tolerance in these plants are regulated in such way that allow a more successful response to salt stress than in other plants (Hasegawa *et al.,* 2000a, b). The halophytic land plants like *M.crystallinum* and *Puccinellia tenuiflora* have been frequently used as model plant in salt tolerance studies (Bohnert *et al.,* 2001; Ying *et al.,* 2014). Identification of gene is the major problem in the use of most halophytes (i.e. searching for salt hypersensitive mutants). For this purpose, the study of *Thellungiella halophila* might be of particular interest in the identification of genes involved in salt tolerance. This halophyte plants can survive at seawater level salinity and its DNA sequence have a similarity of more than 90% of *Arabidopsis* (Zhu, 2001) which can also be easily transformed allowing insertion tag mutagenesis (Bressan *et al.,* 2001).

6.3 Glycophytes

Salinity stress significantly reduces growth and productivity of glycophytes, which are the majority of agricultural products. Of all the glycophytes, undoubtly

Arabidopsis is becoming very useful in the determination of processes involved in salt tolerance (Zhu, 2000). Another glycophyte recently employed in genetic analysis using mutagenesis is tomato (Borsani *et al.,* 2001a). However, in spite of its broad adaptation, production is concentrated in a few warm and rather dry areas (Cuartero and Fernandez-Munoz, 1999) where salinity is a serious problem (Szabolcs, 1994). For this reason, a large number of physiological studies of salt stress have been performed using tomato as a model plant (Cuartero and Fernandez-Munoz, 1999). Unlike *Arabidopsis*, direct studies on salinity, adaptation, and molecular changes in this plant can be assessed also for crop yield.

7. Genetic Engineering for Enhancing Drought and Salt Stress Tolerance

7.1. Improving Response to Drought Stress by Manipulating Single Action Genes

Early attempts to develop transgenic plants resistant to water stress focused on single action genes responsible for the modification of a single metabolite or protein that would confer increased tolerance to drought stress. Genes induced during water-stress conditions are thought to function in protecting cells from water deficit by production of important metabolic proteins and regulation of genes for signal transduction in water-tress response. Recently, a number of droughts responsive genes were cloned and characterized from different plant species (Nepomuceno *et al.,* 2000). Transcription of many of these genes is unregulated by drought stress. Stress-induced proteins with known functions such as water channel proteins, key enzymes for osmolyte (proline, betaine, sugars such as trehalose, and polyamines) biosynthesis, detoxification enzymes, and transport proteins were the initial targets of plant transformation.

7.2. Single Function Genes

7.2.1. *Osmoprotectants*

Osmoregulation is one of the most effective ways evolved by stress-tolerant plants to combat abiotic stress, but most crop plants lack the ability to synthesize the osmoprotectants naturally produced by stress-tolerant plants. Therefore genes concerned with the synthesis of osmoprotectants have been incorporated into transgenic plants to confer stress tolerance (Bhatnagar-Mathur, 2008). In stress-tolerant transgenic plants, many genes involved in the synthesis of osmoprotectants, an organic compounds such as amino acids (e.g. proline), quaternary and other amines (e.g. glycinebetaine and polyamines) and a variety of sugars and sugar alcohols (e.g. mannitol, trehalose and galactinol) that accumulate during osmotic adjustment have been used to date (Vincour and Altman, 2005). Many crops lack the ability to synthesize the special osmoprotectants that are naturally accumulated by stress tolerant organisms. It is believed that osmoregulation would be the best strategy for abiotic stress tolerance, especially if osmoregulatory genes could be triggered in response to different abiotic stresses like drought, salinity and high temperature. Therefore, a widely adopted strategy has been to engineer certain osmolytes or by over expressing it in plants, as a potential route to breed stress-tolerant crops. Various strategies are being employed to modify osmoprotection in plants. Many examples in the literature of increasing compatible solute synthesis as a strategy to improve tolerance to abiotic stress are available. The first step involved in obtaining stress tolerant transgenic plants has been to engineer genes encoding enzymes for the synthesis of specific osmolytes (Bray, 1993). This has resulted in a profusion of reports involving osmoprotectants such as glycine-betaine (Ishitani *et al.*, 1997; Hayashi *et al.*, 1997, 1998; Sakamoto *et al.*, 1998, 2000; McNeil *et al.*, 2000) and proline (Zhu *et al.*, 1998; Yamada *et al.*, 2005). Transformation of chickpea (Cicer arietinum) with the osmoregulatory gene P5CSF129A driven by 35S promoter encoding the mutagenized 1-pyrroline5-carboxylate synthetase (P5CS) for the overp roduction of proline showed significantly higher proline accumulation. However, the transgenic plants resulted only in a modest increase in transpiration efficiency suggesting that enhanced proline had little bearing on the components of yield in chickpea (Bhatnagar-Mathur *et al.*, 2009). Transgenic rice over expressing P5CS showed significantly higher tolerance to salinity and water stress produced in terms of faster growth of shoots and roots (Su and Wu, 2004). Rice plants over expressing the ZFP252 gene, resulted in increased amount of free proline and soluble sugars, elevated the expression of stress defence genes and enhanced tolerance to salt and drought stresses (Xu *et al.*, 2008). Soybean plants expressing 1-pyrroline-5-carboxylate reductase (P5CR) under control of an inducible heat shock promoter were found in greenhouse trials to accumulate proline without deleterious effects and to retain higher relative water content (RWC), and higher glucose and fructose levels than the antisense and control plants (de Ronde *et al.*, 2004). Field trials have been conducted in South Africa with apparent yield advantages for the proline accumulating soybean transgenic plants under reduced watering conditions and heat stress (ARC Research Highlights, 2006). Also, a number of "sugar alcohols" like mannitol, trehalose, myo-inositol and sorbitol have been targeted for the engineering of compatible solute overproduction, thus protecting the membrane and protein complexes during stress (Abebe *et al.*, 2003; Zhao *et al.*, 2000; Garg *et al.*, 2002; Cortina and Culianez 2005; Gao *et al.*, 2000). Similarly, transgenics developed for the over expression of polyamines have also been developed (Kumria and Rajam, 2002; Waie and Rajam, 2003; Capell *et al.*, 2004). As with other compatible solutes discussed above, the concentration of mannitol in the transgenic plants that showed better response to water and salinity stress at the whole-plant level was too small to be osmotically relevant. Mannitol is accumulated as a compatible solute in many plants and organisms of other kingdoms, although its accumulation in celery is often cited, perhaps because in celery up to half of fixed CO_2 is converted to mannitol (Stoop *et al.*, 1996). The over expression of mannitol-1-phosphate dehydrogenase (the Escherichia coli locus mtlD) resulted in the accumulation of a small amount of mannitol and also in the improved tolerance to salinity and drought in Arabidopsis (Thomas *et al.*, 1995). In wheat, where mannitol is normally not synthesized, constitutive expression of the mtlD (under the control of the ZmUbi-1 promoter) improved growth and tolerance to water stress and salinity, although growth in the absence of stress was accompanied with sterility, stunted growth and leaf curling at levels of mannitol higher than 0.7 mmol/gFW (Abebe *et al.*, 2003).

Similarly, glycine betain (GB), a fully N-methyl-substituted derivative of glycine, accumulates in the chloroplasts and plastids of many species such as Poaceae, Amaranthaceae, Asteraceae, Malvaceae and Chenopodiaceae, in response to drought and salinity. In some species, GB accumulates to concentrations that would contribute to cellular osmotic pressure (Munns and Tester, 2008), but in most cases, plants accumulate less than this amount. At lower concentrations, GB stabilizes the quaternary structures of enzymes and complex proteins and protects the photosynthetic machinery via ROS scavenging (Chen and Murata, 2008). Transgenic maize expressing the betA locus of *E. coli*, encoding *choline dehydrogenase*, showed more GB accumulation under drought and salinity in the field (Quan *et al.*, 2004). Under drought stress, imposed at the reproductive stage, transgenic maize lines that showed the highest amounts of GB accumulation (between 5.4 and 5.7 mmol/gFW) also had a 10–23% higher yield than wild-type plants under the same treatment (Quan *et al.*, 2004). Transgenic potato (*Solanum tuberosum* L.) plants, developed via the introduction of the bacterial choline oxidase (codA) gene, expressed under the control of an oxidative stress-inducible SWPA2 promoter and directed to the chloroplast with the addition of a transit peptide at the N-terminus, showed enhanced tolerance to NaCl and drought stress at the whole-plant level (Ahmad *et al.*, 2008). While not yet tested under field conditions, greenhouse testing with transgenic potato plants having relatively low levels of GB (0.9–1.4 mmol/gFW) showed greater dry weight accumulation after recovery from 150 mM NaCl treatment and water withholding stress treatments. Recently, wheat plants over expressing a BADH gene, encoding betaine aldehyde dehydrogenase (BADH), were shown to be more tolerant to drought and heat, by improving the photosynthesis capacity of flag leaves (Wang *et al.*, 2010). There is thought to be a signalling role for trehalose at least in part through its inhibition of SNF-1-related kinase (SnRK1), which results in an up-regulation of biosynthetic reactions supporting photosynthesis and starch synthesis, among others (reviewed by Iturriaga et al., 2009). Transgenic tomatoes (Solanum lycopersicum) over expressing the yeast trehalose-6-phosphate synthase (TPS1) gene (under control of 35S promoter) showed higher tolerance to salt, drought and oxidative stresses (Cortina and Culianez-Macia 2005). The alteration of soluble carbohydrate content suggests that the stress tolerance phenotype in trehalose genetically engineered plants could be partly due to modulation of sugar sensing and carbohydrate metabolism (Fernandez et al., 2010). Incredibly, the transgenic rice accumulated up to 1000 mg/g FW trehalose, which was attributed to the increased efficiency of the fusion protein over two separate enzymes (Jang et al., 2003). Even more surprising was the absence of abnormal developmental and morphological phenotypes, given the high level of trehalose and the occurrence of such deleterious phenotypes in Arabidopsis, potato and tobacco (Goddijn and van Dun, 1999). The results of transgenic modifications of biosynthetic and metabolic pathways in most of the mentioned cases indicate higher stress tolerance and the accumulation of compatible solutes may also protect plants against damage by scavenging of reactive oxygen species (ROS), and by their chaperone like activities in maintaining protein structures and functions (Hare et al., 1998; Diamant et al., 2001). Also, there are also some reports showing a negative effect of osmotic stress on yield potential (Fukai and Cooper, 1995). Osmotin is a stress-responsive multifunctional 24-kDa protein with roles in plant response to fungal pathogens and osmotic tolerance. Over expression of a heterologous osmotin-like protein (under control of 35S) in potato (S. tuberosum) improved tolerance to salinity stress (Evers et al., 1999). The tobacco osmotin gene (driven by the 35S promoter) was transformed into tomato and was reported to enhance tolerance to salt and drought stresses (Goel et al., 2010).

Genetic manipulations of compatible solutes do not always lead to a significant accumulation of the compound (Chen and Murata, 2002), thereby, suggesting that the function of compatible solutes is not restricted to osmotic adjustment and that osmoprotection may not always confer drought tolerance. A recent review of Serraj and Sinclair (2002) shows that virtually none of the studies that tested the effect of osmotic adjustment on yield under water stress showed any benefit at all, since some benefit of osmotic adjustment might be in the ability of plants to maintain root growth under severe stress (Voetberg and Sharp, 1991). Another recent study with chickpea has also shown that osmotic adjustment provided no beneficial effect on yield under drought stress (Turner et al., 2007). The various genes encoding enzymes responsible for synthesizing osmotic and other protectants are given in Table 1.

Table 1: Genes encoding enzymes that synthesize osmotic and other protectants

Gene	Gene Action	Species	Phenotype	Reference
ADC	Arginine decarboxylase	Rice	Reduced chlorophyll loss under drought stress	Capell et al., 1998
ADC	Polyamine synthesis	Rice	Drought resistance	Capell et al., 2004
ADC	Polyamine synthesis	Rice	Drought resistance	Peremarti et al., 2009
ADC	Polyamine accumulation	Rice	Salt resistance in biomass accumulation	Roy and Wu, 2001
ADC2	Putrescine accumulation with no changes in spermidine and spermine content	Arabidopsis	Drought resistance and transpiration reduction	Alcazar et al., 2010
AtHAL3	Phosphoprotein phosphatase	Tobacco	Improved salt, osmotic and Lithium tolerance of cell cultures	Yonamine et al., 2004
AtTPS1	trehalose-6-phosphate synthase	Tobacco	Drought resistance; sustained photosyntehsis	Almeida et al., 2007
BADH-1	Glycine-Betaine production	Carrot	Salinity tolerance	Kumar et al., 2004
BADH-1	Glycine-Betaine production	Maize	Salinity tolerance	Wu et al., 2007
BADH-1	Glycine-Betaine production	Trifoliate orange	Salinity tolerance	Fu et al., 2011
BADH-1	Glycine-Betaine production	Tobacco	Salinity tolerance	Yang et al., 2008
BADH-1	Glycine-Betaine production	Wheat	Sustained photosynthesis under drought and heat stress	Wang et al., 2010
betA	Choline dehydrogenase (glycinebetaine synthesis)	Cotton	Osmotic adjustment and salt tolerance	Zhang et al., 2011
betA	Choline dehydrogenase (glycinebetaine synthesis)	Maize	Drought resistance at seedling stage and high yield after drought	Quang et al., 2004
betA	Choline dehydrogenase (glycinebetaine synthesis)	Tobacco	Increased tolerance to salinity stress	Lilius et al., 1996

[Table Contd.

Contd. Table]

Gene	Gene Action	Species	Phenotype	Reference
betA	Choline dehydrogenase (glycinebetaine synthesis)	Wheat	Seedling drought resistance	He et al., 20111
CHIT33, CHIT42	Endochitinase synthesis	Tobacco	Salt and metal toxicity resistance (& disease)	Dana et al., 2006
CMO	Choline monooxygenase (glycine betaine synthesis)	Tobacco	Better in vitro growth under salinity and osmotic (PEG6000) stress	Yi-Guo Shen et al., 2002
codA	Choline oxidase (glycine betaine synthesis)	Arabidopsis	Salt tolerance in terms of reproduction	Sulpice et al., 2003
codA	Choline oxidase (glycine betaine synthesis)	Rice	Increased tolerance to salinity and cold	Sakamoto et al., 1998
codA	Choline oxidase (glycine betaine synthesis)	Rice	Recovery from a week long salt stress	Mohanty et al., 2003
codA	Choline oxidase (glycine betaine synthesis)	Rice	Drought resistance, antioxidative action	Kathuria et al., 2009
codA	Choline oxidase (glycine betaine synthesis)	Tomato	Salt and drought resistance	Goel et al., 2011
DSM	Control of the xanthophyll cycle	Rice	Drought resistance	Du et al., 2010
COX	Choline oxidase (glycine betaine synthesis)	Rice	Salt and 'stress' tolerance	Su et al., 2006
Ect A... ect C	Edtoin accumulation in chloroplasts	Tobacco	Salt and cold tolerance	Rai et al., 2006
ggpPS	Glucosylglycerol accumulation	Arabidopsis	Salt tolerance	Klähn et al., 2009
GS2	Chloroplastic glutamine synthetase	Rice	Increased salinity resistance and chilling tolerance	Hoshida et al., 2000

[Table Contd.

Contd. Table]

Gene	Gene Action	Species	Phenotype	Reference
IMT1	Myo-inositol o-methyltransferase (D-ononitol synthesis)	Tobacco	Better CO_2 fixation under salinity stress. Better recovery after drought stress.	Sheveleva et al., 1997
M6PR	Mannose-6-phosphate reductase	Arabidopsis	Mannitol accumulation and salt tolerance due to chloroplast protection	Sickler et al., 2007
mt1D	Mannitol-1-phosphate ehydrogenase (mannitol synthesis)	Arabidopsis	Increased germination under salinity stress	Thomas et al., 1995
mt1D	Mannitol-1-phosphate dehydrogenase (mannitol synthesis)	Tobacco	Increased plant height and fresh weight under salinity stress	Tarczynski et al., 1993
mt1D	Mannitol-1-phosphate dehydrogenase (mannitol synthesis)	Wheat	Drought and salinity tolerance of calli and plants	Abebe et al., 2003
mt1D & GutD	Mannitol-1-phosphate dehydrogenase & glucitol-6-phosphate dehydrogenase	loblolly pine	High salt tolerance due to mannitol and glucitol accumulation	Tang et al., 2005
mt1D	Mannitol-1-phosphate dehydrogenase (mannitol synthesis)	Populus tomentosa	Salinity tolerance	Hu et al., 2005
mt1D	Mannitol-1-phosphate dehydrogenase (mannitol synthesis)	Sorghum	Salt and drought tolerance	Maheswari et al., 2010
Osm1 ... Osm4	Osmotin protein accumulation	Tobacco	Drought and salt tolerance in plant water status and proline accumulation	Barthakur et al., 2001
	Osmotin protein accumulation	Mulberry	Drought and salt tolerance	Das et al., 2011

[Table Contd.

Contd. Table]

Gene	Gene Action	Species	Phenotype	Reference
Osmyb4	Cold induced transcription factor	Tomato	Drought tolerance	Vannini et al., 2007
otsA	Trehalose-6-phosphate synthase (trehalose synthesis)	Tobacco	Increased leaf dry weight and photosynthetic activity under drought. Increased carbohydrate accumulation	Pilon-smits et al., 1998
P5CS	Pyrroline carboxylate synthase (proline synthesis)	Bean	Drought, salt and cold resistance	Chen et al.,2009
P5CS	Pyrroline carboxylate synthase (proline synthesis)	Chickpea	Salinity tolerance	Ghanti et al., 2011
P5CS	Pyrroline carboxylate synthase (proline synthesis)	Citrus	Osmotic adjustment and drought tolerance	Molinari et al., 2004
P5CSF 129A	Pyrroline carboxylate synthase (proline synthesis) (tomato)	Citrus (citrumelo)	Drought tolerance via turgor maintenance	De Cavalho et al., 2011
P5CS	Pyrroline carboxylate synthase (proline synthesis)	Petunia	Drought tolerance and high proline	Yamada et al., 2005
P5CS	Pyrroline carboxylate synthase (proline synthesis)	Potato	Salinity tolerance	Hmida-Sayari et al., 2005
P5CS	Pyrroline carboxylate synthase (proline synthesis)	Rice	Increased biomass production under drought and salinity stress	Zhu et al., 1998
P5CS	Pyrroline carboxylate synthase (proline synthesis)	Rice	tolerance to water and sainity stress	Su and Wu, 2004
P5CS	Pyrroline carboxylate synthase (proline synthesis)	Soybean	Drought tolerance, high RWC, high proline	De Ronde et al., 2004
P5CS	Pyrroline carboxylate synthase (proline synthesis) (tomato)	Sugarcane	Drought tolerance via antioxidant role of proline	Molinari et al., 2007

[Table Contd.

Contd. Table]

Gene	Gene Action	Species	Phenotype	Reference
P5CS	Pyrroline carboxylate synthase (proline synthesis)	Tobacco	Increased biomass production and enhance flower development under salinity stress	Kishor et al., 1995
P5CS	Pyrroline carboxylate synthase (proline synthesis)	Tobacco	Various hormonal repercussion under heat and drought	Dobra et al., 2010
P5CS	Pyrroline carboxylate synthase (proline synthesis)	Wheat	Drought resistance due to antioxidative action	Vendruscolo et al., 2007
PPO 2004	Protoporphyrinogen oxidase	Tomato	Drought resistance	Thipyapong et al.,
PPO	Protoporphyrinogen oxidase	Rice	Drought resistance	Phung et al., 2011
SAMDC	S-adenosylmethioninedecar-boxylase (polyamine synthesis)	Rice	Better seedling growth under a 2 day NaCl stress	Roy and Wu, 2002
SAMDC	S-adenosylmethioninede carboxylase (polyamine synthesis)	Tobacco	Drought, salinity, Verticillium and Fusarium wilts resistance	Waie and Rajam, 2003
SAMDC	S-adenosylmethioninede carboxylase (polyamine synthesis)	Rice	Improved recovery from drought	Peremarti et al., 2009
SPE	Spermidine synthase	Arabidopsis	Chilling, freezing, salinity, drought hyperosmosis	Kasukabe et al., 2004
AtSPMS	Spermine synthase	Arabidopsis	salt tolerance	Alet et al, 2012
TPP1	Trehalose synthesis	Rice	Salt and cold tolerance	Ge et al., 2008
TPS; TPP	Trehalose synthesis	Arabidopsis	Drought, freezing, salt and heat tolerance	Miranda et al., 2007
OsTPS1	Trehalose synthesis	Rice	Drought, salt and cold resistance	Li et al., 2011
TPS1	Trehalose synthesis	Alfalfa	Drought, freezing, salt, and heat tolerance.	Suorez et al., 2009

[Table Contd.

Contd. Table]

Gene	Gene Action	Species	Phenotype	Reference
TPS1	Trehalose synthesis	Tomato	Drought, salt and oxidative stress tolerance	Cortina & Culiáñez-Macià, 2005
TPS1	Trehalose synthesis	Potato	Delayed wilting under drought	Stiller et al.,2008
TPS1 & TPS2	Trehalose synthesis	Tobacco	Maintenance of water status under drought stress	Karim et al., 2007
TPSP	Trehalose synthesis	Rice	Drought, salt and cold tolerance expressed by chlorophyll fluorescence	In-Cheol Jang et al., 2003
OsTPS1	Trehalose-6- phosphate synthase (Osmoticadjustment)	Rice	Osmotic adjustment	Li H-W, 2011

Source: (www.plantstress.com)

7.2.2. *Detoxifying Genes*

The production of enzymes catalyzing detoxification of reactive oxygen species is one of the mechanisms controlling shoot tissue tolerance under abiotic stress. To prevent stress injury, cellular ROS need to remain at non toxic levels under drought stress. There are different antioxidants involved in plant strategies to degrade ROS include: 1) enzymes such as: catalase, superoxide dismutase (SOD), ascorbate peroxidase (APX), and glutathione reductase; and 2) nonenzymes such as ascorbate, glutathi-one, carotenoids, and anthocyanins (Wang *et al.*, 2003b). Some proteins, osmolytes, and amphiphilic molecules also have antioxidative functionality (Bowler *et al.*, 1992; Noctor and Foyer 1998). A number of transgenic improvements for abiotic stress tolerance have been achieved through detoxification strategy. These include transgenic plants over expressing enzymes involved in oxidative protection (Zhu *et al.*, 1999; Roxas *et al.*, 1997). As an antioxidant enzyme, glutathione peroxidase (GPX) reduces hydroperoxides in the presence of glutathione to protect cells from oxidative damage, including lipid peroxidation (Maiorino *et al.*, 1995). Gaber *et al.* (2006) generated transgenic Arabidopsis plants over expressing GPX-2 genes in cytosol (AcGPX2) and chloroplasts (ApGPX2). SOD is the first enzyme in the enzymatic antioxidative pathway and halophytic plants, such as mangroves, reported to have a high level of SOD activity. SOD plays a major role in defending mangrove species against severe abiotic stresses. Prashanth *et al.* (2008) further characterized the Sod1 cDNA by transforming it into rice. In plant cells, APXs are directly involved in catalyzing the reduction of H_2O_2 to water, which is facilitated by specific electron donation by ascorbic acid. APXs are ubiquitous in plant cells and are localized in chloroplasts (Takahiro *et al.*, 1995), peroxisomes (Shi *et al.*, 2001), and cytosol (Caldwell *et al.*, 1998). Xu *et al.*, (2008b) transformed Arabidopsis plants with a pAPX gene from barley (HvAPX1). The transgenic line was found to be more tolerant to salt stress than the WT.

Apart from catalase and various peroxidases and peroxiredoxins (Dietz 2003), four enzymes, APX, dehydroascorbate reductase, monodehydro ascorbate reductase and glutathione reductase (GR), are involved in the ascorbate-glutathione cycle, a pathway that allows the scavenging of superoxide radicals and H_2O_2 (Asada 1999). Lee and Jo (2004) introduced BcGR1, a Chinese cabbage gene that encodes cytosolic GR into tobacco plants via *Agrobacterium*-mediated transformation. Homozygous lines containing BcGR1 gene were generated and tested for their acquisition of increased tolerance to oxidative stress. When ten-day old transgenic tobacco seedlings were treated with 5 to 20 mM MV, they showed significantly increased tolerance compared to WT seedlings. In addition, when leaf discs were subjected to MV, the transgenic plants were less damaged than the WT with regard to their electrical conductivity and chlorophyll content.

7.2.3. *Late Embryogenesis Abundant (LEA)*
Proteins Coding Genes

A LEA protein symbolizes another category of proteins with high molecular weight possessing crucial roles in cellular dehydration tolerance preventing protein aggregation during desiccation. These are having antioxidant capacity together with possible role as chaperons (Goyal and Walton, 2005). These proteins accumulate during seed desiccation and in response to water stress (Galau *et al.*, 1987). Amongst the several groups of LEA proteins, those belonging to group 3 are predicted to play a role in sequestering ions that are concentrated during cellular dehydration. These proteins have 11-mer amino acid motifs with the consensus sequence TAQAAKEKAGE repeated as many as 13 times (Dure 1993). The group 1 LEA proteins are predicted to have enhanced water binding capacity, while the group 5 LEA proteins are thought to sequester ions during water loss. Dehydrins are a subfamily of group 2 LEA proteins that accumulate in vegetative tissues subjected to drought, salinity and cold stress (Battaglia *et al.*, 2008). Constitutive over expression of the HVA1, a group 3 LEA protein from barley conferred tolerance to soil water deficit and salt stress in transgenic rice plants (Xu *et al.*, 1996). Similarly, Park *et al.* (2005b) introduced a *B. napus* LEA protein gene, ME-leaN4 into lettuce (*Lactuca sativa* L.) using *Agrobacterium*-mediated transformation (Wakui and Takahata 2002). Transgenic lettuce demonstrated enhanced growth ability compared with WT plants under salt- and water-deficit stress. Brini *et al.* (2007a) analyzed the ectopic expression of dehydrin cDNA in Arabidopsis under salt and osmotic stress. When compared to WT plants, the Dhn-5-expressing transgenic plants exhibited stronger growth under high concentrations of NaCl or water deprivation and showed a faster recovery from mannitol treatment. Leaf area and seed germination rate decreased much more in WT than in transgenic plants subjected to salt stress. Moreover, the water potential was more negative in transgenic than in WT plants and the transgenic lines had higher proline contents and lower water loss rates under water stress. Na^+ and K^+ also accumulated to a greater extent in the leaves of the transgenic plants. The effects of over expression of the HVA1 gene in mulberry under a constitutive promoter was studied by Lal *et al.*, (2008). Over expression of OsLEA3-1 in rice, resulted in improved yields under drought stress without yield penalties under control conditions (Xiao *et al.,* 2007). Multilocation trials with the transgenic maize expressing CspA and CspB showed improved yields (11–21%) under water-stress conditions. Importantly, the improvements in water-limited field trials were not associated with a yield penalty in high-yielding environments (Castiglioni *et al.,* 2008). Whereas LEA, CSPs, and HSP proteins have been repeatedly shown to be involved in abiotic stress response (reviewed by, only limited experiments have used this strategy for engineering abiotic stress tolerant crops (Hand *et al.,* 2011). Similarly, Table 2 depicts various late embryogenesis abundant (LEA) related genes

Table 2: Late embryogenesis abundant (LEA) related genes

Gene	Gene Action	Species	Phenotype	Reference
DQ663481	Lea gene	Tobacco	Drought resistance via cell membrane stability	Wang et al., 2006
AtLEA4	Group 4 LEA proteins	Arabidopsis	Drought resistance	Olvera-Carrillo et al., 2010
HVA1	Group 3 LEA protein gene	Oat	Delayed wilting under drought stress	Maqbool et al., 2002
HVA1	Group 3 LEA protein gene	Oat	Salinity tolerance in yield/plant	Oraby et al., 2005
HVA1	Group 3 LEA protein gene	Rice	Dehydration avoidance and cell membrane stability	Babu et al., 2004
HVA1	Group 3 LEA protein gene	Rice	Drought and salinity tolerance	Rohila et al., 2002
HVA1	Group 3 LEA protein gene	Wheat	Increased biomass and WUE under stress	Sivamani et al., 2000
HVA1	Group 3 LEA protein gene	Wheat	Improved plant water status and yield under field drought conditions	Bahieldin et al., 2005
HVA1	Group 3 LEA protein gene	Wheat	Drought resistance in germination, seedling growth and biomass accumulation	Chauhan and Khurana, 2011
HVA1	Group 3 LEA protein gene	Mulberry	Salinity and drought resistance	Lal et al., 2008
BhLEA1, LEA2	LEA proteins	Tobacco	Dehydration tolerance	Liu et al., 2008
OsLEA3-1	Lea protein	Rice	Drought resistance for yield in the field	Xiao et al., 2007
ME-leaN4	LEA protein	Lettuce	Enhanced growth and delayed wilting under drought. Salt resistance	Park et al., 2005
ME-leaN4	LEA protein	Chinese cabbage	Drought and salt resistance	Park et al., 2005
Rab17	LEA protein	Arabidopsis	Resistance to osmotic and salinity stress	Figueras et al., 2004
SpERD15	Solute accumulation and more	Tobacco	Drought resistance	Ziat et al., 2011

Source: (www.plantstress.com)

which have been transferred in different crops for developing tolerance against drought and salt stress.

7.2.4. *Genes for Ionic Balance*

In most saline soils, Na^+ and Cl^- are the predominant ions in the soil solution. At sufficiently high concentrations, both ions contribute to an unfavorable osmotic gradient between the soil solution and the plant roots. Both ions also cause ion-specific toxicity when accumulated in salt sensitive plants. And while it is clear that the exclusion of Na^+ or Cl^-, or both, is correlated with improved salinity tolerance in some species (and the accumulation of both with others), the state knowledge of Na^+ transport mechanisms is more advanced than that for Cl^- transport (Teakle and Tyerman, 2010). Decreasing Na^+ uptake in both glycophytes and halophytes, the net uptake of sodium into the roots is the sum of sodium influx and efflux. The negative electrical membrane potential difference at the plasma membrane of root cells (40 mV) favors the passive transport of sodium into root cells, and especially so when sodium concentrations increase in the soil solution.

The entry of sodium into root cells is mediated by uniporter or ion channel-type transporters, like HKT, LCT1 and NSCC (reviewed in Plett and Moller, 2010). The reduction of Na^+ uptake might be accomplished by decreasing the number or activity of these transporters in the roots. Reduction of TaHKT2;1 expression in wheat by antisense suppression resulted in lower net sodium uptake of transgenic roots and higher fresh weight of plants grown under salinity stress in controlled growth conditions (Laurie *et al.*, 2002). Similarly, Arabidopsis T-DNA knockout mutants of AtCNGC3, a cyclic nucleotide gated channel which catalyses Na^+ uptake, had lower net influx of Na^+ and were more tolerant to salinity at germination (Gobert *et al.*, 2006). The efflux of sodium from the roots is an active process, which is presumed to be mediated by plasma membrane Na^+/H^+ antiporters. These secondary transporters use the energy of the proton gradient across the plasma membrane to drive the active efflux of sodium from the cytosol to the apoplast. The Na^+/H^+ antiporter, SOS1 (identified in a mutant screen as salt overly sensitive 1), is the only $Na^+/$ efflux protein at the plasma membrane of plants characterized so far. The over expression of AtSOS1, a plasma membrane bound Na^+/H^+ antiporter, improved the ability of the Arabidopsis transgenic plants to grow in the presence of high NaCl concentrations (Shi *et al.*, 2003). The rice orthologue, OsSOS1, is able to complement the Arabidopsis sos1 mutant (Martinez-Atienza *et al.*, 2007). The SOD2 (Sodium2) gene was identified in yeast, *Schizosaccharomyces pombe*, as a Na^+/H^+ antiporter on the plasma membrane involved in salt tolerance. Transformation of rice with the SOD2 gene (under 35S promoter) resulted in accumulation of more K^+, Ca^{2+}, Mg^{2+} and less Na^+ in the shoots compared with wild type (Zhao *et al.*, 2006b). The transgenic rice plants were able to maintain higher photosynthesis level and root proton exportation capacity, whereas reduced ROS generation. Although yield data not reported, the trials were conducted outdoors, which is the closest to field level study of a crop plant for this approach in the literature.

7.2.5. *Decreasing Root to Shoot Translocation of Na$^+$*

The accumulation of sodium in shoots occurs via the translocation of sodium from the roots along the transpirational stream. The removal of sodium from the xylem, which reduces the rate of sodium transfer to the shoot tissue, has been shown to be mediated by members of the HKT gene family (reviewed in Plett and Moller, 2010). AtHKT1;1 in Arabidopsis, OsHKT1;5 in rice, and HKT1;4 in wheat are all critical in reducing Na$^+$ shoot concentrations by transporting Naþ from the xylem into the root stele (reviewed in Hauser and Horie, 2010). One strategy for improving salinity tolerance is to increase the expression of such genes to further reduce sodium concentrations in the xylem (Plett *et al.*, 2010). The over expression of AtHKT1;1 under the control of the constitutive promoter CaMV35S leads to increased salt sensitivity, presumably because Naþ fluxes are increased in inappropriate cells and tissues (Moller *et al.*, 2009). However, when expressed under the control of a promoter directing expression in root epidermal and cortical cells, both in rice and in Arabidopsis, HKT1;1 over expression causes an increase in root cortical sodium, a decrease in shoot sodium and a higher accumulation of fresh weight during the course of the experiment (Plett *et al.*, 2010).

Sequestering Na$^+$ The accumulation of Na$^+$ ions into vacuoles through the operation of a vacuolar Na$^+$/H$^+$ antiporter provided an efficient strategy to avert the deleterious effect of Na$^+$in the cytosol and maintain osmotic balance by using Na$^+$ (and Cl$^-$) accumulated in the vacuole to drive water into the cells (Apse *et al.*, 1999; Apse and Blumwald, 2002). Transgenic plants over expressing an Arabidopsis vacuolar Na$^+$/H$^+$ antiporter, AtNHX1, exhibited improved salt tolerance in *Brassica napus* (Zhang *et al.*, 2001), tomato (Zhang and Blumwald, 2001), cotton (He *et al.*, 2005), wheat (Xue *et al.*, 2004), beet (Yang *et al.*, 2005) and tall fescue (Zhao *et al.*, 2007). The transformation of an orthologue gene (AgNHX1) from halophytic plant Atriplex gmelini into rice improved salt tolerance of the transgenic rice (Ohta *et al.*, 2002). Maize plants over expressing rice OsNHX1 gene accumulated more biomass, under 200 mM NaCl in greenhouse (Chen *et al.*, 2007). Moreover, under field trail conditions, the transgenic maize plants produced higher grain yields than the wild-type plants. Transformation of another Na$^+$/H$^+$antiporter family member, AtNHX3 (from Arabidopsis), in sugar beet (*Beta vulgaris* L.) resulted in increased salt accumulation in leaves, but not in the storage roots, with enhanced constituent soluble sugar contents under salt stress condition (Liu *et al.*, 2008). The introduction of genes associated with the maintenance of ion homeostasis in halotolerant plant into crop plants confirmed salinity tolerance. The yeast gene HAL1 was introduced into tomato (Gisbert *et al.*, 2000), watermelon (*Citrullus lanatus* (Thunb.); Ellul *et al.*, 2003) and melon (*Cucumis melo* L.; Bordas *et al.*, 1997), which confirmed higher level of salt tolerance, with higher cellular K^+ to Na^+ ratio under salt stress. Likewise, the introduction of the yeast HAL2 gene into tomato resulted in improved root growth under NaCl conditions, contributing to improved salt tolerance (Arrillaga *et al.*, 1998). Over expression of HAL3 (*S. cerevisiae*) homologue NtHAL3 in tobacco increased proline biosynthesis and the enhancement of salt and osmotic tolerance in cultured tobacco cells (Yonamine *et al.*, 2004). The electrochemical gradient of protons across the vacuolar membrane is generated by the activity of the vacuolar H^+ translocating enzymes, H^+ATPase and H^+-pyrophosphatase. Increasing vacuolar H^+ pumping might be required to provide the additional driving force for vacuolar accumulation via sodium/proton antiporters. A gene coding for a vacuolar H^+-pyrophosphatase proton pump (AVP1) from Arabidopsis was over expressed in tomato (Park *et al.*, 2005), cotton (Pasapula *et al.*, 2011) and rice (Zhao *et al.*, 2006a) and induced improved growth during drought and salt stress. Interestingly, the over expressed AVP1 resulted in a more robust root system which could possibly improve the plants ability to absorb more water from the soil (Pasapula *et al.*, 2011).

8. Targetting Pathways: Manipulating Regulatory Genes

8.1. Role of Transcription Factors in the Activation of Stress Responsive Genes

A striking strategy for manipulation and gene regulation is the small group of transcription factors that have been identified to bind to promoter regulatory elements in genes that are regulated by abiotic stresses (Shinozaki and Yamaguchi-Shinozaki 1997; Winicov and Bastola 1997). The transcription factors activate cascades of genes that act together in enhancing tolerance towards multiple stresses.

Dozens of transcription factors are involved in the plant response to drought stress (Vincour and Altman 2005; Bartels and Sunkar 2005). Most of these falls into several large transcription factor families, such as AP2/ERF, bZIP, NAC, MYB, MYC, Cys2His2 zinc-finger and WRKY. Individual members of the same family often respond differently to various stress stimuli. On the other hand, some stress responsive genes may share the same transcription factors, as indicated by the significant overlap of the gene expression profiles that are induced in response to different stresses (Seki *et al.*, 2001; Chen and Murata 2002). Transcriptional activation of stress-induced genes has been possible in transgenic plants over expressing one or more transcription factors that recognize promoter regulatory elements of these genes. Two families, bZIP and MYB, are involved in ABA signaling and its gene activation. Many ABA inducible genes share the (C/T) ACGTGGC consensus, cis-acting ABA-responsive element (ABRE) in their promoter regions (Guiltinan *et al.*, 1990; Mundy *et al.*, 1990). Introduction of transcription factors in the ABA signaling pathway can also be a mechanism of genetic improvement of plant stress tolerance. Constitutive expression of ABF3 or ABF4 demonstrated enhanced drought tolerance in Arabidopsis, with altered expression of ABA/stress-responsive genes, e.g. rd29B, rab18, ABI1 and ABI2 (Kagaya *et al.*, 2002). Several ABA-associated phenotypes, such as ABA hypersensitivity and sugar hypersensitivity, were observed in such plants. Moreover, salt hypersensitivity was observed in ABF3- and ABF4-over expressing plants at the germination and young seedling stages indicating the possible participation of ABF3 and ABF4 in response tosalinityat these particular developmental stages. Improved osmotic stress tolerance in35S:At-MYC2/AtMYB2 transgenic plants as judged by an electrolyte-leakage test was reported by (Abebe *et al.*, 2003). Transgenic Arabidopsis plants constitutively over-expressing a cold inducible transcription factor (CBF1; CRT/DRE binding protein) showed tolerance to freezing without any negative effect on the development and growth characteristics (Jaglo-Ottosen *et al.*, 1998). Over expression of Arabidopsis CBF1 (CRT/DRE binding protein) has been shown to activate or homologous genes at non-acclimating temperatures (Jaglo *et al.,* 2001). The CBF1 cDNA when introduced into tomato (*Lycopersicon esculentum*) under the control of a CaMV35S promoter improved tolerance to chilling, drought and salt stress but exhibited dwarf phenotype and reduction in fruit set and seed number (Hsieh *et al.*, 2002).

Another transcriptional regulator, Alfin1, when over expressed in transgenic alfalfa (*Medicago sativa* L.) plants regulated endogenous MsPRP2 (NaCl-inducible gene) mRNA levels, resulting in salinity tolerance, comparable, to a few available salt tolerant plants (Winicov and Bastola 1999). Lee *et al.* (1995) produced thermo-tolerant Arabidopsis plants by de-repressing the activity of ATHSF1, a heat shock transcription factor leading to the constitutive expression of heat shock proteins at normal temperature. Several stress induced cor genes such as rd29A, cor15A, kin1 and cor6.6 are triggered in response to cold treatment, ABA and water deficit stress (Thomashow1998).

The promoters of stress responsive genes have typical *cis*-regulatory elements like DRE/CRT, ABRE, MYCRS/MYBRS and are regulated by various upstream transcriptional factors (Figure 4). These transcription factors fall in the category of early genes and are induced within minutes of stress. The transcriptional activation of some of the genes including RD29A has been well worked out. The promoter of this gene family contains both ABRE as well as DRE/ CRT elements (Stockinger and Gilmour, 1997). Transcription factors, which can bind to these elements were isolated and were found to belong to AP2/EREBP family and were designated as CBF1/ DREB1B, CBF2/DREB1C, and CBF3/DREB1A (Medina *et al.,* 1999). These transcription factors (CBF1, 2 and 3) are cold responsive and in turn bind CRT/DRE elements and activate the transcription of various stress responsive genes. A novel transcription factor responsive to cold as well as ABA was isolated from soybean and termed as SCOF-1 (soybean zinc finger protein). This transcription factor, however, was not responsive to drought or salinity stress. SCOF1 was a zinc finger nuclear localized protein but failed to bind directly to either CRT/DRE or ABRE elements. Yeast 2-hybrid study revealed that SCOF-1 interacted strongly with SGBF-1 (Soybean G-box bind-ing bZip transcription factor) and in vitro DNA binding activity of SGBF-1 to ABRE elements was greatly improved by the presence of SCOF-1. This study supported that protein–protein interaction is essential for the activation of ABRE-mediated cold responsive genes. Transcription factors like DREB2A and DREB2B gets activated in response to dehydration and confer tolerance by induction of genes involved in maintaining the osmotic equilibrium of the cell (Liu *et al.,* 1998). Several basic leucine zipper (bZip) transcription factors (namely ABF/AREB) have been isolated which can specifically bind to ABRE element and activate the expression of stress genes (Choi *et al.,* 2000). These AREB genes (AREB1 and AREB2) are ABA responsive and need ABA for their full activation. These transcription factors exhibited reduced activity in the ABA-deficient mutant aba2 as well as in ABA insensitive mutant aba1-1. Some of the stress responsive genes for example RD22 lack the typical CRT/DRE elements in their promoter indicating their regulation by other mechanisms. Transcription factor RD22BP1 (a MYC transcription factor) and AtMYB2 (a MYB transcription factor) could bind MYCRS (MYC recognition sequence) and MYBRS (MYB recognition sequence) elements, respectively, and could cooperatively activate the expression of RD22 gene (Abe *et al.,* 1997). As cold, salinity and drought stress ultimately impair the osmotic equilibrium of the cell it is likely that these transcription factors as well as the major stress genes may cross talk with each other for their maximal response and help in reinstating the normal physiology of the plant. ABA is an important phytohormone and plays a critical role in response to various stress signals. The application of ABA to plant mimics the effect of a stress condition. As many abiotic stresses ultimately results in desiccation of the cell and osmotic imbalance, there is an overlap in the expression pattern of stress genes after cold, drought, high salt or ABA application. This suggests that various stress signals and ABA share common elements in their signaling pathways and these common elements cross talk with each other, to maintain cellular homeostasis (Finkelstein *et al.*, 2002). Functions of ABA include: 1) ABA causes seed dormancy and delays its germination. 2) ABA promotes stomatal closure.

ABA levels are induced in response to various stress signals. ABA actually helps the seeds to surpass the stress conditions and germinate only when the conditions are conducive for seed germination and growth. ABA also pre-vents the precocious germination of premature embryos. Stomatal closure under drought conditions prevents the intracellular water loss and thus ABA is aptly called as a stress hormone. The main function of ABA seems to be the regulation of plant water balance and osmotic stress tolerance. Several ABA deWcient mutants namely *aba1*, *aba2* and *aba3* have been reported for *Arabidopsis* (Koornneef *et al.*, 1998).

There have been numerous efforts in enhancing tolerance towards multiple stresses such as cold, drought and salt stress in crops other than the model plants like Arabidopsis, tobacco and alfalfa. An increased tolerance to freezing and drought in Arabidopsis was achieved by over expressing CBF4, a close CBF/ DREB1 homolog whose expression is rapidly induced during drought stress and by ABA treatment, but not by cold (Haake *et al.*, 2002). Similarly, a cis-acting element, dehydration responsive element (DRE) identified in *A. thaliana*, is also involved in ABA-independent gene expression under drought, low temperature and high salt stress conditions in many dehydration responsive genes like rd29A that are responsible for dehydration and cold-induced gene expression (Yamaguchi-Shinozaki and Shinozaki 1993; Iwasaki *et al.*, 1997; Nordin *et al.*, 1991). Several cDNAs encoding the DRE binding proteins, DREB1A and DREB2A have been isolated from A. thaliana and shown to specifically bind and activate the transcription of genes containing DRE sequences(Liu *et al.*, 1998). DREB1/CBFs are thought to function in cold-responsive gene expression, whereas DREB2s are involved in drought-responsive gene expression. The transcriptional activation of stress-induced genes has been possible in transgenic plants over-expressing one or more transcription factors that recognize regulatory elements of these genes. In Arabidopsis, the transcription factor DREB1A specifically interacts with the DRE and induces expression of stress tolerance genes (Shinozaki and Yamaguchi-Shinozaki 1997). DREB1A cDNA under the control of CaMV 35S promoter in transgenic plants elicits strong constitutive expression of the stress inducible genes and brings about increased tolerance to freezing, salt and drought stresses (Liu *et al.*, 1998). Strong tolerance to freezing stress was observed in transgenic Arabidopsis plants that over express CBF1 (DREB1B) cDNA under the control of the CaMV 35S promoter (Jaglo-Ottosen *et al.*, 1998). Subsequently, the over expression of DREB1A has been shown to improve the drought and low-temperature stress tolerance in tobacco, wheat and groundnut (Kasuga *et al.*,

Table 3: Various regulatory genes

Gene	Gene Action	Species	Phenotype	Reference
ABF3	Transcription factor	Rice	Drought resistance	Oh et al., 2005
ABP9	ABRE binding protein9	Arabidopsis	Regulation of plant photosynthesis under stress	Zhang et al., 2008
ABP9	ABRE binding protein9	Arabidopsis	drought and salt stress tolerance	Zhang et al., 2011
AlSAP	Transcription factor	Tobacco	Drought, salinity and freezing tolerance	Ben Saad et al., 2010
ALDH3I1 & ALDH7B4	Aldehyde dehydrogenase	Arabidopsis	Salt and dehydration stress tolerance	Kotchoni et al., 2006
ZmALDH22 A1	Aldehyde dehydrogenase	Tobacco	Salt and dehydration tolerance	Huang et al., 2008
Alx8	High APX2 and ABA	Arabidopsis	Drought tolerance	Rossel et al., 2006
AnnAt1	Annexin synthesis	Arabidopsis	Drought tolerance	Konopka-Postupolska et al., 2009
AnnBj1	Annexin synthesis	Cotton	Salt and osmotic resistance	Divya et al., 2010
AP37	Transcription factor	Rice	Drought tolerance in yield	Oh et al., 2009
SlAREB1	ABA-responsive element binding protein	Tomato	Salinity and drought tolerance	Orellana et al., 2010
ASR1	Undetermined	Tobacco	Decreased water loss; salt tolerance	Kalifa et al., 2004
HvCBF4	Induced expression of COR genes	Rice	Drought resistance	Lourenço et al., 2011
AtCML9	Transcription factor	Arabidopsis	Drought and salt tolerance	Magnan et al., 2008
AtCPK6	Calcium-dependent protein kinase	Arabidopsis	Salt and drought tolerance	Xu et al., 2010
OsCPK21	Calcium-dependent protein kinase	Rice	Salt tolerance	Asano et al., 2011
AtGSK1	Homologue of GSK3/ shaggy like protein kinase	Arabidopsis	Salt tolerance in plant and root growth	Piao et al., 2001

[Table Contd.

Contd. Table]

Gene	Gene Action	Species	Phenotype	Reference
ATHB6	Transcription factor	Tomato	Drought resistance	Mishra et al., 2012
AtMYB102	Chimeric repressors	Arabidopsis, rice	Salt tolerance	Mito et al., 2011
GhMT3a	Metallothionein synthesis and ROS scavenging	Tobacco	Drought, salt, cold tolerance	Xue et al., 2009
AtNOA1	Nitric Oxide synthesis	Arabidopsis	Salt tolerance	Zhao et al., 2007
AtNOA1	Nitric Oxide synthesis	Arabidopsis	Salt tolerance	Qiao et al., 2009
OsPR4	Transcription factor	rice	Drought resistance	Wang et al., 2011
AtRabG3e	Intracellular vesicle trafficking	Arabidopsis	Salt and osmotic stress tolerance	Mazel et al., 2004
GhDi19-1 GhDi19-2	Cys2/His2-Type Zinc-Finger Proteins	Arabidopsis	Salt and ABA sensitivity	Li et al., 2010
AtSZF1 & AtSZF2	CCCH-type zinc finger proteins, involved in salt stress responses	Arabidopsis	Salt tolerance	Sun et al., 2007
StZFP1	TFIIIA-type zinc finger protein	Tobacco	Salt tolerance	Tian et al., 2010
At-SRO5	Antioxidative action	Arabidopsis	Salt tolerance	Rabajani et al., 2009
BnPtdIns-PLC2	Phosphatidylinositol-specific phospholipase C	Canola	Drought resistance, early flowering	Georges et al., 2009
CAbZIP1	Plant development (dwarf phenotype)	Arabidopsis	Drought and salt tolerance	Lee et al., 2006
AtbZIP17	Transcription factor	Arabidopsis	Salt tolerance	lu et al., 2008
BAX	BCL2-associated x protein as the pro-PCD factor	Tobacco	Drought, salt and heat tolerance	Isbat et al., 2009

[Table Contd.

Contd. Table]

Gene	Gene Action	Species	Phenotype	Reference
bZIP23	Transcription factor	Rice	ABA sensitivity, drought and salt tolerance	Xiang et al., 2008
ZIP72	Transcription factor	Rice	ABA sensitivity and drought resistance	Lu et al., 2009
ThbZIP1	Transcription factor	Tobacco	Salt tolerance and antioxidant activity	Wang et al., 2010
CAP2	Transcription factor	Tobacco	Drought and salt tolerance	Shukla et al., 2006
CBF3	Transcription factor	Rice	Drought and salt resistance	Oh et al., 2005
CBF4	Transcription factor	Arabidopsis	Drought and freezing tolerance (via activation of C-repeat/dehydration responsive element)	Haake et al., 2002
CBL1	Ca sensing protein	Arabidopsis	Salt and drought tolerance & cold sensitivity	Cheong et al., 2003
CBL1	Ca sensing protein	Arabidopsis	Salt and drought resistance – reduced transpiration	Albrecht et al., 2003
CBL1	Ca sensing protein	Arabidopsis	Salt tolerance	Wang et al., 2007
CBP20	cap binding complex	Arabidopsis	Loss of function (recessive) induces drought resistance	Papp et al., 2004
CcHyPRP	A hybrid-proline-rich protein encoding gene	Arabidopsis	Heat, salt and osmotic resistance	Priyanka et al., 2010
CpMYB10	Glucose sensitive and ABA hypersensitive	Arabidopsis	Desiccation and salinity tolerance	Villalobos et al., 2004
DREB	Transcription factor	Arabidopsis	Increased tolerance to cold, drought and salinity	Kasuga et al., 1999
DREB	Transcription factor	Arabidopsis	Sality tolerance	Xu et al., 2009
CgDREBa	Transcription factor	Chrysanthemum	Drought and salinity tolerance	Chen et al.,
GhDREB1	Transcription factor	Arabidopsis	Salt and osmotic tolerance	Huang et al., 2009

[Table Contd.

Contd. Table]

Gene	Gene Action	Species	Phenotype	Reference
DREB1A	Transcription factor	Paspalum grass	Salinity and dehydration tolerance	James et al., 2008
DREB1 or OsDREB1	Transcription factor	Rice	Drought, salt and cold tolerance with reduced growth under non-stress	Ito et al., 2006
DREB1A	Transcription factor	Tobacco	Drought and cold tolerance	Kasuga et al., 2004
DREB1A	Transcription factor	Tobacco	Salinity tolerance and dwarfing	Cong et al., 2008
DREB1A	Transcription factor	wheat	Delayed wilting under drought stress	Pellegrineschi et al., 2004
DREB1A; DREB2A	Transcription factor	Arabidopsis	Drought-cold tolerance	Maruyama et al., 2009
DREB2A	Transcription factor	Arabidopsis	Drought tolerance	Sakuma et al., 2006
DREB2	Transcription factor	Rice	Improve yield under limited water	Bihani et al., 2011
MbDREB1	Transcription factor	Arabidopsis	Drought and salt tolerance	Yang et al., 2011
SiDREB2	Transcription factor	Foxtail millet	Salinity and osmotic stress resistance	Lata et al., 2011
TaDREB2-3	Transcription factor	Wheat, barley	Drought and frost resistance	Morran et al., 2011
ERA1	Farnesyltransferase	Canola	When down regulated promotes drought tolerance	Wang et al., 2009
FAD3 & FAD8	Increased fatty acid desaturation	Tobacco	Drought tolerance	Zhang et al., 2005
FLD	Flavodoxin overexpression	Medicago truncatula	Salinity tolerance	Peña et al., 2010
FTL1/DDF1	Transcription factor	Arabidopsis	Resistance to cold, drought and heat	Kang et al., 2011
LeGPAT	glycerol-3-phosphate acyltransferase of chloroplasts	Tomato	Salt tolerance	Sun et al., 2010

[Table Contd.

Contd. Table]

Gene	Gene Action	Species	Phenotype	Reference
GsCBRLK	Calcium/calmodulin-independent kinase	Arabidopsis	Salt and ABA tolerance	Yang et al., 2010
HAL1	Promote K+/Na+ selectivity	Tomato	Salt tolerance in growth and fruit production	Rus et al., 2001
HAL1	Promote K+/Na+ selectivity	Watermelon	Salt tolerance in growth	Ellul et al., 2003
HAL2	Promote K+/Na+ selectivity	Tomato	Salt tolerance in calli and rooting	Arrillaga et al., 1998
HAL I or HAL II	Promote K+/Na+ selectivity	Tomato	Salt tolerance	Safdar et al., 2011
Hardy	AP2/ERF (APETALA2/ethylene responsive element binding factors) transcription factor	Clover	Drought and salt tolerance	Abogadallah et al., 2011
HOT2	Encode a chitinase-like protein	Arabidopsis	Salt tolerance	Kwon et al., 2007
Hrf1	Harpin protein	Rice	Drought tolerance via ABA signaling and antioxidants	Zhang et al., 2011
HvCBF4	Transcription factor	Rice	Drought, salt chilling tolerance	Oh et al., 2007
InsP3	Human type Inositol-(1,4,5)-trisphosphate	Tomato	Drought resistance	Khodakovskaya et al., 2010
ITN1	Transcription factor	Arabidopsis	Salt tolerance	Sakamoto et al., 2008
GmERF3	Jasmonate and ethylene-responsive factor 3	Tobacco	Drought, salt and disease resistance	Zhang et al., 2009
JERF3	Jasmonate and ethylene-responsive factor 3	Tobacco	Salinity tolerance	Wang et al., 2004
SodERF3	Ethylene-responsive factor 3	Tobacco	Drought tolerance	Trujillo et al., 2008

[Table Contd.

Gene	Gene Action	Species	Phenotype	Reference
JERF1	Jasmonate and ethylene-responsive factor 1	Tobacco	Salinity tolerance	Zhang et al., 2004
JERF1	Jasmonate and ethylene-responsive factor 1	Tobacco	Salt and cold tolerance	Wu et al., 2007
JERF1	Jasmonate and ethylene-responsive factor 1	Rice	Drought tolerance	Zhang et al., 2010
JERF1	Jasmonate and ethylene-responsive factor 1	Wheat	Multiple stress tolerance	Xu et al., 2007
JERF3	Jasmonate and ethylene-responsive factor 3	Tobacco	Drought, salt and freezing tolerance	Wu et al., 2008
TSRF1	Ethylene-responsive factor 1	Rice	Drought tolerance	Quan et al., 2010
KAPP	Kinase-associated protein phosphatase	Arabidopsis	Salt (Na+) tolerance	Manabe et al., 2008
lew2	Wilting allele; cellulose synthesis complex	Arabidopsis	Drought tolerance	Chen et al., 2005
LOS5	Regulates ABA biosynthesis	Tobacco	Drought tolerance	Yue et al., 2011
LOS5	Molybdenum cofactor sulfurase (Metabolism of abscisic acid)	Soybean	Drought tolerance	Li et al., 2013.
MCM6	Transcription factor	Tobacco	Salinity tolerance	Dang et al., 2011
NADP-ME 2	NADP-malic enzyme	Arabidopsis	Salt tolerance	Liu et al., 2007
MH1	DNA helicase	Arabidopsis	Drought and salt tolerance due to antioxidative action	Luo et al., 2009
MKK9	MAP Kinase	Arabidopsis	Salt tolerance in germination	Alzwiya et al., 2007

[Table Contd.

Contd. Table]

Gene	Gene Action	Species	Phenotype	Reference
GhMPK2	MAP Kinase	Tobacco	Salt and drought tolerance	Zhang et al., 2011
PtrMAPK	MAP Kinase	Tobacco	Drought tolerance	Huang et al., 2011
ZmMKK4	MAP Kinase	Arabidopsis	Salt and cold resistance	Kong et al., 2011
MsPRP2	Transcription factor	Alfalfa	Increased salinity tolerance	Winicov and Bastola 1999
GmNAC11; GmNAC20	Transcription factors	Arabidopsis	Salt and cold tolerance	Hao et al., 2011
NDPK1	Nucleoside diphosphate kinase 2	Potato	Multiple stress tolerance	Tang et al., 2008
NEK6	NIMA-related kinase	Arabidopsis	Salinity tolerance	Zhang et al., 2011
NFYB2	Transcription factor	Maize	Drought resistance	Nelson et al., 2007
NahG	Salicylate hydroxylase expression	Arabidopsis	Reduced leaf necrosis under salt stress	Borsani et al., 2001
NPK1	Mitogen-activated protein kinase	Maize	Drought resistance of photosynthesis	Shou et al., 2004
OsCDPK7	Transcription factor	Rice	Increased cold salinity and drought tolerance	Saijo et al., 2000
OsCIPK01– OsCIPK30	Calcineurin B-like protein-interacting protein kinases	Rice	Salt and drought tolerance	Xiang et al., 2007
OsCIPK03	Calcineurin B-like protein-interacting protein kinase	Rice	Salt tolerance	Rao et al., 2011
CIPK6	Calcineurin B-like protein-interacting protein kinase	Tobacco	Salt tolerance	Tripathi et al., 2009

[Table Contd.

Contd. Table]

Gene	Gene Action	Species	Phenotype	Reference
OrbHLH2	helix-loop-helix (bHLH) ncoding gene	Arabidopsis	Salt and osmotic tolerance	Zhu et al., 2009
OsCOIN	RING finger protein	Rice	Cold, salt and drought tolerance and overexpression of P5CS	Liu et al., 2007
OCPI1	Transcription factor	Rice	Drought resistance in yield	Huang et al., 2007
ocp3	Transcription factor	Arabidopsis	Drought resistance	Ramirez et al., 2009
OPBP1	Transcription factor	Tobacco	Salinity and disease tolerance	Guo et al., 2004
OsSbp	Calvin cycle enzyme sedoheptulose-1, 7- bisphosphatase	Rice	Tolerance of photosynthesis to salt	Feng et al., 2007
OsDREB1A	Transcription factor	Arabidopsis	Drought, salt, freezing tolerance	Dubouz et al., 2003
OsDREB2A	Transcription factor	Rice	Drought tolerance	Cui et al.,2011
OsMYB3R-2	MYB homeodomain, and zinc finger proteins	Arabidopsis	Drought, salt, freezing tolerance	Dai et al., 2007
OsiSAP8	Stress/zinc finger protein	Rice	Salt drought and cold tolerance	Kanneganti and Gupta, 200
OsNAC5	Transcription factor	Rice	Salt tolerance	Song et al., 2011
OsNAC10	Transcription factor	Rice	Drought tolerance in the field	Jeong et al., 2010
PARP1; PARP2	Poly(ADP-ribose) polymerase	Arabidopsis; Brassica	Silencing induces drought and heat tolerance	Block et al., 2004
PDH45	DNA helicase 45	Pea	Salinity tolerance in yield	Sanan-Mishra et al., 2005; Sahoo et al., 2012

[Table Contd.

Contd. Table]

Gene	Gene Action	Species	Phenotype	Reference
PeSCL7	Transcription factor	Arabidopsis	Salt and drought tolerance	Ma et al., 2010
PLD	Phospholipase D	Arabidopsis	Salt tolerance	Bargmann et al., 2009
RGS1	Regulation of G-protein signalling	Arabidopsis	ABA mediated root elongation and drought tolerance	Chen et al., 2006
SCABP8	Interacts with SOS2	Arabidopsis	Salt tolerance	Quan et al., 2007
smGTP	Encode small guanosine triphosphate binding protein	Lolium temulentum	Salt & dehydration tolerance	Dombrowski et al., 2008
SINAGS1	Ornithine accumulation	Arabidopsis	Drought and salt tolerance	Kalamaki et al., 2009
SIZ1	SUMO E3 ligase	Arabidopsis	Salt tolerance	Miura et al., 2011
SIZ1	SUMO E3 ligase	Arabidopsis	Heat tolerance	Chen et al., 2011
SNAC1	Transcription factor	Rice	Drought and salt tolerance	Hu et al., 2006
TaSnRK2.7	Transcription factor	Arabidopsis	Multi-abiotic stress tolerance	Zhang et al., 2011
ONAC063	Transcription factor	Arabidopsis	Salt tolerance	Yokotani et al., 2009
SQE1	Squalene epoxidase enzyme	Arabidopsis	Root sterol biosynthesis and drought tolerance	Posé et al., 2009
OsRDCP1	Transcription factor	Rice	Drought tolerance	Bae et al., 2011
SRK2C	Protein kinase	Arabidopsis	Osmotic stress/drought tolerance	Umezawa et al.,2004
OsSDIR1	RING-finger containing E3 ligase	Rice	Drought tolerance	Gao et al., 2011
StMYB1R-1	MYB-Like Domain Transcription Factor	Potato	Drought tolerance via reduced water loss	Shin et al., 2011
STO	Protein binds to a Myb transcription factor	Arabidopsis	Salt tolerance	Ngaoka and Takano, 20

[Table Contd.

Contd. Table]

Gene	Gene Action	Species	Phenotype	Reference
Sto1	Reduced ABA accumulation	Arabidopsis	Better growth under salt stress	Ruggiero et al., 2004
TaABC1	Protein kinase	Arabidopsis	Drought salt and cold tolerance	Wang et al., 2011
TaCHP	Cysteine, histidine, and proline rich zinc finger protei	Arabidopsis	Promotion of CBF3 and DREB2A expression and salt tolerance	Li et al., 2010
TaPP2Ac-1	catalytic subunit (c) of protein phosphatase 2A	Tobacco	Drought resistance; maintain RWC and membrane stability	Xu et al., 2007
TaSTK	serine/threonine protein kinase	Wheat	Salt tolerance	Ge et al., 2007
TaSrg6	Transcription factor	Arabidopsis	Drought tolerance	Tong et al., 2007
TERF1	ERF transcription activator	Tobacco	ABA sensitivity and drought tolerance	Zhang et al., 2005
ThIPK2	Inositol polyphosphate kinase	Brassica	Salt and drought tolerance	Zhu et al., 2009
Tsi1	Transcription factor	Tobacco	Increase osmotic stress tolerance	Park et al., 2001
VuNCED1	Involved in ABA biosynthesis	Creepingbent grass	Salinity and drought tolerance	Aswath et al., 2005
WAB15	Transcription factor	Tobacco	Freezing, osmotic and salt tolerance	Kobayashi et al., 2008
WIN1/SHN1	Wax inducer	Arabidopsis	Epicuticular wax, stomata number and drought tolerance	Yang et al., 2011
GmWNK1	(With No Lysine K) serine—threonine kinase	Arabidopsis	Seedling salt tolerance	Wang et al., 2011
WRKY25 & WRKY33	Transcription factor	Arabidopsis	Salt tolerance	Jiang and Deyholos, 2009
WRKY45	Transcription factor	Arabidopsis	Drought resistance	Qiu and Yu, 2009
OsWRKY45	Transcription factor	Rice	Drought and cold resistance	Tao et al., 2011

[Table Contd.

Contd. Table]

Gene	Gene Action	Species	Phenotype	Reference
AtWRKY63	Transcription factor	Arabidopsis	ABA response and drought tolerance	Ren et al., 2010
WXP1	Epicuticular wax accumulation	Alfalfa	Drought resistance in maintained leaf water status and delayed wilting	Zhang et al., 2005
WXP1	Epicuticular wax accumulation	White clover	Drought resistance	Jiang et al., 2010
WXP1;WXP2	Epicuticular wax accumulation	Arabidopsis	Drought and freezing tolerance	Zhang et al., 2007
WRSI5	Protease inhibitors	Arabidopsis	Salt tolerance	Shan et al., 2008
	Protease inhibitors	Tobacco	Salt tolerance	Srinivasan et al., 2009
ZmDREB2A	Encodes HSP &LEA proteins	Arabidopsis	Drought and heat tolerance	Qin et al., 2007
ThZFL	zinc finger protein	Tobacco	Salinity tolerance	An et al., 2011
MtZpt2	zinc finger protein	Medicao	Recover Root growth under salt stress	Merchan et al., 2007

Source: (www.plantstress.com)

Table 4: Genes encoding proton pumps, antiporters, ion transporters and aquaporins

Gene	Gene Action	Species	Phenotype	Reference
AtABCG36/ AtPDR8	ATP-binding cassette (ABC) transporter	Arabidopsis	Salt tolerance due to sodium exclusion	Kim et al., 2010
Atchx21	Putative Na+/H+ antiporter	Arabidopsis	Sodium concentrations in plant, root growth, plant size	Hall et al., 2006
AtCNGC10	Plasma membrane cation transport	Arabidopsis	Salt tolerance	Guo et al., 2008
AtCLC	Chloride channel	Arabidopsis	Salt tolerance	Jossier et al., 2010
AtHKT1	Reduction in Sodium in root	Arabidopsis	Salt tolerance	Horie et al., 2006
AtHKT1	Sodium and Potassium transporter	cells	Reduced sodium accumulation	Sunarpi et al., 2005
GmHKT1	Sodium and Potassium transporter	Tobacco	Salinity tolerance	Chen et al., 2011
HKT2;1	Sodium and Potassium transporter	Barley	Salinity tolerance	Mian et al., 2011
AtMRP4	Stomatal guard cell plasma membrane ABCC-type ABC transporter,	Arabidopsis	Drought susceptibility due to loss of stomatal control	Markus et al., 2004
AtNHX1	Vacuolar Na$^+$/H$^+$ antiporter	Arabidopsis	Salt tolerance	Yokoi et al., 2002
AtNHX1	Vacuolar Na$^+$/H$^+$ antiporter	Brassicanapus	Salt tolerance, growth, seed yield and seed oil quality	Zhang et al., 2001
AtNHX1	Vacuolar Na$^+$/H$^+$ antiporter	Buckwheat	Salt tolerance	Chen et al., 2008
AtNHX1	Vacuolar Na$^+$/H$^+$ antiporter	Cotton	Salt tolerance in photosynthesis and yield	He et al., 2005
AtNHX1	Vacuolar Na$^+$/H$^+$ antiporter	Tall fwscue	Salt tolerance	Zhao et al., 2007

[Table Contd.

Contd. Table]

Gene	Gene Action	Species	Phenotype	Reference
AtNHX1	Vacuolar Na$^+$/H$^+$ antiporter	Tomato	Salt tolerance, growth, fruit yield	Apse *et al.*, 1999
AtNHX1	Vacuolar Na$^+$/H$^+$ antiporter	Wheat	Salt tolerance for grain yield in the field	Xue *et al.*, 2004
AtNHX1	Vacuolar Na$^+$/H$^+$ antiporter	Sugar beet	Salt tolerance, sugar accumulation	Liu *et al.*, 2008
LnNHX2	Vacuolar Na$^+$/H$^+$ antiporter	Arabidopsis	K+ accumulation, salt tolerance	Rodriguez-Rosales *et al.*, 2008
AtNHX5	Vacuolar Na$^+$/H$^+$ antiporter	Torenia	Salt tolerance	Shi *et al.*, 2008
AtNHX5	Vacuolar Na$^+$/H$^+$ antiporter	Paper Mulberry	Salt and drought resistance	Li *et al.*, 2011
AtNHX2 AtNHX5	Vacuolar Na$^+$/H$^+$ antiporter	Arabidopsis	Salt tolerance	Yokoi *et al.*, 2002
AVP1	Vacuolar H$^+$-pyrophosphatase (H$^+$-PPase) gene	Arabidopsis	Salt tolerance in growth and sustained plant water status	Gaxiola *et al.*, 2001
AVP1	Vacuolar H$^+$-pyrophosphatase (H$^+$-PPase) gene	Alfalfa	Drought and salt tolerance	Bao *et al.*, 2009
AVP1	Vacuolar H$^+$-pyrophosphatase (H$^+$-PPase) gene	Agrostis stolonifera L	Salt tolerance	Li *et al.*, 2010
AVP1	Vacuolar H$^+$-pyrophosphatase (H$^+$-PPase) gene	Cotton	Drought and salt resistance and yield in the field	Pasapula *et al.*, 2011
MdVHP1	Vacuolar H$^+$-pyrophosphatase (H$^+$-PPase) gene	Tomato	Drought and salinity resistance	Dong *et al.*, 2011
AVP1 + PgNHX1	H$^+$-PPase + Vacuolar Na$^+$/H$^+$ antiporter	Tomato	Salinity tolerance	Bhaskaran & Savithramma 2011
TsVP	H$^+$-PPase gene	Cotton	Drought resistance (yield)	Lv *et al.*, 2009
GhNHX1	Vacuolar Na$^+$/H$^+$ antiporter	Arabidopsis (cotton)	Salt tolerance	Wu *et al.*, 2004

[Table Contd.

Contd. Table]

Gene	Gene Action	Species	Phenotype	Reference
GmCAX1	Cation/proton antiporter	Arabidopsis	Salt tolerance	Luo et al., 2005
HKT1	Potassium transporter	Wheat	Salt tolerance in growth and improved K^+/Na^+ ratio	Laurie et al., 2002
NtAQP1	PIP1 plasma membrane aquaporin	Tobacco	High root hydraulic conductance and reduced plant water deficit under drought stress	Siefritz et al., 2002
OsARP	Antiporter-regulating protein	Tobacco	Salt tolerance by Na^+ compartmentation	Uddin et al., 2006
OsNHX1	Vacuolar Na^+/H^+ antiporter	Rice	Salt tolerance	Fukuda et al., 2004
OsSOS1	Plasma membrane Na^+/H^+ exchanger	Rice	Salt tolerance et al., 2007	Martinez-Atienza
SlSOS1	Plasma membrane Na^+/H^+ exchanger	Tomato	Salt tolerance	Olias et al., 2009
PcSrp	Serine rich protein (enhancing ion homeostasis?)	Finger millet	Salt tolerance	Mahalakshmi et al., 2006
PgTIP1	Tonoplast intrinsic protein	Arabidopsis	Salt tolerance; root dependant drought tolerance	Peng et al., 2007
PpENA1	Plasma membrane Na^+ pumping ATPase	Rice	Salt tolerance	Jacobs et al., 2011
PIP1 (VfPIP1)	Plasma membrane aquaporin overexression	Arabidopsis	Faster growth, stomatal closure under drought stress	Cui et al., 2008
PIP1;4 & PIP2;5	Plasma membrane aquaporin overexression	Tobacco	Excessive water loss and retarded seedling growth under drought stress	Jang et al., 2007

Table Contd.

Contd. Table]

Gene	Gene Action	Species	Phenotype	Reference
TdPIP1;1 or TdPIP2;1	Plasma membrane aquaporin overexression	Tobacco	Salt and osmotic tolerance	Ayadi et al., 2011
SAT32	Enhanced vacuolar H^+-pyrophosphatase (H^+-PPase)	Arabidopsis	Salt tolerance	Park et al., 2009
SlTIP2;2	Tonoplast aquaporin	Tomato	Enhanced transpiration and f yield under drought and control	Sade et al., 2009
SOD2	Vacuolar Na^+/H^+ antiporter	Arabidopsis	Salt tolerance; higher plant K/Na ratio	Gao et al., 2004
SOD2	Vacuolar Na^+/H^+ antiporter	Rice	Salt tolerance	Zhao et al., 2006
SOS1	Na^+/H^+ antiporter	Arabidopsis	Protect K+ permeability during salt stress	Qi and Spalding, 2004
SOS3	Sodium accumulation in roots	Arabidopsis	Salt tolerance	Horie et al., 2006
SOS4	Involved in the synthesis of pyridoxal-5-phosphate which modulates ion transporters	Arabidopsis	Salt tolerance through Na+/K+ homeostasis	Shi et al.,2002
SsNHX1	Vacuolar Na^+/H^+ antiporter	Rice	Salt tolerance	Zhao et al., 2006
SsVP-2	Vacuolar Na^+/H^+ antiporter	Arabidopsis	Salt tolerance	Guo et al., 2006
TNHX1 and H+-PPase TVP1	Vacuolar Na^+/H^+ antiporter	Arabidopsis	Salt tolerance	Brini et al., 2007
TaVB	H^+-ATPases (V-ATPase) subunit B	Arabidopsis	Salt tolerance	Wang et al., 2011
TsVP	Vacuolar Na^+/H^+ antiporter	Tobacco	Salt tolerance	Gao et al., 2006
TsVP	Vacuolar Na^+/H^+ antiporter	Cotton	Salt tolerance	Lv et al., 2008
YCF1	Sequester glutathione-chelates of heavy metals into vacuoles	Arabidopsis	Heavy metal and salt tolerance	Koh et al., 2006

Source: (www.plantstress.com)

2004; Pellegrineschi *et al.*, 2004; Behnam *et al.*, 2006; Bhatnagar-Mathur *et al.*, 2004, 2006). The use of stress inducible rd29A promoter minimized the negative effects on plant growth in these crop species. However, over expression of DREB2 in transgenic plants did not improve stress tolerance, suggesting involvement of post-translational activation of DREB2 proteins (Liu *et al.*, 1998). Recently, an active form of DREB2 was shown to trans activate target stress-inducible genes and improve drought tolerance in transgenic Arabidopsis (Sakuma *et al.*, 2006). The DREB2 protein is expressed under normal growth conditions and activated by osmotic stress through post-translational modification in the early stages of the osmotic stress response. Another ABA-independent, stress-responsive and senescence-activated gene expression involves ERD gene, the promoter analysis of which further identified two different novel cis acting elements involved with dehydration stress induction and in dark-induced senescence (Simpson *et al.*, 2003). Similarly, transgenic plants developed by expressing a drought-responsive AP2-type TF, SHN1-3orWXP1, induced several wax-related genes resulting in enhanced cuticular wax accumulation and increased drought tolerance (Aharoni *et al.*, 2004; Zhang *et al.*, 2005). Thus, clearly, the over expression of some drought-responsive transcription factors can lead to the expression of downstream genes and the enhancement of abiotic stress tolerance in plants (see review, Zhang *et al.*, 2004). The regulatory genes/factors reported so far not only playa significant role in drought and salinity stresses, but also in submergence tolerance. More recently, an ethylene response-factor-like gene Sub1A, one of the cluster of three genes at the Sub1 locus have been identified in rice and the over expression of Sub1A-1 in a submergence-intolerant variety conferred enhanced submergence tolerance to the plants (Xu *et al.*,2006),thus confirming the role of this gene in submergence tolerance in rice.

8.2. Role of Helicases inImparting Tolerance to Abiotic Stress

Abiotic stress condition often affects the cellular gene expression machinery. Therefore, the molecules that are involved in the processing of nucleic acids including helicases are also likely to be affected. Multiple DNA helicases are present in the cell and are involved in gene regulation at various developmental stages as well as in stress conditions. These DNA unwinding enzymes may have different substrates as well as structural requirements (Matson *et al.*, 1994; Tuteja and Tuteja, 1996). Though a number of different helicases have been reported from *E. coli*, bacteriophages, viruses, yeast, calf thymus and humans the biological role of only a few DNA helicases have been explored (Lohman and Bjornson, 1996; Tuteja and Tuteja, 2004). Helicase genes are now being reported as powerful gene for developing stress tolerant crops. Most helicases are members of DEAD-box protein super-family and play essential roles in basic cellular processes such as replication, repair, recombination, transcription, ribosome biogenesis and translation initiation. Transient opening of the stable duplex DNA is an essential prerequisite step in many biological processes such as DNA replication, repair, recombination and transcription. DNA helicases catalyses the unwinding of energetically stable duplex DNA (DNA helicase) or inter and intra molecular base -paired duplex RNA (RNA helicase) structures by disrupting the hydrogen bonds between the two strands and thereby plays an important role in all aspects of nucleic acid metabolisms. Gene *pdh45*, the first plant DNA helicase has been cloned, over-expressed and characterized in detail (Hoi *et al.,* 2008). The potential role of PDH45 (pea DNA helicase 45) in overcoming salinity stress was explored (Sanan-Mishra *et al.,* 2005). They have proved that *PDH45* over expressing transgenic lines showed high salinity tolerance and the T_1 transgenic plants were able to grow to maturity and set normal viable seeds under continuous salinity stress without any reduction in plant yield in terms of seed weight. The authors have proposed a dual mode of action for PDH45 (Table 3).

There are various reports published on the isolation of a pea DNA helicase 45 (PDH45) and its novel role in abiotic stress tolerance in model plant tobacco (Plant J, 24, 1-13; PNAS, USA 102, 509-514). The exact mechanism of helicase-mediated salt tolerance is not yet understood. However, based on the properties of PDH45 that were studied earlier (Rocak and Linder, 2004) and those known for DEAD-box proteins (Hasegawa, 2000; Tuteja and Tuteja, 2004). It is revealed that two sites of action of this dual helicase: 1) it may act at the translation level to enhance or stabilize protein synthesis; or 2) it may associate with DNA multi-subunit protein complexes to alter gene expression. After observing the proof of concept in model plants, the PDH45 has been used to transform the bacteria and to different varieties of rice and groundnut crops. The results show that PDH45 also provide the salt tolerance in bacteria, rice and groundnut. Interestingly, there were no yield losses.

8.3. Signal Transduction Genes

Signaling pathway is complex as it involves the coordinated action of various genes in a single pathway or diverse pathways. Calcium is a prime candidate, which functions as a central node in mediating the coordination and synchronization of diverse stimuli into specific cellular responses. Thus, the proteins, which sense cytoplasmic Ca^{2+} perturbations and relay this information to downstream molecules, serve as an important component of signaling. In plant cell many calcium sensors have been recognized which include calmodulin (CaM) and calmodulin-related proteins (Luan *et al.,* 2002), Ca^{2+}-dependent protein kinases (CDPKs) (Sanders *et al.,* 2002) and the relatively recently discovered sensor CBL (calcineurin B-like) protein (Liu and Zhu, 1998). CBLs are characterized by 4 helix-loop-helix calcium binding domains termed as EF hands. Currently, 10 isoforms of CBL have been dis-covered in *Arabidopsis* and named as CBL due to their sig-niWcant sequence similarity to animal calcineurin B. Despite this sequence similarity, *Arabidopsis* lacks calcineurin in its data bank (Gong *et al.*, 2004). Various isoforms of CBL are up-regulated in stress condition. CBLs specially interact with a class of kinases known as CBL-interacting protein kinase (CIPKs) to transduce the signal via phosphorylation of downstream signaling components.

One of the merits for the manipulation of signaling factors is that they can control a broad range of downstream events that can result in superior tolerance for multiple aspects (Umezawa *et al.*, 2006). Alteration of these signal transduction components is an approach to reduce the sensitivity of cells to stress conditions, or such that a low level of constitutive expression of stress genes is induced (Grover *et al.*, 1999). Over expression of functionally conserved At-DBF2 (homolog of yeast DBf2 kinase) showed striking multiple stress tolerance in Arabidopsis plants (Lee *et al.*, 1999). Pardo *et al.* (1998) also achieved salt stress-tolerant transgenic plants by over expressing calcineurin (a Ca^{2+} Calmodulin dependent protein phosphatase), a protein phosphatase known to be involved in salt-stress signal transduction in yeast. Transgenic tobacco plants produced by altering stress signaling through functional reconstitution of activated yeast calcineurin not only opened-up new routes for study of stress signaling, but also for engineering transgenic crops with enhanced stress tolerance (Grover *et al.*, 1999). Over expression of an osmotic-stress-activated protein kinase, SRK2C resulted in a higher drought tolerance in *A. thaliana*, which coincided with the upregulation of stress-responsive genes (Umezawa *et al.*, 2004). Similarly, a truncated tobacco mitogen-activated protein kinase kinase kinase (MAPKKK), NPK1, activated an oxidative signal cascade resulting in cold, heat, salinity and drought tolerance in transgenic plants (Table 3; Kovtun *et al.*, 2000; Shou *et al.*, 2004). However, suppression of signaling factors could also effectively enhance tolerance to abiotic stress (Wang *et al.*, 2005). This hypothesis was based on previous reports indicating that a and b subunits of farnesyltransferase ERA1 functions as a negative regulator of ABA signaling (Cutler *et al.*, 1996; Pei *et al.*, 1998). Conditional antisense downregulation of a or b subunits of protein farnesyl transferase, resulted in enhanced drought tolerance of Arabidopsis and canola plants.

Presently the direction of research is more towards isolation of master switches, which can control these stress genes. As cytosolic calcium up-regulation is more or less a universal phenomenon associated with stress signaling, thus the calcium sensors, which decode these Ca^{2+} signatures and relay the information down stream, may act as master switches in controlling various stress genes. Moreover, mutations in these calcium sensors like *AtCBL1* and their interacting protein kinases have been shown to cause aberrations in the expression of some of the major stress responsive genes like *RD29A*, *KIN1*, *KIN2* and *RD22* indicating their immense significance in stress signaling (Pandey *et al.*, 2004).

8.4. Targeting Pathways: Tandem Expression of Genes

Under natural field conditions plants have to cope with different stress combinations at different developmental stages and for varying duration. Tolerance to abiotic stress is a consequence of genetic and environmental interactions through a complex network that implies physiological, molecular and biochemical responses. Modifying the expression of different components simultaneously has the potential to generate responses apt to the complexity of a combination of stresses. There are only few examples where the simultaneous co-expression of different components of the same pathway has been tried. Increase in biosynthesis of proline was achieved by co-expression of *E. coli* P5C biosynthetic enzymes gamma-glutamyl kinase 74 (GK74) and gamma-glutamylphosphate reductase (GPR) and the antisense transcription of proline dehydrogenase (ProDH) in Arabidopsis and tobacco (Stein *et al.*, 2011). The transgenic plants displayed improved tolerance to heat stress associated with the accumulation of cell wall proline-rich proteins (Stein *et al.*, 2011). Simultaneous co-expression of dehydroascorbate reductase (DHAR), glutathione reductase (GR) or glutathione-S-transferase (GST) and glutathione reductase (GR) in tobacco plants also resulted in the increased tolerance of the transgenic plants to a variety of abiotic stresses (Martret *et al.*, 2011). In tobacco seeds, higher antioxidant enzymes activity driven by the simultaneous over expression of the CuZnSOD and APX genes in plastids, allowed the increase of germination rates and longevity of long-term stored seeds under combined stress conditions (Lee *et al.*, 2010), demonstrating the enormous potential of simultaneous gene expression in plant engineering (Table 3).

8.5. Modifying Function: Engineering C_4 Photosynthetic Pathway into C_3 Crops

Abiotic stress is the major factor limiting photosynthetic activity, resulting in growth and yield reduction. The photosynthesis machinery also affects metabolic processes such as carbon and nitrogen partitioning (Ainsworth and Bush, 2011) and oxidative stress regulation (Foyer and Shigeoka, 2011). The projected effects of climate change in rising ambient temperatures and CO_2 concentrations will have influence plant CO_2 assimilation (and yield), and photorespiration. Research efforts are focused on obtaining Kranz anatomy (Hibberd *et al.*, 2008), especially in rice which have an intermediate anatomical characteristics between C_3 and C_4 plants (Sage and Sage, 2009). While most genes controlling bundle density in C_4 plants are still unknown, it has been postulated that about 20 genes are involved (reviewed by Peterhansel, 2011).The ability of the C_4 photosynthetic pathway to suppress ribulose 1,5-bisphosphate (RuBP) oxygenation and photorespiration represents the most efficient form of photosyn-thesis on Earth (Sage 2004). In recent years, efforts have been given to engineer C_4 photosynthesis into C_3 crops (Sage and Zhu, 2011). The expression of genes encoding enzymes such as phosphoenol pyruvate carboxylase (PEPC), the chloroplastic pyruvate orthophosphate dikinase (PPDK), and NADP-malic enzyme (NADP-ME) into rice (Ku *et al.,* 2007), tobacco (Häusler *et al.,* 2002) and potato (Rademacher *et al.,* 2002) improved photosynthetic rate and yield. Although considerable efforts have been made, the over expression of either single or multiple C_4-enzyme related genes in C_3 plants have resulted in contradictory results (Shao *et al.,* 2011). Thus, in order to obtain C_4 crops, new transformation methods together with additional efforts to better understand the function of C_4 enzymes in a proper leaf anatomy (Furbank *et al.,* 2009) are needed. Another important aspect that has to be addressed is source/sink relationships. From an evolutionary perspective C_3 plants have modified their sink size proportionally to the source size (i.e. photosynthesis organs). Thus, more efficient carbon fixation via C_4 pathway in the transformed plants would require adapting the sinks to attain efficient harvest index (Murchie *et al.,* 2009).

9. Epigenetic and Post-Transcriptional Control

Epigenetic processes such as DNA methylation, histone modifica-tions, generation of small RNAs (sRNA) molecules and transposable element activity, play essential roles in modulating gene activity in response to environmental stimuli (Henderson and Jacobsen *et al.,* 2007; Feng *et al.,* 2010). While most mechanisms involved in epigenetic and its heritance have not yet indentified, they play a major role in gene silencing on one hand and as a target for manipulation on the other. Abiotic stress can induce changes in gene expression through hypomethylation or hypermethylation of DNA which are related with stress tolerance. The stress-induced-specific CpHpG-hypermethylation in the halophyte Mesembryanthemum crys-tallinum L. induced the switch in photosynthesis mode from C_3 to CAM, contributing to the adaptation to salt stress (Dyachenko *et al.,* 2006). In wheat, the use of the methylation inhibitor 5-azacytidine resulted in the increased tolerance to salt stress at the seedling stage. Decrease levels of histone acetylation levels (antisense) in tomato resulted in higher photosynthetic rates under water-stress (Scippa *et al.,* 2004). The control of methylation and histone patterns is emerging as a potential tool for improving tolerance to abiotic stress in crops, however, little is known about how to control the effect of post transcriptional manipulation.

The small RNAs (sRNAs) are the bioregulators of plant stress response, regulates by transcriptional gene silencing (TGS), controlling mRNA stability and translation, or targeting epigenetic modifications (Khraiwesh *et al.*, 2011). Abiotic stress can induce both the over- or under-expression of specific sRNAs that are involved in pathways that contribute to re-program complex processes of metabolism and physiology. Several reports have recently indicated the possible use of these sRNAs as targets for the genetic manipulation of crops. The over expression of miR398 in Arabidopsis, which targets two closely related Cu/Zn superoxide dismutases (cytosolic CSD1 and chloroplastic CSD2) resulted in increased tolerance to oxidative stress (Sunkar *et al.*, 2006).

Transgenic tomatoes expressing Sly-miR169c displayed decreased stomata opening, a decrease in leaf water loss and enhanced drought tolerance (Zhang *et al.*, 2011). Several miRNAs regulating drought and salinity stress have been discovered till the date (Table 5). Transgenic rice constitutively expressing osa-MIR396c showed increased sensitivity to salt stress (Gao *et al.*, 2010). The identification and characterization of the role(s) of sRNAs in the regulation of gene expression, together with the development of artificial miRNA methodologies open new avenues for the generation of transgenic stress tolerant crops (Schwab *et al.*, 2005).

10. Choice of Promoters: When and How Much to Express

An important aspect of transgenic technology is the regulated expression of transgenes. Tissue specificity of transgene expression is also an important consideration while deciding on the choice of the promoter so as to increase the level of expression of the gene. Thus, the strength of the promoter and the possibility of using stress inducible, developmental stage, or tissue-specific promoters have also proved to be critical for tailoring plant response to these stresses (Bajaj *et al.*, 1999). Some gene products are needed in large amounts, such as LEA3, thereby necessitating the need for a very strong promoter. With other gene products, such as enzymes for polyamine biosynthesis, it may be better to use an inducible promoter of moderate strength. The promoters that have been most commonly used in the production of abiotic stress tolerant plants so far, include the CaMV 35S, ubiquitin 1 and actin promoters.

These promoters being constitutive in nature, by and large express the downstream transgenes in all organs and at all the stages. However, constitutive over production of molecules, such as trehalose (Romero *et al.*, 1997) or polyamines (Capell *et al.*, 1998) causes abnormalities in plants grown under normal conditions. Also, the production of the above-described molecules can be metabolically expensive. In these cases, the use of astress inducible promoter may be more

Table 5: The potential miRNA biomarkers identified in drought and salinity stress conditions

Type of Stress	miRNA biomarkers	Up/Down regulated	References
Drought Stress	miR393, miR319, miR397	↓↑	Sunkar and Zhu 2004
	miR169g, miR171a	↓↑	Zhao et al., 2007, Jian et al., 2010
	miR156, miR159, miR168, miR170, miR171, miR172, miR319, miR396, miR397, miR408, miR529, miR896, miR1030, miR159,	↓	Zhou et al., 2010
	miR169, miR171, miR319, miR395, miR474, miR845, miR851, miR854, miR896, miR901, miR903, miR1026, miR1125	↓↑	
	miR1035, miR1050, miR1088, miR1126,	↓↑	Arenas-Huertero et al., 2009
	miR159.2, miR393, miR2118, miR398 a, b, miR408	↓↑	Trindade et al., 2010
Salinity Stress	miR169g, miR169n, miR169o, miR393	↓↑	Zhao et al., 2009, Gao et al., 2011
	miR156, miR158, miR159, miR165, miR167, miR168, miR169, miR171, miR319, miR393, miR394, miR396, miR397	↓↑	Liu et al., 2008
	miR398	↓	

Table 6: Various promoters used for targeted expression of transgenes in crop plants to develop tolerance against drought and salinity stress

Promoter:: Transgene	Plant species	Stress application	Increase in enzyme activity	Output	Reference
35S::*Cucurbita ficifofia* SF'DS	*Arabidopsis thaliana*	Drought	5–6-fold (SPDS)	Increased tolerance to salinity and drought	Kasukabe *et al.,* 2004
35S::*Cucurbita ficifolia*SPDS	*Lpomoea batatus*	Salt (NaCl; 114 day from planting)	NA	Enhanced tolerance to Salt and drought stress	Kasukabe *et al.,*2006
35S::*Malus sylvestrisvar.* domestica SPDS	*Pyrus communis* L. "Ballad"	Salt (250 mM NaCl for 10 day), high osmoticum (300 mM mannitol for 10 day)	NA	Greater tolerance to salt stress	Wen *et al.,* 2008, 2009, 2010
35S::*Malus sylvestrisvar.* domestica SPDS1	*Lycopersicon esculentum*	Salt (100 or 250 mM NaCl; 4 week-old plants for 60–65 day)	NA	Enhanced tolerance to salt stress	Neily *et al.,* 2011
35S::Mouse ODC	*Nicotiana tabacum* var. xanthi	NaCl (200 mM; up to 4 week from germination or 15 day old seedlings subjected to 300 mM NaCl for 4 week)	Very high (mouse ODC; native ODC or ADC activity was lower in the transgenics)	Greater tolerance to salt stress	Kumria and Rajam, 2002
ABA-inducible:: *Avena sativa* ADC	*Oryza sativa*	NaCl (150 mM; 2-day in 10-day old seedlings)	3–4-fold	Increased tolerance to salinity stress	Roy and Wu, 2001

[Table Contd.

Contd. Table]

Promoter:: Transgene	Plant species	Stress application	Increase in enzyme activity	Output	Reference
35S::*Datura stramonium* ADC	*Oryza sativa*	Drought (60-day old plants; 6 day in 20% PEG followed re-watering for 3 day)	NA	High tolerance to drought	Capell *et al.,* 2004
35S::*Avena sativa* ADC	*Solanum melongena*	Salinity (150–200 mM NaCl; 8–10 day), drought (7.5–10% PEG; 8–10 day), low temperature (6–8æ%C; 10 day), high temperature (45æ%C for 3 h), cadmium (0.5–2 mM for 1 month) in 8–10 day old seedlings	3–4-fold (ADC, DAO), ~2-fold (ODC)	Enhanced tolerance tomultiple stresses	Prabhavathi and Rajam, 2007
35S::*Arabidopsis thaliana* ADC2	*Arabidopsis thaliana*	Drought (4 week-old plants for 14 day followed by 7 dayrecovery)	NA	Increased tolerance to Drought stress	Alcazar *et al.,* 2010
35S: *Poncirus trifoliate* ADC	*Arabidopsis thaliana*	High osmoticum, drought, and low temperature (up to14-day from germination,1–18 day in 3–4 week-old plants)	NA	Enhanced resistance to long-term drought	Wang *et al.,* 2011
pRD29A::*Avena sativa* ADC	*Lotus tenuis*	Drought (6–8 week-old plants exposed to soil water potential of –2 MPa)	~2.2-fold (drought)	Increased tolerance to drought	Espasandin *et al.,* 2014
ABA inducible:: *Tritordeum* SAMDC	*Oryza sativa*	NaCl (150 mM; 11 day-old seedlings for 2 day)	NA	Enhanced salt tolerance	Roy and Wu, 2002

[Table Contd.

Contd. Table]

Promoter:: Transgene	Plant species	Stress application	Increase in enzyme activity	Output	Reference
35S::Homo sapiens SAMDC	*Nicotiana tabacum* var. xanthi	NaCl (250 mM), PEG (20%) up to 2 months from sowing	~1.3–5-fold (overall SAMDC), ~2-fold (DAO)	Greater tolerance to salt and drought	Waie and Rajam, 2003
35S::*Dianthus caryophyllus* SAMDC	*Nicotiana tabacum*	Salt (NaCl; 0–400 mM from sowing through 8 week) Low temperature (5 week-old plants for 24 h at OoG)	2-fold	Increased tolerance to salt and other stresses	Wi et al., 2006
Ubi::*Datura stramonium* SAMDC	*Oryza sativa* L. subsp. *Japonica* cv. EYI105	Osmoticum (PEG; 60 day-old plants for 6 day followed by 20 day recovery period)	NA	Greater tolerance to high osmoticum induced drought and better recovery	Peremarti et al., 2009
35S:: *Capsicum annuum* SAMDC	*Arabidopsis thaliana*	Drought (2 week old plants for 6 h or 3 week-old plants for 11 day followed by 3 day recovery)	1.4–1.6-fold (total SAMDC)	Increased drought tolerance	Wi et al., 2014

Fold increases of PAs in transgenic plants are from the basal level unless otherwise stated (NA, not available) desirable. In plants, various types of abiotic stresses induce a large number of well characterized and useful promoters. An ideal inducible promoter should not only be devoid of any basal level of gene expression in the absence of inducing agents, but the expression should be reversible and dose-dependent.

The transcriptional regulatory regions of the drought-induced and cold-induced genes have been analyzed to identify several cis-acting and trans-acting elements involved in the gene expression that is induced by abiotic stress (Shinwari, 1999). Most of the stress promoters contain an array of stress-specific cis-acting elements that are recognized by the requisite transcription factors; for example, the transcriptional regulation of hsp genes is mediated by the core "heat shock element" (HSE) located in the promoter region of these genes, 5' of the TATA box. All the plant hsp genes sequenced so far have been shown to contain partly overlapping multiple HSEs proximal to TATA motif. Apart from these hsp promoters, rd29 and adh gene promoters induced by osmotic stress and anaerobic stress, respectively, have also been studied. The Arabidopsis rd29A and rd29B are stress responsive genes, but are differentially induced under abiotic stress conditions. The rd29A promoter includes both DRE and ABRE elements, where dehydration, high salinity and low temperatures induce the gene, while the rd29B promoter includes only ABREs and the induction is ABA-dependent. Over expression of DREB1A transcription factors under the control of stress inducible promoter from rd29A showed a better phenotypic growth of the transgenic plants than the ones obtained using the constitutive CaMV 35S promoter (Kasuga *et al.,* 1999). A stress inducible expression of Arabidopsis CBF1 in transgenic tomato was achieved using the ABRC1 promoter from barley HAV22 (Lee *et al.,* 2003). Gene expression is induced by the binding of DREB1A, which in itself is induced by cold and water stress, to a cisacting DRE element in the promoters of genes such as rd29A, rd17, cor 6.6, cor 15A, erd 10, and kin1, thereby, initiating synthesis of gene products imparting tolerance to low temperatures and water stress in plants. The regions of respiratory alcohol dehydrogenase adh1 gene promoter in maize and rice that are required for anaerobic induction include a string of bases called anoxia response element (ARE) with the consensus sequence of its core element as TGGTTT. Besides, other stress-responsive cis-acting promoter sequences like low temperature responsive elements (LTRD) with a consensus sequence of A/GCCGAC have been identified in genes such as Cor 6.6, Cor 15 and Cor 78 These basic findings on stress promoters have led to a major shift in the paradigm for genetically engineering stress tolerant crops (Katiyar-Aggarwal *et al.,* 1999).

11. Applications of Chloroplast Engineering for Abiotic Stress Tolerence

Abiotic stresses Chloroplast engineering had been successfully applied for the development of plants with tolerance to salt, drought and low temperature. Previous research has shown that over-expression of enzymes for Glycine betaine (GlyBet) biosynthesis in transgenic plants improved tolerance to various abiotic stresses (Rhodes and Hanson, 1993). Choline monooxygenase (BvCMO) from beet (Beta vulgaris), the enzyme that catalyzes the conversion of choline into betaine aldehyde, has been recently transferred into the plastid genome of tobacco. Transplastomic plants demonstrated that higher photosynthetic rate and apparent quantum yield of photosynthesis in the presence of 150 mmol/L NaCl. Salt stress caused no significant change on the maximal efficiency of PSII photochemistry (Fv/Fm) in both wild type and transplastomic plants (Zhang *et al.*, 2008). Transplastomic carrot plants expressing BADH could be grown in the presence of high concentrations of NaCl (up to 400 mmol/L), the highest level of salt tolerance reported so far among genetically modified crop plants (Kumar *et al.*, 2004b). Trehalose has been found to accumulate under stress conditions such as freezing, heat, salt, or drought, so that it is thought to play a role in protecting cells against damage caused by these stresses. In contrast to nuclear transgenic plants that exhibited pleiotropic effects even at low levels of TPS1 expression, chloroplast transgenic plants grew normally and accumulated trehalose 25-fold higher (Lee *et al.*, 2003). Chloroplast transgenic plants showed a high degree of drought tolerance by remaining green and healthy in 6 percent PEG, whereas wildtype plants were completely bleached (Lee *et al.*, 2003). The unsaturation level of fatty acids (FA) in plant lipids has several implications for the stress tolerance of higher plants as well as for their nutritional value and industrial utilisation. Ä9 desaturase gene, an important gene in lipid metabolic pathways, was transformed into tobacco chloroplast. The transplastomic plants demonstrated the feasibility of using plastid transformation to engineer lipid component in both vegetative and reproductive tissues for increasing cold tolerance (Craig *et al.*, 2008). The feasibility to use chloroplast genetic engineering for weed control has been explored in several studies that aimed at producing glyphosate-tolerant tobacco plants (Daniell *et al.*, 1998; Lutz *et al.*, 2001). Plastid expression of the bar gene encoding the herbicide-inactivating phosphinothricin acetyltransferase (PAT) enzyme led to high-level enzyme accumulation (up to > 7% of TSP) and conferred field-level tolerance to glufosinate (Lutz *et al.*, 2001). These cases demonstrate that transplastomic technology might be particularly useful to develop plants resistant to abiotic and biotic stresses. The plastids engineered to have an adequate expression of resistance genes provide effective plant protection in the field.

12. Means of Stress Impositions, Growth Conditions, and Evaluations

Stress conditions used to evaluate the transgenic material in most of the reports so far, are usually too severe (Nanjo *et al.*, 1999a; Shinwari *et al.*, 1998; Garg *et al.*, 2002) as plants are very unlikely to undergo such stresses under field conditions. Also, the means of evaluation are often significantly different from natural conditions. For example, Pellegrineschi *et al.* (2004) compared the performance of initial events of DREB1A transgenic wheat to the wild parent by with holding water to 2-week-old seedlings grown in 5 cm 9 5 cm pots, and then re-watering until maturity when they were evaluated. Untransformed plants were nearly dead within 10–15 days of stress imposition, likely because of a different pattern of water use, whereas transgenic plants survived in these small pots and "passed" the evaluation successfully; such conditions would obviously not occur in the field. Besides, the type of systems used to assess plant performance; one would expect the evaluation to be made, at least, on the basis of biomass accumulated during the stress. While the use of PEG (polyethylene glycol) in hydroponics can be useful to test certain response of plants under a given osmotic potential as reported by Pilon-Smits *et al.* (1996, 1999), it offers relatively different conditions than in the soil where the water reservoir is by definition finite. Here, the observation on improved growth was explained by an increased water uptake under the water potential applied, due to osmolyte production by the transgenic plant. This is quite possible in such a system because the water reservoir is unlimited in hydroponics, and because the water potential is constant. Under soil conditions, however, the volume of soil surrounding the root where water can be extracted is limited, and the water potential of that soil quickly declines upon water uptake by roots, reaching soil water potential where even the enhanced osmolyte production of the transgenics would be unable to extract any significant additional amount of water. A more realistic test of the ability to take up water using osmotic potential-enhanced transgenics would be to compare their capacity to extract water from a soil system rather than a hydroponic system. A recent study by Sivamani *et al.* (2000) reported an increased WUE in the transgenic wheat. Unfortunately there was no control over the soil evaporation that probably accounts for most of the water loss and explained the very low values of WUE observed. Besides, investigating drought responses by using fresh weight (Sun *et al.*, 2001) and other indirect estimates of performance like growth rate, stem elongation (Pilon-Smits *et al.*, 1995; Lee *et al.*, 2003), or survival are likely to give inconsistent results (Pardo *et al.*, 1998). While applying a drought stress, it is important to know the stages of drought stress that the plants are exposed to, for which, a detailed description of growth conditions, plant size, container size, water availability, and transpiration is needed. It is also crucial to report the dry weight of tested plants, possibly before and after the stress period. Similarly, often the stress imposed has been modified from 2 days, to 2 weeks, and even 4 weeks using the same experimental conditions (Lee *et al.*, 2003), without indicating the water holding capacity of the potting mixture used as well as the plant density. This obviously leads to different types of stresses, where the plants exposed for 2 days of water stress may well have remained in stage I when water is abundant (see below), while plants exposed to 4 weeks stress may have spent most of the time under stage III where roots may have exhausted all the available water. Also there are cases where a given quantity of water is applied to the plants on alternate days from 2 to 10 weeks (Sivamani *et al.*, 2000), thereby, disregarding the fact that the water requirements increase dramatically during the period, and probably exposing their plants to an initial flooding before a severe stress.

13. Conclusion and Future Perspectives

Developing drought and salinity tolerance crop plants using conventional plant breeding methods had limited success during the past century. New technologies are providing opportunities to address the challenging problem of maintaining high yielding crop production under stressful environmental conditions and changing climates. The last century of breeding effort and crop physiology studies have led to increases in the economic yield of most major crop species and have elucidated many traits that are associated with plant adaptability to drought prone environments. However, many attempts to improve drought tolerance through breeding have been associated with reduced yield potential. Nevertheless, enormous advances should have been making in the understanding of the physiological and molecular responses of plants to water deficit through breeding and physiological studies and scientists in these areas will continue to have a fundamental role in the development of transgenic drought tolerant crops. One of the greatest limitations in drought stress tolerance breeding has been the fact that drought takes many varied forms. Depending on the crop and the season, water stress may be experienced in early vegetative stages, during transition to flowering and even post anthesis. In some cases, this may be a terminal stress or, more commonly, a recurring event broken by sporadic rainfall precipitating a recovery of the whole plant, sometimes with reduced yield potential as a result of conservative plant cell survival responses. Transgenic approaches for increasing plant salt tolerance are feasible. So far, results obtained with many genes are encouraging, and recent results obtained in transgenic plants harbouring genes encoding an Na^+ / H^+ antiporter or a transcription factor show the possibility of increasing the salt tolerance. However, it still needs to be tested whether or not this is the case for all crops since may diverge in the mechanisms of stress responses. Therefore, the basic understanding of the mechanisms underlying the functioning of stress genes is important for the development of transgenic plants. Each stress is a multigenic trait and therefore their manipulation may result in alteration of a large number of genes as well as their products. A deeper understanding of the transcription factors regulating these genes, the products of the major stress responsive genes and cross talk between different signaling components should remain an area of intense research activity in future.

The recent efforts to improve abiotic stress tolerance in crop plants by employing some of the stress-related genes and transcription factors that have been cloned and characterized. The following general conclusions emerge from this chapter:

1. The use of transgenes to improve the tolerance of crops to abiotic stresses remains an attractive option and options targeting multiple gene regulation appear better than targeting single genes.

2. An important issue to address is how the tolerance to specific abiotic stress is assessed and whether the achieved tolerance compares to existing tolerance. The biological cost of production of different metabolites to cope with stress and their effect on yield should be properly evaluated.

3. A well focused approach combining the molecular, physiological and metabolic aspects of abiotic stress tolerance is required for bridging the knowledge gaps between short and long term effects of the genes and their products and between the molecular or cellular expression of the genes and the whole plant phenotype under stress.

4. Thorough understanding of the underlying physiological processes in response to different abiotic stresses can efficiently drive the choice of a given promoter or transcription factor to be used for transformation.

14. References

1. Abebe, T.; Guenzi, A.C.; Martin, B.; Cushman J.C. Tolerance of mannitol-accumulating transgenic wheat to water stress and salinity. *Plant Physiol.***2003**, 131, 1748–1755.

2. Adams, R.P.; Kendall, E.; Kartha, K.K. Comparison of free sugars in growing and desiccated plants of Selaginella lepidophylla. *Biochem. Syst. Ecol.***1990**, 18, 107–110.

3. Allen, R. Dissection of oxidative stress tolerance using transgenic plants. *Plant Physiol.* **1995**, 107, 1049–1054.

4. Alcazar, R.; Planas, J.; Saxena, T.; Zarza, X.; Bortolotti, C.; Cuevas, J.; Bitriαn, M.; Tiburcio, A. F.; Altabella, T. Putrescine accumulation confers

drought tolerance in transgenic Arabidopsis plants over-expressing the homologous Arginine decarboxylase 2 gene. *Plant Physiol. Biochem.* **2010**, 48, 547–552.

5. Apse, M.P.; Blumwald, E. Engineering salt tolerance in plants. *Curr. Opin. Plant Biol.* **2002,** 13, 146–150.

6. Apse, M.P.; Aharon, G.S.; Snedden, W.A.; Blumwald, E. Salt tolerance conferred by over expression of a vacuolar Na^+/H^+ antiport in Arabidopsis. *Science.* **1999**, 285, 1256–1258.

7. Arenas-Huertero, C.; Pérez, B.; Rabanal, F.; Blanco-Melo, D.; De la Rosa, C.; Estrada-Navarrete, G.; Sanchez, F.; Co-varrubias, A.; and Reyes, J. Conserved and Novel miRNAs in the Legume *Phaseolus vulgaris* in Response to Stress. *Plant Mol. Biol.* **2009**, 70, 385-401.

8. Bajaj, S.; Targoli, J.; Liu, L-F.; Ho, T. H.;Wu, R.Transgenic approaches to increase dehydration-stress in plants. *Mol. Breed.* **1999**, 5, 493–503.

9. Bartels, D.; Sunkar, R. Drought and salt tolerance in plants. *Crit. Rev. Plant Sci.* **2005,** 24, 23–58.

10. Battaglia, M.; Olvera-Carrillo, Y.; Garciarrubio, A.; Campos, F.; Covarrubias, A.A. The enigmatic LEA proteins and other hydrophilins. *Plant Physiol.* **2008**, 148, 6–24.

11. Bhatnagar-Mathur, P.; Vadez, V.; Jyostna Devi, M.; Lavanya, M.; Vani, G.; Sharma, K. Genetic engineering of chickpea (*Cicer arietinum* L.) with the P5CSF129A gene for osmoregulation with implications on drought tolerance. *Mol. Breed.* **2009**, 23, 591–606.

12. Bhora, J. S.; Dooffling, H.; Dooffling, K. Salinity tolerance of rice with reference to endogenous and exogenous abscised acid. *J. Agron. Crop. Sci.* **1995**, 174, 79–86.

13. Bohnert, H.; Ayoubi, P.; Borchert, C.; Bressan, R. A genomics approach towards salt stress tolerance. *Plant Physiol. Biochem.* **2001**,39, 295–311.

14. Bordas, M.; Montesinos, C.; Dabauza, M.; Salvador, A.; Roig, L.A.; Serrano, R.; Moreno, V. Transfer of the yeast salt tolerance gene *HAL1* to *Cucumis melo* L. cultivars and *in vitro* evaluation of salt tolerance. *Trans. Res.* **1997**, 6, 41–50.

15. Borsani, O.; Cuartero, J.; Fernandez, J. A.; Valpuesta, V.; Botella, M. A. Identification of two loci in tomato reveals distinct mechanisms for salt tolerance.*Plant Cell.* **2001a**, 13, 873–888.

16. Borsani, O.; Valpuesta, V.; Botella, M. A. Evidence for a role of salicylic acid in the oxidative damage generated by NaCl and osmotic stress in *Arabidopsis* seedlings. *Plant Physiol.* **2001b**, 126, 1024–1030.

17. Bray, E. A. Molecular responses to water deficit. *Plant Physiol.* **1993**, 103, 1035–1040.

18. Bray, E.A. Plant responses to water deficit. *Trends Plant Sci.* **1997**, 2,48–54.

19. Bressan, R. A.; Zhang, C.; Zhang, H.; Hasegawa, P. M.; Bohnert, H. J.; Zhu, J. K. Learning from the Arabidopsis experience. The next gene search paradigm. *Plant Physiol.* **2001**, 127, 1354–1360.

20. Brewster, J. L.; de Valoir, T.; Dwyer, N. D.; Winter, E.; Gustin, M. C. An osmosensing signal transduction pathway in yeast. *Science.* **1993**, 259, 1760–1763.

21. Capell, T.; Bassie, L.; Christou, P. Modulation of the polyamine biosynthetic pathway in transgenic rice confers tolerance to drought stress. *Proc. Natl. Acad. Sci. U. S. A.* **2004**, 101, 9909–9914.

22. Chen, M.; Chen, Q. J.; Niu, X.-G.; Zhang, R.; Lin, H. Q.; Xu, C. Y.; Wang, X. C.; Wang, G. Y.; Chen J. Expression of OsNHX1 gene in maize confers salt tolerance and promotes plant growth in the field. *Plant, Soil Environ.* **2007**, 53, 490–498.

23. Cortina, C.; Culianez-Macia, F. A. Tomato abiotic stress enhanced tolerance by trehalose biosynthesis. *Plant Sci.* **2005**, 169, 75-82.

24. Cuartero, J.; Fernandez-Munoz, R. Tomato and salinity. *Sci. Hort.* **1999**, 78, 83–125.

25. de Ronde, J. A.; Laurie, R.N.; Caetano, T.; Greyling, M.M.; Kerepesi, I. Comparative study between transgenic and non-transgenic soybean lines proved transgenic lines to be more drought tolerant. *Euphytica.* **2004**, 138, 123–132.

26. Dyachenko, O.; Zakharchenko, N.; Shevchuk, T.; Bohnert, H.; Cushman, J., Buryanov, Y. Effect of hyper methylation of CCWGG sequences in DNA of *Mesembryanthemum crystallinum* plants on their adaptation to salt stress. *Biochemistry (Mosc.).* **2006**, 71, 461–465.

27. Espasandin, F. D.; Maiale, S. J.; Calzadilla, P.; Ruiz, O. A.; Sansberro, P. A. Transcriptional regulation of 9-cis-epoxycarotenoid dioxygenase (NCED) gene by putrescine accumulation positively modulates ABA synthesis and drought tolerance in Lotus tenuis plants. *Plant Physiol. Biochem.* **2014**, 76, 29–35.

28. Feng, S.; Jacobsen, S.E.; Reik, W. Epigenetic reprogramming in plant and animal development. *Science.* **2010**, 330, 622–627.

29. Flowers, T. J.; Yeo, A. R. Breeding for salinity resistance in crop plants: where next? *Aust. J. Plant Physiol.* **1995**, 22, 875-884.

30. Flowers, T. J.; Koyama, M. L.; Flowers, S. A.; Chinta Sudhakar, K. P.; Shing, K. P.; Yeo, A. R. QTL: their place in engineering tolerance of rice to salinity. *J. Exp. Bot.* **2000**, 51, 99-106.

31. Foolad, M. R.; Lin, G. Y. Genetic potential for salt tolerance during germination in Lycopersicon species. *Hort. Sci.* **1997**, 32, 296–300.

32. Foyer, C. H.; Shigeoka, S. Understanding oxidative stress and antioxidant functions to enhance photosynthesis. *Plant Physiol.* **2011**, 155, 93–100.

33. Furbank, R.T.; von Caemmerer, S.; Sheehy, J.; Edwards, G. C_4 rice: a challenge for plant phenomics. *Funct. Plant Biol.* **2009**, 36, 845–856.

34. Gao, P.; Bai, X.; Yang, L.; Lv, D.; Li, Y.; Cai, H.; Ji, W.; Guo, D.; Zhu, Y. Over-expression of osa-MIR396c decreases salt and alkali stress tolerance. *Planta.* **2010**, 231, 991–1001.

35. Gao, P.; Bai, X.; Yang, L.; Lv, D.; Pan, X.; Li, Y.; Cai, H.; Ji, W.; Chen, Q.; Zhu Y. M. *osa-MIR393*: A salinity and Alkaline Stress-Related MicroRNA Gene. *Mol. Biol. Rep.* **2011**, 38, 237-242.

36. Garg, A. K.; Kim, J. K.; Owens, T. G.; Ranwala, A. P.; Choi, Y. D.; Kochian, L. V.; Wu, R. J. Trehalose accumulation in rice plants confers high tolerance levels to different abiotic stresses, Proc. *Natl. Acad. Sci. U. S. A.* **2002**, 99, 15898–15903.

37. Gisbert, C.; Rus, A. M.; Bolarin, M. C.; Lopez-Coronado, J. M.; Arrillaga, I.; Montesinos, C.; Caro, M.; Serrano, R.; Moreno, V. The yeast *HAL1* gene improves salt tolerance of transgenic tomato. *Plant Physiol.* **2000**, 123, 393–402.

38. Glenn, E.P., Brown, J. J., and Blumwald E. (1999). Salt tolerance and crop potential of halophytes. *Crit. Rev. Plant Sci.* 18, 227–255.

39. Goddijn, O.; Verwoerd, T. C.; Voogd, E.; Krutwagen, R.; de Graff, P.; Poels, V.; van Dun, K.; Ponstein, A. S.; Damm, B.; Pen, J. Inhibition of trehalose activity enhances trehalose accumulation in transgenic plants. *Plant Physiol.* **1997**, 113, 181–190.

40. Goyal, K.; Walton, L. J.; and Tunnacliffe, A. LEA proteins prevent protein aggregation due to water stress. *Biochem. J.* **2005**, 388, 151–157.

41. Hand, S. C.; Menze, M. A.; Toner, M.; Boswell, L.; Moore, D. LEA proteins during water stress: not just for plants anymore. *Annu. Rev. Physiol.* **2011**, 73, 115–134.

42. Hare, P. D.; Cress, W. A.;Van Staden, J. Dissecting the roles of osmolyte accumulation during stress. *Plant Cell Environ.* **1998**, 21, 535–554.

43. Hasegawa, M.; Bressan, R.; Pardo, J. M. The dawn of plant salt tolerance genetics. *Trends Plant Sci.* **2000a**, 5, 317–319.

44. Hasegawa, M.; Bressan, R.; Zhu, J. K.; Bhonert, H. Plant cellular and molecular responses to high salinity. *Ann. Rev. Plant Physiol.* **2000b**, 51, 493–499.

45. Hauser, F.; Horie, T. A conserved primary salt tolerance mechanism mediated by HKT transporters: a mechanism for sodium exclusion and maintenance of high K^+/Na^+ ratio in leaves during salinity stress. *Plant Cell Environ.* **2010**, 33, 552-565.

46. Hausler, R. E.; Hirsch, H. J.; Kreuzaler, F.; Peterhänsel, C. Over expression of C_4-cycle enzymes in transgenic C_3 plants: a biotechnological approach to improve C_3-photosynthesis. *J. Exp. Bot.* **2002**, 53, 591–607.

47. Hayashi, H.; Alia Mustardy, L.; Deshnium, P.; Ida, M.; Murata, N. Transformation of *Arabidopsis thaliana* with the *codA* gene for choline oxidase; accumulation of glycinebetaine and enhanced tolerance to salt and cold stress. *Plant J.* **1997**, 12, 133–142.

48. He, C.; Yan, J.; Shen, G.; Fu, L.; Holaday, A. S.; Auld, D.; Blumwald, E.; Zhang, H. Expression of an Arabidopsis vacuolar sodium/proton antiporter gene in cotton improves photosynthetic performance under salt conditions and increases fiber yield in the field. *Plant Cell Physiol.* **2005**, 46, 1848–1854.

49. Hibberd, J. M.; Sheehy, J. E.; Langdale, J. A. Using C_4 photosynthesis to increase the yield of rice rationale and feasibility. *Curr. Opin. Plant Biol.* **2008**, 11, 228–231.

50. Hoi, P. X.; Ham, L. H.; Tuteja,N. A nuclear DNA helicase from pea (Pisum sativum L.) That translocates in the 3' to 5' direction.Journal of Biology. **2008**, 30 (2), 50-55.

51. Hsieh, T.H.; Lee, J. T.; Charng, Y. Y.; Chan, M. T. Tomato plants ectopically expressing Arabidopsis CBF1 show enhanced resistance to water deficit stress. *Plant Physiol.* **2002**, 130, 618–626.

52. Ingram, J.; Bartels, D. The molecular basis of dehydration tolerance in plants. *Ann. Rev. Plant Physiol. Plant Mol. Biol.* **1996**, 47, 377–403.

53. Ishitani, M.; Xiong, L.; Stevenson, B.; Zhu, J. K. Genetic analysis of osmotic and cold stress signal transduction in Arabidopsis: interactions and convergence of abscisic acid-de-pendent and abscisic acid-independent pathways. *Plant Cell.* **1997**, 9, 1935–1949.

54. Jaglo, K. R.; Kleff, S.; Amundsen, K. L.; Zhang, X.; Haake, V.; Zhang, J. Z.; Deits, T.; Thomashow, M. F. Components of the Arabidopsis C-Repeat/ dehydration-responsive element binding factor cold-response pathway are conserved in Brassica napus and other plant species. *Plant Physiol.* **2001**, 127, 910–917.

55. Jaglo-Ottosen, K. R.; Gilmour, S. J.; Zarka, D. G.; Schabenberger, O.; Thomashow, M. F. Arabidopsis CBF1 over expression induces COR genes and enhances freezing tolerance. *Science.* **1998**, 280, 104–106.

56. James, R. A.; Blake, C.; Zwart, A. B.; Hare, R. A.; Rathjen, A. J.; Munns, R. Impact of ancestral wheat sodium exclusion genes Nax1 and Nax2 on grain yield of durum wheat on saline soils. *Funct. Plant Biol.* **2012**, 39,609-618.

57. Jang, I. C.; Oh, S. J.; Seo, J. S.; Choi, W. B.; Song, S. I.; Kim, C.H.; Kim, Y. S.; Seo, H. S.; Choi, Y. D.; Nahm, B. H.; and Kim, J. K. Expression of a bifunctional fusion of the Escherichia coli genes for trehalose-6-phosphate synthase and trehalose-6-phosphate phosphatase in transgenic rice plants increases trehalose accumulation and abiotic stress tolerance without stunting growth. *Plant Physiol.* **2003**,131, 516–524.

58. Jagadeeswaran, G.; Saini, A.; Sunkar, R. Biotic and abiotic stress down-regulate miR398 expression in arabidopsis. *Planta.* **2009**, 229, 1009-1014.

59. Jian, X.; Zhang, L.; Li, G.; Zhang, L.; Wang, X.; Cao, X.; Fang, X. H.; Chen, F. Identification of Novel Stress Regulated MicroRNAs from *Oryza sativa* L. *Genomics.* **2010**, 95, 47-55.

60. Johnson, D.; Smith, S.; Dobrenz, A. Genetic and phenotypic relationships in response to NaCl at different developmental stages in alfalfa. *Theor. Appl. Gen.* **1992**, 83, 833–838.

61. Kasuga, M.; Liu, Q.; Miura, S. Yamaguchi-Shinozaki K, Shinozaki K: Improving plant drought, salt, and freezing tolerance by gene transfer of a single stress-inducible transcription factor. *Nat Biotechnol.* **1999**, 17, 287-291.

62. Kasukabe, Y.; He, L.; Nada, K.; Misawa, S.; Ihara, I.; Tachibana, S. Over expression of spermidine synthase enhances tolerance to multiple environmental stresses and up-regulates the expression of various stress-regulated genes in transgenic Arabidopsis thaliana. *Plant Cell Physiol.* **2004**, 45, 712–722.

63. Khraiwesh, B.; Zhu, J. K.; and Zhu, J. Role of miRNAs and siRNAs in biotic and abiotic stress responses of plants, BBA. *Gene Regul. Mech.* **2011**, 1819, 137–148.

64. Koornneef, M.; Leon-Kloosterziel, K. M.; Scwartz, S. H.; Zeevart J. A. D. The genetic and molecular dissection of abscisic acid biosynthesis and signal transduction in *Arabidopsis. Plant Physiol. Biochem.* **1998**, 36: 83–89.

65. Kruszka, K.; Pieczynskia, M.; Windelsb, D.; Bielewicza, D.; Jarmolowskia, A.; Kulinskaa, Z. S.; Vazquezb, F. Role of MicroRNAs and Other sRNAs of Plants in Their Changing Environments. *J. Plant Physiol.* **2012**, 169, 1664-1672.

66. Ku, M. S. B.; Cho, D.; Li, X.; Jiao, D. M.; Pinto, M.; Miyao, M.; Matsuoka, M. Introduction of Genes Encoding C_4 Photosynthesis Enzymes into Rice Plants: Physiological Consequences, in: J.A. Goode, D. Chadwick (Eds.),

Novartis Foundation Symposium 236 — Rice Biotechnology: Improving Yield, Stress Tolerance and Grain Quality, John Wiley & Sons, Ltd., Chichester, UK. **2007**, pp. 100–116.

67. Kudla, J.; Xu, Q.; Harter, K.; Gruissem, W.; Luan, S. Genes for calcineurin B-like proteins in Arabidopsis are differentially regulated by stress signals. *Proc. Natl. Acad. Sci. USA.* **1999**, 96, 4718–4723.

68. Kumria, R.; Rajam, M. V. Ornithine decarboxylase transgene in tobacco affects polyamines, in vitro-morphogenesis and response to salt stress. *J. Plant Physiol.* **2002,** 159, 983–990.

69. Lee, Y. P.; Baek, K. H.; Lee, H. S.; Kwak, S. S.; Bang, J. W.; Kwon, S. Y. Tobacco seeds simultaneously over-expressing Cu/Zn-superoxide dismutase and ascorbate peroxidase display enhanced seed longevity and germination rates under stress conditions. *J. Exp. Bot.* **2010**, 61, 2499–2506.

70. Liu, H.; Wang, Q.; Yu, M.; Zhang, Y.; Wu, Y.; Zhang, H. Transgenic salt-tolerant sugar beet (*Beta vulgaris* L.) constitutively expressing an Arabidopsis thaliana vacuolar Na^+/H^+ antiporter gene, AtNHX3, accumulates more soluble sugar but less salt in storage roots. *Funct. Plant Biol.* **2008**, 31, 1325–1334.

71. Liu, H. H.; Tian, X.; Li, Y. J.; Wu, C. A.; Zheng, C. C. Microarray Based Analysis of Stress-Regulated MicroRNAs in Arabidopsis thaliana. *RNA.* **2008**,14, 836-843.

72. Liu, J.; Ishitani, M.; Halfter, U.; Kim, C. S.; Zhu, J. K. The *Arabidopsis thaliana SOS2* gene encodes a protein kinase that isrequired for salt tolerance. Proc. *Natl. Acad. Sci. USA.* **2000**, 97, 3730– 3734.

73. Liu, Q.; Kasuga, M.; Sakuma, Y.; Abe, H.; Miura, S.; Yamaguchi-Shinozaki, K.; Shinozaki, K. Two transcription factors, DREB1 and DREB2, with an EREBP/AP2 DNA binding domain separate two cellular signal transduction pathways in drought-and low-temperature-responsive gene expression, respectively, in *Arabidopsis. Plant Cell.* **1998,** 10, 1391–1406.

74. Lohman, T. M.; Bjornson, K. P. Mechanism of helicase catalyzed DNA unwinding. *Ann. Rev. Biochem.* **1996**, 65,169–214.

75. Luan, S.; Li, W.; Rusnak, F.; Assmann, S. M.; Schreiber, S. L. Immunosuppressant implicate protein phosphatase-regulation of K^+ channels in guard cells. *Proc. Natl. Acad. Sci. USA.* **1993**, 90, 2202–2206.

76. Maeda, T.; Wurgler-Murphy, S. M.; Saito, H. A two-component system that regulates an osmosensing MAP kinase cascade in yeast. *Nature.* **1994**, 369, 242–245.

77. Maeda, T.; Takekawa, M.; Saito H. Activation of yeast PBS2 MAPKK by

MAPKKKs or by binding of an SH3-containing osmosensor. *Science.* **1995**, 269, 554–558.

78. Maischak, H.; Zimmermann, M.R.; Felle, H. H.; Boland, W.; Mithofer, A. Alamethicin induced electrical long distance signaling in plants. *Plant Signal Behav.* **2010**, 5, 988-990.

79. Matson, S. W.; Bean, D.; and George, J. W. DNA helicases; enzymes with essential roles in all aspects of DNA metabolism. *BioEssays.* **1994**, 16, 13–21.

80. McCourt, P. Genetic analysis of hormone signaling. *Ann. Rev. Plant Physiol. Plant Mol. Biol.* **1999**, 50, 219–243.

81. Mendoza, I.; Rubio, F.; Rodriguez-Navarro, A.; Pardo, J. M. The protein phosphatase calcineurin is essential for NaCl tolerance of *Saccharomyces cerevisiae. J. Biol. Chem.* **1994**, 269, 8792–8796.

82. Mittler, R.; Vanderauwera, S.; Suzuki, N.; Miller, G.; Tognetti, V. B.; Vandepoele, K.; Gollery, M.; Shulaev, V.;Van Breusegem, F. ROS signaling: the new wave? *Trends Plant Sci.* **2011**, 16,300-309.

83. Moller, I. S.; Gilliham, M.; Jha, D.; Mayo, G. M.; Roy, S. J.; Coates, J. C.; Haseloff, J.; Tester, M. Shoot Na+ exclusion and increased salinity tolerance engineered by cell type-specific alteration of Na+ transport in Arabidopsis. *Plant Cell.* **2009**, 21, 2163–2178.

84. Muuns, R. Comparative physiology of salt and water stress. *Plant Cell Environ.* **2002**, 25, 239–250.

85. Munns, R.; Tester, M. Mechanisms of salinity tolerance. *Annu. Rev. Plant Biol.* **2008**, 59, 651-681.

86. Murchie, E. H.; Pinto, M.; Horton, P. Agriculture and the new challenges for photosynthesis research. *New Phytol.* **2009,** 181, 532–552.

87. Nakamura, T.; Liu, Y.; Hirata, D.; Namba, H.; Harada, S.; Hirokawa, T.; Miyakawa, T. Protein phosphatase type 2B (calcineurin)-mediated, FK506-sensitive regulation of intracellular ions in yeast is an important determinant for adaptation to high salt stress conditions. *EMBO J.* **1993**, 12, 4063–4071.

88. Nanjo, T.; Kobayashi, M.; Yoshiba, Y.; Kakubari, Y.; Yamaguchi-Shinozaki, K.; Shinozaki, K. Antisense suppression of proline degradation improves tolerance to freezing and salinity in Arabidopsis thaliana. *FEBS Lett.* **1999**, 461, 205–210.

89. Neily, M. H.; Baldet, P.; Arfaoui, I.; Saito, T.; Li, Q. L.; Asamizu, E. Over expression of apple spermidine synthase 1 (MdSPDS1) leads to significant salt tolerance in tomato plants. *Plant Biotechnol. J.* **2011**, 28, 33–42.

90. Niu, X.; Bressan, R. A.; Hasegawa, P. M.; Pardo, J. M. Ion homeostasis in NaCl stress enviroments. *Plant Physiol.* **1995**, 109, 735–742.

91. Pardo, J. M. Biotechnology of water and salinity stress tolerance, *Curr. Opin. Plant Biol.* **2010**, 21, 185–196.

92. Pasapula, V.; Shen, G.; Kuppu, S.; Paez-Valencia, J.; Mendoza, M.; Hou, P.; Chen, J.; Qiu, X.; Zhu, L.; Zhang, X. Expression of an Arabidopsis vacuolar H^+ pyrophosphatase gene (AVP1) in cotton improves drought- and salt tolerance and increases fibre yield in the field conditions. *Plant Biotechnol. J.* **2011**, 9, 88-99.

93. Peleg, Z.; Reguera, M.; Tumimbang, E.; Walia, H.; Blumwald, E. Cytokinin mediated source/sink modifications improve drought tolerance and increases grain yield in rice under water stress. *Plant Biotechnol. J.* **2011**, 9, 747–758.

94. Peleg, Z.; Apse, M. P.; Blumwald, E. Engineering salinity and water-stress tolerance in crop plants: getting closer to the field, *Adv. Bot. Res.* **2011**, 57, 405–443.

95. Peremarti, A.; Bassie, L.; Christou, P.; Capell, T. Spermine facilitates recovery from drought but does not confer drought tolerance in transgenic rice plants expressing Datura stramonium S-adenosylmethionine decarboxylase. *Plant Mol. Biol.* **2009**, 70, 253–264.

96. Peterhansel, C. Best practice procedures for the establishment of a C_4 cycle in transgenic C_3 plants. *J. Exp. Bot.* **2011**, 62, 3011–3019.

97. Piao, H. L.; Pih, K. T.; Lim, J. H.; Kang, S. G.; Jin, J. B.; Kim, S. H.; Hwang, I. An *Arabidopsis GSK3 /shaggy*-like gene that comple-ments yeast salt stress-sensitive mutants is induced by NaCl and abscisic acid. *Plant Physiol.* **1999**, 119, 1527–1534.

98. Plett, D. C.; Moller, I. S. Na^+ transport in glycophytic plants: what we know and would like to know. *Funct. Plant Biol.* **2010**, 33, 612–626.

99. Popping, B.; Gibbons, T.; Watson, M. D. The *Pisum sativum* MAP kinase homologue (*PsMAPK*) rescues the *Saccharomycescerevisiae* hog1 deletion mutant under conditions of high os-motic stress. *Plant Mol. Biol.* **1996**, 31, 355–363.

100. Prabhavathi, V. R.; Rajam, M. V. Polyamine accumulation in transgenic eggplant enhances tolerance to multiple abiotic stresses and fungal resistance. *Plant Biotechnol.***2007**, 24, 273–282.

101. Price, A.; Hendry, G. Iron-catalyzed oxygen radical forma-tion and its possible contribution to drought damage in nine native grasses and three cereals. *Plant Cell Environ.* **1991**, 14, 477–484.

102. Quan, R.; Shang, M.; Zhang, H.; Zhao, Y.; Zhang, J. Engineering of enhanced glycine betaine synthesis improves drought tolerance in maize. *Plant Biotechnol. J.* **2004**, 2, 477–486.

103. Rademacher, T.; Hausler, R. E.; Hirsch, H. J.; Zhang, L.; Lipka, V.; Weier, D.; Kreuzaler, F.; Peterhänsel, C. An engineered phosphoenolpyruvate carboxylase redirects carbon and nitrogen flow in transgenic potato plants. *Plant J.* **2002**, 32, 25–39.

104. Rhodes, D.; Hanson, A. D. Quaternary ammonium and tertiary sulfonium compounds in higher plants. Ann. Rev. Plant Physiol. *Plant Mol. Biol.* **1993**, 44, 357–384.

105. Rivero, R. M.; Gimeno, J.; Van Deynze, A.; Walia, H.; Blumwald, E. Enhanced cytokinin synthesis in tobacco plants expressing P_{SARK}::IPT prevents the degradation of photosynthetic protein complexes during drought. *Plant Cell Physiol.* **2010**, 51, 1929–1941.

106. Rivero, R. M.; Kojima, M.; Gepstein, A.; Sakakibara, H.; Mittler, R.; Gepstein, S.; Blumwald, E. Delayed leaf senescence induces extreme drought tolerance in a flowering plant. *Proc. Natl. Acad. Sci. USA.* **2007**, 104, 19631–19636.

107. Roberts, S. K. Regulation of K^+ channels in maize roots by water stress and abscisic acid. *Plant Physiol.* **1998**, 116, 145–153.

108. Roberts, S. K.; Snowman, B. N. The effects of ABA on channel-mediated K^+transport across higher plant roots. *J. Exp. Bot.* **2000**, 51, 1585–1594.

109. Rocak, S.; Linder, P. DEAD-box proteins: the driving forces behind RNA metabolism. *Nat. Rev. Mol. Cell Biol.* **2004**, 5, 232-241.

110. Rock, C. D. Pathways to abscisic acid-regulated gene expression. *New Phytol.* **2000**, 148, 357–396.

111. Rodriguez-Navarro, A. Potassium transport in fungi and plants. *Biochem. Biophys. Acta.* **2000**, 1469, 1–30.

112. Romero, C.; Belles, J.; Vaya, J.; Serrano, R.; Culianez-Macia, F. Expression of the yeast trehalose-6-phosphate synthase gene in transgenic tobacco plants: pleiotropic phenotypes include drought tolerance. *Planta.* **1997**, 201, 293-297.

113. Roxas, V. P.; Smith, R. K.; Allen, E. R.; Allen, R. D. Over expression of glutathione S-transferase/glutathione peroxidase en-hances the growth of transgenic tobacco seedlings during stress. *Nat. Biotechnol.* **1997**, 15, 988–991.

114. Roy, M.; Wu, R. Over expression of S-adenosylmethionine decarboxylase gene in rice increases polyamine level and enhances sodium chloride-stress tolerance. *Plant Sci.* **2002**, 163, 987–992.

115. Roy, M., and Wu, R. Arginine decarboxylase transgene expression and analysis of environmental stress tolerance in transgenic rice. *Plant Sci.* **2001**, 160, 869–875.

116. Sahoo, R. K.; Gill, S. S.; Tuteja, N. Pea DNA helicase 45 promotes salinity stress tolerance in IR64 rice with improved yield. *Plant Sign. Behav.* **2012**, 7.

117. Sage, R. F.; Zhu, X. G. Exploiting the engine of C$_4$ photosynthesis. *J. Exp. Bot.* **2011**, 62, 2989–3000.

118. Sage, R. F. The evolution of C$_4$ photosynthesis. *New Phytol.* **2004**,161, 341–370.

119. Sage, T. L.; Sage, R. F. The functional anatomy of rice leaves: implications for refixation of photorespiratory CO$_2$ and efforts to engineer C$_4$ photosynthesis into rice. *Plant Cell Physiol.* **2009**, 50, 756–772.

120. Schwab, R.; Palatnik, J. F.; Riester, M.; Schommer, C.; Schmid, M.; Weigel, D. Specific effects of microRNAs on the plant transcriptome. *Dev. Cell.* **2005**, 8, 517–527.

121. Scippa, G. S.; Di Michele, M.; Onelli, E.; Patrignani, G.; Chiatante, D.; Bray, E. A. The histone-like protein H1-S and the response of tomato leaves to water deficit. *J. Exp. Bot.* **2004**, 55, 99–109.

122. Serrano, R. Salt tolerance in plants and microorganisms: toxicity targets and defense responses. *Int. Rev. Cytol.* **1996**, 165, 1–52.

123. Shi, H.; Ishitani, M.; Kim, C.; Zhu, J. K. The *Arabidopsisthaliana* salt tolerance gene *SOS1* encodes a putative Na^1/H^{1}antiporter. *Proc. Natl. Acad. Sci. USA.* **2000**, 97, 6896–6901.

124. Shi, H.; Lee, B. H.; Wu, S. J.;Zhu, J. K. Over expression of a plasma membrane Na$^+$/H$^+$ antiporter gene improves salt tolerance in Arabidopsis thaliana. *Nat. Biotechnol.* **2003**, 21, 81-85.

125. Shi, H.; Quintero, F. J.; Pardo, J. M.;Zhu, J. K. The putative plasma membrane Na$^+$/H$^+$ antiporter SOS1 controls long-distance Na$^+$ transport in plants. *Plant Cell.* **2002a**, 14, 456–477.

126. Shi, H.; Xiong, L.; Stevenson, B.; Lu, T.; Zhu, J. K. The Arabidopsis salt overly sensitive 4 mutants uncover a critical role for vitamin B6 in plant salt tolerance. *Plant Cell.* **2002b**, 14, 575–588.

127. Smirnoff, N. The role of active oxygen in the response of plants to water deficit and desiccation. *New Phytol.* **1993**, 27–58.

128. Stewart, C. R.; Voetberg, G. Abscisic acid accumulation is not required for proline accumulation in wilted leaves. *Plant Physiol.* **1987**, 83, 747–749.

129. Stein, H.; Honig, A.; Miller, G.; Erster, O.; Eilenberg, H.; Csonka, L. N.; Szabados, L .; Koncz, C.; Zilberstein, A. Elevation of free proline and proline-rich protein levels by simultaneous manipulations of proline biosynthesis and degradation in plants. *Plant Sci.* **2011**, 181, 140–150.

130. Su, J.; Wu, R. Stress-inducible synthesis of proline in transgenic rice confers faster growth under stress conditions than that with constitutive synthesis. *Plant Sci.* **2004**, 166, 941-948.

131. Sunkar, R.; Kapoor, A.; Zhu, J. K. Posttranscriptional induction of two Cu/Zn superoxide dismutase genes in Arabidopsis is mediated by downregulation of miR398 and important for oxidative stress tolerance. *Plant Cell.* **2006**,18, 2051–2065.

132. Sunkar, R. MicroRNAs with macro-effects on plant stress responses. *Semin. Cell Dev. Biol.* **2010**, 21, 805–811.

133. Sunkar, R.; Zhu, J. K. Novel stress-regulated microRNAs and other small RNAs from Arabidopsis. *PlantCell.* **2004**,16, 2001-2019.

134. Szabolcs, I. Soil salinization. In: Pressarkli, M (ed) Hand-book of plant crop stress. **1994**, pp. 3–11.

135. Taylor, I. B.; Burbidage, A.; Thompson, A. J. Control of abscisic acid synthesis. *J. Exp. Bot.***2000**, 51, 1563–1574.

136. Thomas, J. C.; Smigocki, A. C.; Bohnert, H. J. Light-induced expression of ipt from *Agrobacterium tumefaciens* results in cytokinin accumulation and osmotic stress symptoms in trans-genic tobacco. *Plant Mol. Biol.***1995**, 27, 225–235.

137. Toone, W. M.; Jones, N. Stress-activated signaling pathways in yeast. *Genes Cells.* **1998**, 3, 485–498.

138. Trindade, I.; Capitao, C.; Dalmay, T.; Fevereiro, M. P.; Santos, D. M. miR398 and miR408 are up regulated in response to water deficit in *Medicago truncatula. Planta.* **2010**, 231, 705-716.

139. Tuteja, N.; Tuteja, R. Prokaryotic and eukaryotic DNA helicases: Essential molecular motor proteins for cellular machinery. *Eur. J. Biochem.* **2004**, 271(10), 1835-48.

140. Tuteja, N.; Tuteja, R. DNA helicases: the long unwinding road. *Nat. Genet.* **1996**, 13, 11–12.

141. Umezawa, T.; Fujita, M.; Fujita, Y.; Yamaguchi-Shinozaki, K.; Shinozaki, K. Engineering drought tolerance in plants: discovering and tailoring genes to unlock the future. *Curr. Opin. Plant Biol.* **2006**,17, 113–122.

142. Vinocur, B.; Altman, A. Recent advances in engineering plant tolerance to abiotic stress: achievements and limitations. *Curr. Opin. Plant Biol.* **2005**, 16 123–132.

143. Waie, B.; Rajam, M. V. Effect of increased polyamine biosynthesis on stress responses in transgenic tobacco by introduction of human S-adenosylmethionine gene. *Plant Sci.* **2003**, 164, 727–734.

144. Wang, G. P.; Hui, Z.; Li, F.; Zhao, M. R.; Zhang, J.; Wang, W. Improvement of heat and drought photosynthetic tolerance in wheat by over accumulation of glycinebetaine. *Plant Biotechnol. Rep.* **2010**, 4, 213–222.

145. Wang, Y.; Ying, J.; Kuzma, M.; Chalifoux, M.; Sample, A.; McArthur, C.; Uchacz, T.; Sarvas, C.; Wan, J.; Dennis, D. T.; McCourt, P.; Huang, Y. Molecular tailoring of farnesylation for plant drought tolerance and yield protection. *Plant J.* **2005,** 43:413–424.

146. Wen, X. P.; Pang, X. M.; Matsuda, N.; Kita, M.; Inoue, H.; Hao, Y. J. Over-expression of the apple spermidine synthase gene in pear confers multiple abiotic stress tolerance by altering polyamine titers. *Transgenic Res.* **2008**, 17, 251–263.

147. Wen, X. P.; Ban, Y.; Inoue, H.; Matsuda, N.; Moriguchi, T. Aluminum tolerance in a spermidine synthase-over expressing transgenic European pear is correlated with the enhanced level of spermidine via alleviating oxidative status. *Environ. Exp. Bot.***2009**,66, 471–478.

148. Wen, X. P.; Ban, Y.; Inoue, H.; Matsuda, N.; Moriguchi, T. Spermidine levels are implicated in heavy metal tolerance in a spermidine synthase over expressing transgenic European pear by exerting antioxidant activities. *Transgenic Res.* **2010**,19, 91–103.

149. Wang, J.; Sun, P. P.; Chen, C. L.; Wang, Y.; Fu, X. Z.; Liu, J. H. An arginine decarboxylase gene PtADC from Poncirus trifoliata confers abiotic stress tolerance and promotes primary root growth in Arabidopsis. *J. Exp. Bot.* **2011**, 62, 2899–2914.

150. Wi, S.J.; Kim, W.T.;Park, K.Y. Over expression of carnation S adenosyl-methionine decarboxylase gene generates a broad-spectrum tolerance to abiotic stresses in transgenic tobacco plants. *Plant Cell Rep.* **2006**, 25, 1111–1121.

151. Wi, S. J.; Kim, S. J.; Kim, W. T.; Park, K. Y. Constitutive S adenosylmethionine decarboxylase gene expression increases drought tolerance through inhibition of reactive oxygen species accumulation in Arabidopsis. *Planta.* **2014**, 30.

152. Xu, D.; Duan, X.; Wang, B.; Hong, B.; Ho, T.; Wu, R. Expression of a late embryogenesis abundant protein gene, HVA1, from barley confers tolerance to water deficit and salt stress in transgenic rice. *Plant Physiol.* **1996**,110, 249–257.

153. Xue, Z. Y.; Zhi, D.Y.; Xue, G.P.; Zhang, H.; Zhao, Y. X.; Xia, G. M. Enhanced salt tolerance of transgenic wheat (*Tritivum aestivum* L.) expressing a vacuolar Na^+/H^+ antiporter gene with improved grain yields in saline soils in the field and a reduced level of leaf Na^+. *Plant Sci.* **2004**, 167, 849–859.

154. Yamasaki, H.; Abdel-Ghany, S. E.; Cohu, C. M.; Kobayashi, Y.; Shikanai, T.; Pilon, M. Regulation of Cop-per Homeostasis by Micro-RNA in Arabidopsis. *J. Biol. Chem.* **2007**, 282, 16369-16378.

155. Yamaguchi-Shinozaki, K.; Shinozaki. K. Characterization of the expression of a desiccation-responsive rd29 gene of Arabidopsis thaliana and analysis of its promoter in transgenic plants. *Mol. Gen. Genet.* **1993a**, 236, 331-340.

156. Yang, A. F.; Duan, X. G.; Gu, X. F.; Gao, F.;Zhang J. R. Efficient transformation of beet (Beta vulgaris) and production of plants with improved salt-tolerance. *Plant Cell Tissue Org.* **2005**, 83, 259–270.

157. Yeo, A. Molecular biology of salt tolerance in the context of whole-plant physiology. *J. Exp. Bot.* **1998**,49, 915–929.

158. Zhang, H.; Hodson, J.; Williams, J. P.; Blumwald, E. Engineering salt tolerant *Brassica* plants: Characterization of yield and seed oil quality in transgenic plants with increased vacuolar sodium accumulation. *Proc. Natl. Acad. Sci. USA.* **2001**, 98, 12832– 12836.

159. Zhang, L.; Xiao, S. W.; Li, W.; Feng, J.; Li, Z.; Wu, X.; Gao, F.; Liu, M. of a Harpin encoding gene hrf1 in rice enhances drought tolerance. *J. Exp. Bot.* **2011**, 62, 4229–4238.

160. Zhao, F.; Zhang, H. Salt and paraquat stress tolerance results from co-expression of the Suaeda salsa glutathione S-transferase and catalase in transgenic rice. *Plant Cell Tissue Org.* **2006**, 86, 349–358.

161. Zhao, F.; Guo, S.; Zhang, H.; Zhao, Y. Expression of yeast SOD2 in transgenic rice results in increased salt tolerance. *Plant Sci.* **2006**, 170, 216–224.

162. Zhao, J.; Zhi, D.; Xue, Z.; Liu, H.; Xia, G. Enhanced salt tolerance of transgenic progeny of tall fescue (*Festuca arundinacea*) expressing a vacuolar Na^+/H^+ antiporter gene from Arabidopsis. *J. Plant Physiol.* **2007**, 164, 1377–1383.

163. Zhao, B. T.; Liang, R. Q.; Ge, L. F.; Li, W.; Xiao, H. S.; Lin, H. X.; Ruan, K. C.; Jin, Y. X. Identification of Drought-Induced MicroRNAs in Rice. *Biochem. Bioph. Res. Co.* **2007**, 354, 585-590.

164. Zhao, B.T.; Ge, L. F.; Liang, R. Q.; Li, W.; Ruan, K. C.; Lin, H. X.; Jin, Y. X. Members of miR-169 Family Are Induced by High Salinity and Transiently Inhibit the NF-YA Transcription Factor. *BMC Mol. Biol.* **2009**, 10, 29.

165. Zhou, L.; Y, Liu.; Z, Liu.; D, Kong.; M, Duan.; L Luo. Genome-wide identification and analysis of drought responsive microRNAs in *Oryza sativa*. *Journal of Experimental Botany*. **2010**, 61, 4157-4168.

166. Zhu, B. C.; Su, J.; Chan, M. C.; Verma, D. P. S.; Fan, Y. L.; Wu, R. Over expression of a Delta (1)-pyrroline-5-carboxylate synthetase gene and analysis of tolerance to water- and salt-stress in transgenic rice. *Plant Sci.* **1998,** 139, 41-48.

167. Zhu, J. K. Genetic analysis of plant salt tolerance using *Arabidopsis. Plant Physiol.* **2000**, 124, 941–948.

168. Zhu, J. K. Plant salt tolerance. *Trends Plant Sci.* **2001**, 6, 66–71.

169. Zhu, J. K. Cell signaling under salt, water and cold stress. *Curr. Opin. Plant Biol.* **2001**, 4, 401-406.

170. Zhu, J. K. Salt and drought stress signal transduction in plants. *Annu. Rev. Plant Biol.* **2002**, 53, 247-273.

171. Zhu, J. K.; Hasegawa, P. M.; Bressan, R. A. Molecular aspects of osmotic stress. *Crit. Rev. Plant Sci.* **1997**, 16, 253–277.

172. Zhu, L.; Tang, G. S.; Hazen, S. P.; Kim, H. S.; Ward, R. W. RFLP-based genetic diversity and its development in Shaanxi wheat lines. *Acta Bot. Boreali Occident Sins.* **1993,** 19, 13.

Abiotic Stress Tolerance Mechanisms in Plants, Pages 85–126
Edited by: Gyanendra K. Rai, Ranjeet Ranjan Kumar and Sreshti Bagati
Copyright © 2018, Narendra Publishing House, Delhi, India

2

BIOTECHNOLOGICAL APPROACHES FOR THE DEVELOPMENT OF HEAT STRESS TOLERANCE IN CROP PLANTS

Sreshti Bagati[1], Gyanendra K. Rai[1], Diksha Bhadwal[1] and Mukesh Kumar Berwal[2]

[1]*School of Biotechnology,*
Sher-e- Kashmir University of Agricultural Sciences and Technology of Jammu-180009 (J&K)
[2]*Division of Crop Production, ICAR-Central Institute for Arid Horticulture, Bikaner*
E-mail: bagati.sreshti@gmail.com

Abstract

Abiotic stresses including water deficit conditions (drought), salinity, extreme temperatures (heat, cold), light intensities beyond those saturating for photosynthesis, and radiation (UV B, C) pose a serious threat to agricultural production and quality worldwide. On an average crop losses across the globe account for millions of US dollars. Hence, prevention of such losses has become necessary and this can be achieved by the development of stress tolerant crops. In this context, breeding strategies along with molecular markers can be used to identify resistant germplasm and transfer the desirable attributes to sensitive varieties, but this approach is a timely process. Introduction of genes that can improve stress tolerance in crops against heat, drought, and salinity is relatively a more effective technology. Since a number of critical genes, particularly transcription factors that regulate gene expression in response to environmental stresses have been identified and validated. In the recent times, post transcriptional and post translational regulation mechanisms of the abiotic stress response, like micro RNAs and ubiquitination, emerged as promising tools to develop abiotic stress-tolerant plants and contribute towards crop improvement for feeding the growing masses.

Keywords: Agriculture, Climate change, Environmental stress, Transcription

1. Heat Stress

Global warming due to the elevated levels of carbon dioxide and other greenhouse gases in the environment has been an issue of concern since long. Since the environment around us is changing constantly, it creates a number of unfavorable conditions (stressful conditions), thereby making the survival of the living organisms difficult. Among all the living organisms, crop plants (particularly) being sessile (unable to move) cannot escape the stress emerged due to changing environment by travelling to favorable environments; hence their growth and developmental processes are badly affected (Lobell and Asner, 2003; Lobell and Field, 2007). While the constituents of the environment have been constantly changing, the ambient temperature is found to increase constantly. This substantial increase in the temperature is detrimental for the crop plants. According to IPCC, the global air temperature will increase at the rate of 0.2°C per decade and in 2100 the temperature will be 1.8–4.0°C higher than the current level (IPCC, 2007). The constantly rising temperature often acts lethal to the crops (Lobell and Asner, 2003; Lobell and Field, 2007). High temperature (HT) stress or heat stress occurs when the environmental temperatures rises beyond a threshold level. A threshold temperature is the temperature beyond which there is an irreversible and detectable reduction in the growth and development of plants. The threshold temperatures (both upper and lower limit) for various crop plants have been determined via a number of laboratory experiments and field trials. The growth and the development of the plant species cease both below and above the threshold temperature. Generally, when the temperature of a place goes up by 10-15°C above normal, the temperature is expected to create heat stress/ heat shock. The heat stress is a combined function of various factors i.e. intensity (temperature in degrees), duration, as well as the rate of increase in temperature and the stress intensity depends upon the probability and duration of high temperatures occurring during the day and/or the night (Wahid *et al.,* 2007). Continuous studies during the past few years have suggested that the continuous and frequent changes in the temperature, creating extreme conditions will be more crucial in future. The summer heat wave in Europe in the year 2003 proved the researchers right as it rose the temperature by 5°C above normal, creating extreme and critical conditions which prevailed throughout summers (Rennenberg *et al.,* 2006). As the threat of the rapidly changing climate is accelerating, it will lead to a decelerating effect on the agricultural production throughout the world. Heat waves generated due to the increased temperature do considerable damage to crops with respect to their yield, thereby creating a threat to the global food security in near future (Christensen and Christensen, 2007). Approximately 50% of the people living in the developing countries count totally on agriculture and will be most critically affected by the global food insecurity caused by HT stress (Bita and Gerats, 2013).The immediate effects of heat stress are manifold leading to antagonistic effects on the growth, development, yield as well as the physiological processes (photosynthesis and respiration) of the crop plants (Hasanuzzaman *et al.,* 2012 and Hasanuzzaman *et al.,* 2013). The heat stress never arrives alone and is often associated with other stresses like increased solar irradiance, drought, and wind; all of them combined together infuriate the damage caused to plants due to heat stress (Paulsen, 1994). The damage caused by the increased temperature is frequent at certain target sites in crop plants like the CO_2 fixation system, photophosphorylation, the electron transportchain, and the oxygen- evolving complex (OEC) (Nash *et al.*, 1985; Feller *et al.*, 1998; Bukhov and Mohanty, 1999; Carpentier 1999; Sharkey, 2005). In several plant species the consequences of the HT stress are more conspicuous during the reproductive phase bringing about pollen infertility, henceforth an unusual drop in the yield is seen (Young *et al.*, 2004; Zinn *et al.*, 2010). It has been reported that in the coming days amid the growing season, the tropical and the subtropical regions will experience very high temperatures surpassing the most intense seasonal temperatures till date, which will expedite the mechanism of land degradation and shoot up the decline in the crop yield (Battisti and Naylor, 2009; Varshney *et al.*, 2011). At present, near about 15% of the world wheat production occurs in the Indian lowlands but it has been foreseen that the frequently observed changes in the climate will convert these lowlands into a HT-stressed environment with shortened production season. Likewise, in Africa, Asia and Middle East the ambient temperature is expected to increase by 3–4°C which could lead to a decline in the crop yields by 15–35% and by 25–35% respectively (Ortiz *et al.*, 2008). Concisely, the increase in temperature beyond threshold (heat stress) acts antagonistically on various physiological processes (photosynthesis, lipid biosynthesis, hormone signaling and the synthesis of primary and secondary metabolites) and it disorganizes the membrane stability, stability of several proteins as well as leads to the disruption of the cytoskeleton structures; hence creating a negative effect on the growth and development of plants (Bita and Gerats, 2013). The concept of mathematical modeling enabled the researchers to understand that if novel strategies were not developed to combat the detrimental effects of heat stress, the production of cereals in Southeast Asia and Southern Africa is probably expected to decline (Nelson, 2009; Fischer and Edmeades, 2010). Harshness of the HT- induced catastrophe relies on several parameters like tested systems, capability of plants to withstand heat, temperature gradient and the manner in which the heat is applied (Bukhov and Mohanty, 1999). As soon as a plant encounters heat stress, certain changes take place in the thylakoid membrane fluidity, which acts as an indication for the plant to undergo few functional changes in order to withstand the stress (Horvath *et al.*, 1998; Los and Murata 2004). Apart from the changes in the membrane fluidity, heat stress stimulates the aggregation of HSPs (heat shock proteins) which hinder the protein degradation pathways. Apart from preventing the protein degradation, HSP's also disturb the metabolic balance by enhancing the release of toxic by-products like ROS (reactive oxygen species) which in turn hamper the vegetative and reproductive growth of plants, leading to decline in the fruit set and yield (Bita and Gerats, 2013). During Heat stress, the up or down regulation of the gene expression leads to certain alterations in the physiological and the biochemical processes for the development of resistance/ tolerance towards heat, thereby helping the plant undergoing stress to acclimatize or adapt in the prevailing conditions (Moreno and Orellana, 2011; Hasanuzzaman *et al.*, 2010). Recently, it has been revealed that the information about the lower threshold temperature plays an important role while carrying out research on physiological parameters and in crop production (Wahid *et al.*, 2007). The external spray of a number of biomolecules, usually known as protectants also plays a crucial role in soothening the damage caused due to elevated temperature. The commonly used biomolecules include osmoprotectants (e.g. proline, gylcine betaine, trehalose etc.), phytohormones or plant growth regulators (PGR's) like abscisic acid, gibberllic acid, brassinosteroids, salicyclic acid etc., signaling molecules (e.g., nitric oxide, NO), polyamines (putrescine, spermidine and spermine), trace elements like selenium and silicon as well as macronutrients like nitrogen, potassium, calcium, phosphorus etc. (Hasanuzzaman *et al.*, 2010, 2012, 2012, 2013, 2013; Waraich *et al.*, 2012; Barnabas *et al.*, 2008). On the basis of the factors like severity and extent of heat stress, plant species, ratio of the other constituents present in the surrounding environment, each and every plant species has its own dynamic way of exhibiting an active response towards the interim stress but the recognition and the validation of the features that provoke this response for tolerance against heat stress are yet ambiguous (Wahid *et al.*, 2007; Rodriguez *et al.*, 2005). In order to meet the future demands for food and prevent the extensive damage to crop plants especially food grains and legumes, the development and introduction of novel HT tolerant crop cultivars is the matter of principal concern among plant scientists (Moreno and Orellana, 2011; Hasanuzzaman *et al.*, 2010, 2010, 2012, 2012, 2013, 2013; Waraich *et al.*, 2012; Barnabas *et al.*, 2008; Zhang *et al.*, 2006). The scientists working on HT stress are aspiring to uncover the plant responses which contribute to heat stress tolerance so that the management of plants in the high temperature prone areas becomes easy. In recent times, a better knowledge of the fundamental molecular mechanisms behind the heat stress tolerance will be of much help. The utilization of the lately developed molecular techniques and approaches along with the currently introduced "Omics" techniques as well as the transgenic technology which involves alteration of the target genes (Kosova *et al.*, 2011; Duque *et al.*, 2013; Schoffl *et al.*, 1999) will prove beneficial for the development and the cultivation of agriculturally important crop varieties resistant to HT- stress. The present chapter focuses on the effects of the rapidly rising temperature (heat stress) on crop plants and the currently used biotechnological and the molecular approaches for the development of HT- stress tolerance in agronomically important crop plants.

2. Effects of Heat (HT- induced) Stress on Crop Plants

Exposure to the continuously increasing temperature (HT) has become a regular event for the agricultural crops (Sarkar *et al.,* 2009; Perez *et al.,* 2009). Due to the rapidly changing climate across the globe, high temperature (HT) has emerged as an acute factor for the growth and productivity of the crop plants and hence high temperature (HT) or heat stress is being looked upon as one of the major abiotic stresses which hampers the agricultural crop production (Hasanuzzaman *et al.,* 2012). Among all the abiotic stresses that a plant goes through, HT- stress acts autonomously on the physiological and the metabolic activities of the plant cells (Barnabas *et al.,* 2008; Sakata and Higashitani, 2008). High temperature has disastrous effects on the growth and metabolism as these processes are temperature bound and work efficiently between the optimal temperature limits, which vary from species to species (Zrobek-sokolnik, 2012). Commonly, high temperature stress never arrives alone, and is usually accompanied by other stress creating companions like drought and salinity stress. While the plant growth and development occur via a cascade of innumerable number of biochemical reactions which are perceptive to temperature (Zrobek-sokolnik, 2012), it becomes necessary to dissect the sovereign role of temperature and its effects on the biology of crops in order to mitigate the consequences of the mixed stresses. Sensitivity to high temperature and its effects on crop plants depends on various factors like the developmental stage in which the plant is, species and genotypes (with enormous inter and intra specific variations) respectively (Barnabas *et al.,* 2008; Sakata and Higashitani, 2008). In order to overcome the heat stress generated due to increased temperature, the plants show some response towards it which varies with various determinants like the range to which the temperature has increased, the duration till which the high temperature prevails and the type of the plant under stress (Mittler, 2006). Whenever the temperature reaches to extreme and the conditions become unfavorable for the growing crops, cell destruction and cellular injury take place within a fraction of time, because of which the catastrophic shutdown of the cellular machinery occurs (Ahuja *et al.,* 2010). High temperature (HT) or Heat stress has widespread affects and effects almost all the possible facets of the plant processes including germination, growth, development, reproduction and yield (Hasanuzzaman *et al.,* 2010; Mittler and Blumwald, 2010; Lobell *et al.,* 2011; Mc Clung *et al.,* 2010). Apart from remarkably affecting the stable behaviour of various proteins, membranes, RNA species and cytoskeleton structures, the HT-stress also metamorphoses the effectiveness of the enzymatic reactions, hence heat stress acts a serious hindrance to the primary physiological processes of the plant system, leading to metabolic imbalance (Ruelland and Zachowski, 2010; Suzuki *et al.,* 2011, 2012; Pagamas and Nawata, 2008). Some of the most common effects of HT- stress on crop plants with respect to different parameters are discussed below and are represented in Figure 1.

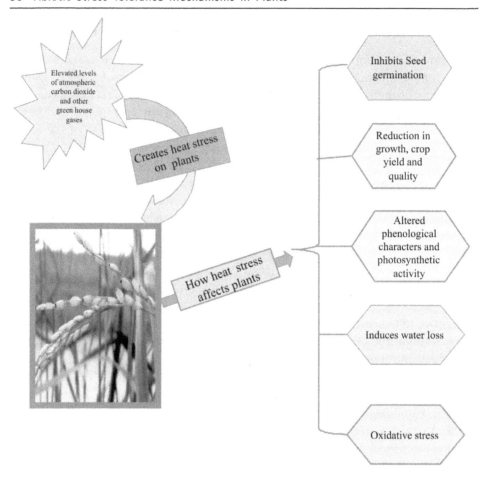

Fig. 1: Substantial effects of heat stress on crop plants

2.1. Morphological, Anatomical and Phenological Effects

2.1.1. *Morphological Effects*

Extremely high temperature leads to a number of changes in the morphology of the plants exposed to it. Due to the continuous introduction of crop plants to heat stress, noticeable pre and post harvest harm occurs to them (Guilioni *et al.,* 1997; Ismail and Hall, 1999; Vollenweider and Gunthardt- Goerg, 2005). The most remarkable morphological effects are the scorching and sunburns of leaves and twigs, branches and stems, leaf senescence and abscission, inhibition of shoot and root growth, discoloration and deterioration of fruits as well as huge reduction in yield (Rodriguez *et al.,* 2005; Wahid, 2007). Apart from affecting the shoot and the root growth, a frequent decrease in the root number and the root diameter is also observed in the plants undergoing heat stress (Porter and Gawith, 1999). The effects of the HT- stress on the plants and their response towards it varies with the phenological state, genotype and the species of the plants.For example: The continuous exposure of sugarcane to high temperature leads to the rolling and drying of leaves, necrosis along with prominent blemishes to the leaf- tips and margins (Omae *et al.*, 2012). Introduction of the wheat crop to acute HT-stress slowed down the rate of stem growth and restrained the plant height from 66.4- 97.3 cm (normal) to 55.7- 82.3 cm (heat stress exposed plant) (Prasad and Allen, 2006; Rahman, 2004; Ahamed *et al.*, 2010).Although the leaf area and the number of tillers per plant dropped off intensely when the wheat plants were introduced to intense HT (30/25 °C, day/night) (Djanaguiraman *et al.*, 2010; Johkan *et al.*, 2011), yet a slight improvement was seen in the shoot elongation (Kumar *et al.*, 2011).

In spite of the fact that the morphological symptoms due to heat stress in common bean (*Phaseolus vulgaris*) were found similar to that of wheat and included extremely altered shoot growth and elongation along with a severe depreciation in the total biomass, yet there are certain exceptions like when the plant species were raised up in high temperature of about 28 to 29°C, an unusual stem elongation and leaf enlargement i.e. hyponasty was observed (Koini *et al.*, 2009; Patel and Franklin, 2009; Savin *et al.*, 1997). The periodically changing temperature affected the coleoptile growth in maize like at 40°C the growth of the coleoptiles decreased and an increase in the temperature by 5° (i.e. 45°C), refrained the coleoptile growth (Weaich *et al.*, 1996). The large scale consequences of the HT-stress on the crops like common bean, ground nut, wheat, maize, pearl millet, sugarcane, tomato and cereals include decreased biomass and relative growth rate (Ashraf and Hafeez, 2004; Wahid, 2007; Ebrahim *et al.*, 1998), curtailment of the intermodal length (resulting in premature death) (Hall, 1992), extended span for grain filling with reduced number of kernels and loss of kernel density in spring wheat (Guilioni *et al.*, 2003; Ferris *et al.*, 1998), loss of yield in ground nut (Rainey and Griffiths, 2005) and beans (Vara Prasad *et al.*, 1999), meager fruit and seed set (Kinet and Peet, 1997) and unusual elongation of style which prevents self-pollination in tomato.

2.1.2. *Anatomical Effects*

The anatomy (internal morphology) of plants undergoes changes because of the harshness of the high temperature. Plants going through HT- stress have a natural propensity of reducing their cell size, closing the stomata to prevent water loss, increasing both the stomatal and the trichomatous densities along with an enlargement of the xylem vessels of both roots and shoots (Wahid, 2007; Anon *et al.*, 2004). The alterations seen in the anatomy of plants during heat stress were found to be somewhat similar to those seen during the drought stress (Wahid, 2007). The observed effects of heat stress on plant anatomy include the fragmentation of the ultra-structure which is associated to several factors viz. the decrease in the stomatal density, loosening of the mesophyll cells, enlarged stomatal area having a wider opening, thinning of leaf area, underdeveloped vascular bundles (xylem and phloem) and the instability among the organelle structures. A comparative microscopic study of the ultra-structural properties of the mesophyll cells present in the flag cells of both HT- susceptible and resistant rice genotypes was carried out under heat stress (37/30°C), an increase in the permeability of membranes was seen in both the cases whereas the closely packed arrangement of flag leaf mesophyll cells, completed closed stomata and well matured vascular tissues were identified as the characters that contributed to the tolerance for HT- stress in the resistant genotypes. On the other hand, the similar characters were missing in the susceptible genotypes, leading to heat injury in them (Zhang *et al.*, 2009).The research carried out in the recent past has revealed that the photosynthetic membranes are much affected by heat stress because of the swelling of grana or due to the failure of grana to form stacks (Zhang *et al.*, 2005). The alteration of the structural arrangement of thylakoids brings about extensive changes at the sub-cellular level which in turn affect the photosynthesis (Karim *et al.*, 1997). The immediate effects of the HT- stress on the grape plants (*Vitis vinifera*) were listed as, extensively disrupted mesophyll cells, swollen stroma lamellae, round chloroplasts, jumbled vacuolar elements, disorganized cristae, hollow mitochondria and development of PSII (photosystem-II) without antenna; all these factors contributed to the decline in the photosynthetic and the respiratory processes (Zhang *et al.*, 2005). *Zygophyllum qatarense* developed polymorphic leaves and bimodal stomata to reduce the water loss during transpiration while undergoing heat stress (Sayed, 1996). Taking into consideration, the above mentioned impacts of heat stress, it becomes quite clear that the heat stress effects on the anatomy of plants are not limited to the tissue and the cellular levels, but extend up to the sub-cellular level. The combined effects of all these changes eventually lead to the underprivileged growth and productivity of plants (Wahid, 2007).

2.1.3. *Phenological Effects*

Estimation of the harm caused by HT- stress to the crop plants can be predicted by studying the stage of development at which the stress was imposed (Wahid, 2007). Surveillance of the alterations occurring in the phenology of the crops in response to HT- stress will disclose a higher quality picture of the relationships between the plants and the surrounding stress (Wahid, 2007). Since a number of inter and intra specific variations are present among the species and the genotypes, hence the susceptibility of the various phenological phases towards high temperature varies with them (Wollenweber *et al.,* 2003; Howarth, 2005).Almost all the stages or the steps involved in the vegetative and the reproductive development of the plants are vulnerable to HT- stress. For example, an acute impairment in the gaseous exchange properties of leaves can happen due to increased temperature during the day and on the other hand, even a small duration exposure of the plant to heat stress during reproductive phase can lead to remarkable enlargement in the flower buds along with the abortion of completely opened flowers (Wahid, 2007; Guilioni *et al.,* 1997; Young *et al.,* 2004). Apart from these, other phenological effects like deterioration of the pollen, improper anther development responsible for reduced fruit set are seen in some plants which vary from moderate to high temperatures (Peet *et al.,* 1998; Sato *et al.,* 2006). Although it has been observed that for the maintenance of more number of green leaves during anthesis stage and lesser loss to the yield, premature heading has proved much beneficial, yet elevated temperature during the grain filling stage may alter the quality of the flour as well as bread generated from them, along with certain modifications in the protein content of the flour as well as the physical and the chemical properties of the cereal crops like rice and wheat (Tewolde *et al.,* 2006; Perrotta *et al.,* 1998; Wardlaw *et al.,* 2002). Major staple food crops (cereal crops) are unable to withstand the wider temperature ranges and when the temperature exceeds above their normal range, acute harm occurs to the fertilization and the seed production, leading to loss in the crop yield (Porter, 2005). Hence for the prevention of heat stress induced damage and the reduction in huge losses occurring to the agriculturally important crops, it is the need of the hour to get a better insight to the heat sensitive stages in the developmental (both vegetative and the reproductive) processes. A complete and adequate knowledge of the high temperature susceptible phenological stages will help the researchers and plant scientists for the determination and development of heat stress tolerance in crop plants.

2.2. Physiological Effects

Other than morphology and anatomy, physiology of the crop plants is found to be extremely affected by high temperature. Development of tolerance towards continuously elevating temperature is a complex process driven by the factors like the surrounding environment and the genetic potential of the plants under stress (Zrobek, 2012). Respiration and photosynthesis, the two most important physiological processes are susceptible to the harm caused by high temperature. Temperature has a direct impact on the respiratory processes leading to a rapid increase in them but after certain threshold there is an extreme drop in the respiration cycle. Unlike respiration, photosynthesis is relatively less susceptible to high temperature, but the fall in the photosynthetic pathway is much similar to that seen in the respiratory one. It has been reported that with every 10°C rise in the surrounding temperature, a bifold increase is seen in the pace of the enzyme catalyzed reactions (Zrobek, 2012). Since every process involved in the life cycle of crop plants works efficiently within its desired range of temperature, above or below which these processes do not work efficiently and may cease, leading to a long lasting injury to the cell structure and function and in many cases such conditions mark the onset of the end of plant life.

2.2.1. *Water Relations*

One of the most significant components of plant physiology which should be considered during HT- stress is the water content in the plants and this character is found to be quite fluctuating under changing temperatures (Mazorra *et al.*, 2002). The continuous long time exposure of sugarcane plants and tomato cultivars to heat stress have a huge impact on the hydraulic conductance of roots (Morales *et al.*, 2003)and brings about remarkable changes in the leaf water potential and its constituents, inspite of the favorable humidity in the surrounding air and the presence of water rich soil (Wahid and Close, 2007). When studied under field conditions, it has been observed that the HT- stress is often linked to minimal soil water availability (Simoes-Araujo *et al.*, 2003). In plants like lotus (*Lotus creticus)*, an amenable increase in the night temperature is often associated with preeminent deceleration in the leaf water potential (Anon *et al.*, 2004). Generally, increased day temperature accelerates the rate of transpiration and generates water deficit plants having altered physiological processes (Tsukaguchi *et al.*, 2003). Substantially, the day time HT- stress is predicted to stimulate the water deficit in plants at a much faster rate than the HT- stress during night (Wahid, 2007).

2.2.2. *Photosynthesis*

Photosynthetic machinery is an integral part of the plant system and it often falls prey to the high temperature stress. Being highly perceptive to the increased temperature (Crafts-Brandner and Salvucci, 2002), the photosynthetic efficiency of a plant is frequently considered to be temperature dependent. In green plants, the immediate target sites for heat stress injury are the thylakoid lamellae (site for photochemical reactions) and the stroma (site for carbon metabolism) (Wise *et al.*, 2004). Plastids (chloroplasts), the prime location for the photosynthetic machinery undergo extensive modifications under HT- stress such as: disruption of the thylakoid structure, inability of grana to form stacks and Major alterations occur in chloroplasts like altered structural organization of thylakoids, inflammation of grana and its inability to form stacks (Rodriguez *et al.*, 2005; Ashraf and Hafeez,

2004). The antagonistic behaviour of high temperature towards photosynthesis is accelerated by secondary factors like HT- induced reduction in the leaf water potential and leaf area along with the early leaf senescence as well as the exhaustion of the carbohydrate resources (Greer and Weedon, 2012; Young *et al.*, 2005; Djanaguiraman *et al.*, 2009). Since the elevated temperature modifies and ceases the activity of the enzymes involved in carbon metabolism (especially rubisco) (Haldimann *et al.*, 2004) along with oxygen involving enzymes in the PS II, which hinders the electron transport and effects the pace of the RuBP synthesis leading to the conclusion that among the photosynthetic efficiency of C_3 and the C_4 plants, photosynthesis in C_3 plants is more sensitive to temperature (Salvucci and Crafts-Brandner, 2004; Yang *et al.*, 2006). Closing of stomata to prevent water loss, decrease in the amount and activity of certain proteins and enzymes like soluble proteins, rubisco binding proteins (RBP), sucrose phosphate synthase (Chaitanya *et al.*, 2001), ADP-glucose pyrophosphorylase, and invertase (Vu *et al.*, 2001) as well as reduction in the rate of biochemical reactions are certain other factors that affect the photosynthesis of theplants under heat stress (Sumesh *et al.*, 2008; Rodriguez *et al.*, 2005; Djanaguiraman *et al.*, 2009; Ashraf and Hafeez, 2004; Nakamoto and Hiyama, 1999). Apart from curtailing the number of photosynthetic pigments (Marchand *et al.*, 2005) increase in temperature influences the thylakoid structure and the working of the thermolabile photosystem II (PSII) (Morales *et al.*, 2003; Bukhov *et al.*, 1999; Camejo *et al.*, 2005; Mcdonald and Paulsen, 1997). While studying the effect of HT- stress (38/28 °C) on the cultivated soybean crop, 18% reduction was seen in the total chlorophyll content, 7% and 3% decline in chlorophyll a and chlorophyll a/b ratio respectively. The sucrose content dropped to 9 % while as an exceptional increase of 47% and 36% was observed in the reducing sugar content and in the quantity of leaf soluble sugars present (Tan *et al.*, 2011). The two varieties of rice Shuanggui 1 and T219 showed a reduced rate of photosynthesis by 16% and 15% respectively, when exposed to heat (Hurkman *et al.*, 2009). Keeping in consideration the increasing CO_2 content in the atmosphere, the resulting high temperature may have harmful effects on the photosynthetic machinery, yield and productivity of crop plants (Wahid, 2007). Concisely, the photosynthetic efficiency of plants has been found positively correlated to high temperature (Schuster and Monson, 2002). Hence, development of the strategies to prevent the alarming effects of heat stress on photosynthesis of crops is the need of the hour.

2.3. Yield

Continuous increase in the concentration of the atmospheric CO_2 is facilitating the temperature elevation which has been categorized as a huge threat to the crop yield and productivity, in turn hampering the food security in near future (Savin *et al.,* 1997). Sharing a negative correlation with the vital plant characteristics like growth, Dry matter partitioning, reproduction, photosynthesis establishment, growth, DM partitioning, reproductive growth and photosynthesis, this rapidly upraising temperature stands as a major hitch for the crop yield (Hasanuzzaman *et al.,* 2013). Like all other heat susceptible characters discussed above, the effects of HT- stress on the yield of plants is highly dependent upon the variables like crop genotypes and the intensity as well as the length of action of the temperature (Hasanuzzaman *et al.,* 2013) and its outcome is so horrifying that even a petty small increase in temperature above ambient (approximately 1.5 °C above normal) can lead to the remarkable loss in the expected crop yields (Warland *et al.,* 2006). A number of plant varieties whose yield and other yield attributing factors come under the terrible effects of heat have been recorded, these include cereals (e.g., rice, wheat, barley, sorghum, maize), pulse (e.g., chickpea, cowpea), oil yielding crops (mustard, canola) and many more. The susceptible ones in them are the most critically affected ones (Zhang *et al.,* 2013; Foolad *et al.,* 2005; Ahamed *et al.,* 2010; Tubeillo *et al.,* 2007; Kalra *et al.,* 2008; Hatfield *et al.,* 2011; Wang *et al.,* 2012). After effects of the HT- stress on the cereal grains has been well observed, and lead to a conclusion that 1 °C rise in the normal temperature reduced the cereal grain yield by 4.1% to 10.0% (Wang *et al.,* 2012). Heat stress in cereals has been often related to the elevation in the kernel growth rate along with a reduction in the grain length and breadth (Morita *et al.,* 2005; Zakaria *et al.,* 2002). In maize heat stress (33–40 °C) lead to approximately 79 to 95% decrease in the grain dry weight (Commuri and Jones, 2001), hindered the development of endosperm, leaving a negative effect on the RUE, biomass, harvest index, henceforth a reduction in the yield was seen (Monjardino *et al.,* 2005; Zhang *et al.,* 2013). Considering the consequences of heat stress in rice huge reduction in yield (approximately 90% in general) (35.3% to 39.5% and (21.7% to 24.5%) in the susceptible and the tolerant varieties was observed respectively along with 2% reduction in the grain length and width, 61% increase in spikelet fertility and the nitrogen content grains increased by 44% was recorded (Suwa *et al.,* 2010; Ahamed *et al.,* 2010). Wheat, *T. aestivum* being highly susceptible to heat, was found to undergo 20% decrease in the grain weight accompanied by 90% grain deficit in the plants which was associated with the formation of immature seeds, leading to yield loss during high temperature (Mohammed and Tarpley, 2010; Johkan *et al.,* 2011). In recent past, when spring wheat plants were exposed to HT- stress, remarkable observations like 50% decrease in the spike number, 20% fall in the total dry weight, loss of yield by 39% were recorded (Prasad *et al.,* 2011). In summary, the HT- induced yield loss in wheat and maize was found to be a cumulative outcome of the alterations like reduction in photosynthesis and assimilation ability of plants, increased respiration, decrease in the RUE (radiation use efficiency) etc. (Gan *et al.,* 2004; Zhang *et al.,* 2006; Wang and Li, 2006;

Rahman *et al.*, 2009; Lin *et al.*, 2010). Concisely, an unusual elevation in the surrounding temperature has noticeable influence on the performance (yield and yield attributes) as well as on the crop quality and nutritional parameters.

2.4. Effects on Reproduction and Development

Like growth, the processes of reproduction as well as development of crop plants are unable to escape the harshness of high temperature, because of the extreme susceptibility of the reproductive tissue and flowering time towards heat (Mckee *et al.*, 1998; Lobell, 2011). A small duration exposure of susceptible plant species to high temperature was found to exaggerate the flower and floral bud abortion (Guilioni *et al.*, 1997; Young *et al.*, 2004; Sato *et al.*, 2006). Seed sterility is one of the important consequences of the HT- stress, as temperatures induces an erupted meiosis in both male and female organs, abnormal pollen germination as well as pollen tube growth, decrease in the viability of ovules, reduced ability of the stigma to retain pollen, impaired fertilization and hindered endosperm, proembryo as well as unfertilized embryo growth (Maheswari *et al.*, 2012; Foolad, 2005); Cao *et al.*, 2008; Peet *et al.*, 1998; Sato *et al.*, 2006); all these altered parameters contribute to the reduction in the fruit set. As observed in rice, heat stress affects the spikelet fertility, and its intensity is determined by the developmental stage of the spikelet (Zakaria *et al.*, 2002). In soybean, elevated temperature (37/27 °C) during flowering is followed by the reduction in the viability of pollen, lesser pod set, flowerabscission and abortion, reduced seed number, reduced size ovaries and bisexual flowers, (Tubiello *et al.*, 2007; Tan *et al.*, 2011; Kitano *et al.*, 2006; Johkan *et al.*, 2011). Pre –anthesis stage (Porch and Jahn, 2001) in common bean (*Phaseolus vulgaris)* was found to be sensitive to heat and the elevated temperature lead to impaired anther and pollen as well reduced fruit set in both common bean (Gross and Kigel, 1994) and peach (Kozai *et al.*, 2004). Heat stress tends to reduce anther dehiscence, the rate of pollen fertility and the spikelet fertility, formation of sterile seeds in rice Hurkman *et al.*, 2009; Ahamed *et al.*, 2010) and ceases the fruit set in tomato (Abdul- Bakki and Stommel, 1995). The overall effects of high temperature stress on certain important crops are mentioned in Table 1. Larcher (Larcher, 1995) classified all the plant species into three groups i.e. Psychrophiles, Mesophiles and Thermophiles (Figure 2). Development of the ability to withstand the harshness of the high temperature isn't easy.

It may either arise due to the phenological and the morphological alternations that may have occurred in the plant while undergoing evolution or may be due to the interim avoidance and the tolerance mechanisms developed in them (Wahid, 2007; Hasanuzzaman *et al.*, 2013). Being sessile in nature, the extent to which a plant can reverberate to the HT- stress is circumscribed; hence the cellular and

Table 1: Effects of HT- stress on some important plant species

No.	Species	Temperature	Stage/Phase	Effects	References
1.	Wheat (*Triticumaestivum*)	32 and 24 °C, during day and night respectively for 24 hours	Towards the end of the spikelet initiation stage	Leads to spikelet sterility and loss of grain yield	Saitoh, 2008 Mohammed and Tarpley, 2010
2.	Sorghum (*Hordeumvulgare*)	40°C day temperature and 30°C during night	65 days after sowing (DAS) till maturity	Yield reduction, injured thylakoid membrane, increase in the ROS content, reduction in the amount of chlorophyll, impaired photosystem II activity	
3.	Maize(*Zea mays*)	35°C day temperature and 27 °C night temperature for 14 days	Reproductive phase	Reduced photosynthase supply leading to impaired hemicelluloses and cellulose synthesis, suppressed cob extensibility	Suwa *et al.,* 2010
4.	Chili pepper (*Capsicumannuum*)	38 and 30°C during day and night respectively	Reproduction phase Maturity stage and harvesting phase	Decrease in the width and weight of fruits, more number of abnormal seeds/ fruit are produced	Hasanuz zaman *et al.,* 2012
5.	Wheat (*Triticumaestivum*)	37°C during day and 28°C night temperature for 20 days	Grain fill and maturity phase	Reduced yield and weight of kernels, reduced time span of grain filling and maturity	Rahman *et al.,* 2009
6.	Rice (*Oryza sativa*)	Greater than 33°C	Heading stage	Drop in the pollen and spikelet fertility percentage	Hurkman *et al.,* 2009
7.	Rice (*Oryza sativa*)	Night temperature of 32°C	Reproductive phase	Reduction in yield, grain length- breadth and grain weight, increase in the spikelet fertility	Yin *et al.,* 2010

[Table Contd.

Contd. Table]

No.	Crop	Temperature	Stage	Effect	Reference
8.	Wheat (*Triticum aestivum*)	30°C during day and 25°C night temperature	60 days after sowing till the maturity is achieved	Drastic reduction in grain size, grain number/spike, leaf size and yield; reduced time span for booting, heading, anthesis and maturity	Djanaguiraman et al., 2010
9.	Okra (*Abelmoschus esculentus*)	32 and 34 °C	Every stage involved in the growing period	Yield loss, Calcium-pectate deterioration and reduced fibre content, hampered pod quality	Gunawardhana and de Silva, 2011
10.	Tobacco (*Nicotiana tabacum*)	High temperature of about 43°C for 2 hours	Initial stages of growth	Malfunctioning of antioxidant enzymes, fall in the net photosynthetic rate, stomatal conductance and caboxylation efficiency	Tan et al., 2011
11.	Wheat (*Triticum aestivum*)	38°C for 24 and 48 hours	Seedling stage	Ceased antioxidant activity, reduction in relative water content and amount of chlorophyll	Hasanuzzaman et al., 2013
12.	Soybean (*Glycine max*)	14 days exposure to 38°C day temperature and 28 °C during night	Flowering phase	Injured chloroplast, mitochondria, thylakoid, plasma membranes, destroyed cristae and matrix, more thick palisade and spongy layers	Djanaguiraman et al., 2011
13.	Maize (*Zea mays*)	15 days exposure to 33–40°C temperature	Pre-anthesis stage and post silking phase	Plant and ear growth rates are drastically effected	Edreira and Otegui, 2012

the physiological changes promoting the avoidance and the tolerance mechanisms are of keen importance. Just like the effects of HT- stress in plants vary from species to species as well as from tissue to tissue, response of plants towards HT-stress follows the same path (Queitsch *et al.,* 2000). Figure 3 (A and B) represent various ways by which plants adapt to heat stress and the mechanism of the development of stress tolerance.

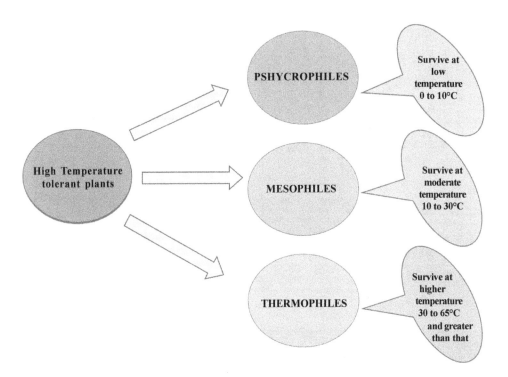

Fig. 2: Larcher's Classification of plant species on the basis of their heat tolerance ability

3. How Plants Respond to the Heat Stress

The rapidly occurring alternations in the surrounding environment are posing a huge threat to the food security in near future by accelerating the yield loss. An overview of the most emphasized plant stress tolerance mechanisms i.e. Avoidance and Tolerance is given below:

3.1. Mechanism of Tolerance

A plant is said to be heat tolerant when it has or it acquires the potential to grow, develop and generate economic yield on exposure to heat stress. A number of heat tolerance mechanisms are present with in various plant systems.Important tolerance mechanisms against the long term exposure of the plants to high temperature include, ion transporters, osmoprotectants, free-radical scavengers, late embryogenesis abundant (LEA) proteins and signaling cascades and transcription factors (Wang *et al.,* 2004; Rodriguez *et al.,* 2005; Wang *et al.,*2004) whereas in case of the abrupt heat stress, temporary changes in leaf orientation, transpirational cooling and membrane lipid composition are beneficial for plant survival (Rodriguez *et al.,* 2005; Radin *et al.,* 1994). As soon as the plant senses

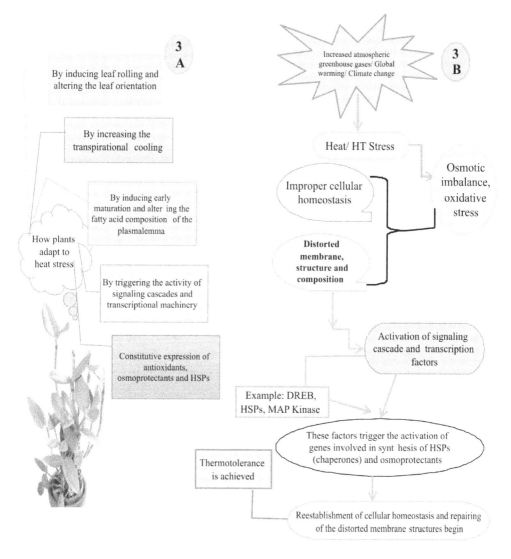

Fig 3: 3(A): Different adaptation strategies of plants towards heat stress
3 (B): Mechanism of the development of heat stress tolerance in plants.

the heat, an initial cascade of stress signals gets activated, which include osmotic or ionic effects, or changes in temperature or membrane fluidity, thereby stimulating the downstream signaling processes and transcription controls, and hence thestress-responsive mechanisms are activated. The immediate consequences of the heat stress are on cytoskeletal organization and the fluidity of the plasmalemma, which faces an influx of Ca^{2+} influx (Bohnert et al., 2006); both these factors contribute in the reestablishment of homeostasis as well as protection and repairmen of the impaired membranes and proteins (Vinocur and Altman, 2005). Several other heat tolerance mechanisms in plants are the activation of antioxidants (Maestri et al., 2002) and the Heat shock proteins (HSPs) orchaperones, both these lead to transduction of the stress signal (Nollen and Morimoto, 2002). Inspite of all the tolerance mechanisms in plants, sometimes improper responses are generated while signal transduction and gene activation, which proves to be critical for theplant as irreversible damage occurs to the cellular homeostasis as well as to the proteins (both structural and functional) bringing about the cell death (Vinocur and Altman, 2005;Bohnert et al., 2006). Heat tolerance being a highly trait specific character is acquired by plants at a number of levels like plasma membrane, ongoing biochemical pathways in the cytosol or cytoplasmic organelles (Sung et al., 2003) and a much better insight to all such mechanisms present within the plant systems is of huge practical importance.

3.2. Avoidance Mechanism

Avoidance mechanisms or acclimation mechanisms are temporary short lived responses of plants towards high temperature. These include visible alterations in the orientation of leaves and the composition of the lipid bilayer, induction of transpirational cooling, stomata closure, reduction in water loss, increased stomatal and trichomatous densities, and enlarged xylem vessels (Srivastava et al., 2012). Another heat stress avoidance strategy in plants is the reduced absorption of the solar radiations which is attributed to the presence of small hairs (tomentose) a thick coat on the leaf surface, cuticles, a protective waxy covering and rolling leaf blades. The presence of small sized leaves with smaller air boundary layer resistance also contribute to heatstress avoidance by expelling the heat at a faster rate as compared to their large counterparts. The adaptive/ avoidance mechanisms in both heat stressed and drought stressed plants have been found to be somewhat similar.

With evolution, certain plants species have developed the ability to survive even during the hottest period of the year by undergoing leaf abscission, having heat resistant buds, or by completing the entire reproductive cycle during the cooler months as in desert annuals (Fitter and Hay, 2002). Adaptation towards heat stress also varies from species to species, as some plants species become more tolerant to heat during summers while others acquire the same in the winter season.Considering the development of avoidance for heat in dormant plants, it has been found that they acquire stress resistance once they reach that stage of development which is extremely hampered by high temperature.

Other important traditionally followed methods to avoid heat stress in crop plants are selecting proper sowing methods, choice of sowing date, cultivars, irrigation methods, *etc.* (Hall, 2011). Since the effects of all the stresses like water deficit, heat etc. are quite similar making it difficult to differentiate the responses created for any of them by the plants. In the recent past, it has been concluded that only those adaptations which bring about either avoidance or tolerance for the stresses can be considered as the potent ones (Fitter and Hay, 2002).

4. Heat stress and Biotechnology

With the help of recent studies, it has been found that all the tolerance mechanisms which provide thermo tolerance to the plants during high temperature are under the control of many genes, making thermotolerance a multigenic trait. Being a quantitative trait, understanding the molecular basis for the abiotic stress tolerance has been quite challenging (Yadav *et al.,* 2013; Howarth, 2005; Bohnert *et al.,* 2006). Although the advent of biotechnology isn't that old, yet it has proved significantly important for getting a better insight to the genetic basis of thermotolerance (Liu *et al.,*2006; Sun *et al.,* 2006; Momcilovic and Ristic, 2007). Apart from genes and genetics, metabolic pathways and the enzymes involved in them have been utilized for the betterment of crop plants. All these characters/ traits can undergo manipulations much easily as compared to structural and developmental ones and hence have been remarkably used for the development of improved stress tolerant crop varieties (Bhatnagar- Mathur *et al.,* 2008). During the past two decades, biotechnological approaches have proved potentially important towards the exploitation of genetic and metabolic (biochemical) basis involved in stress tolerance for the induction of enhanced stress tolerance in agriculturally important crop plants (Foolad, 2005). The most commonly used biotechnological approaches involved in the development of increased stress tolerance approaches in plants are: Marker assisted selection (MAS), genetic transformation as well as the recently discovered omics (especially comparative proteomics) approaches (Foolad, 2005). The diagrammatic representation of how these biotechnological interventions can be directed towards the development of heat stress resistant plants is shown in Figure 4. Inspite of the complexity involved in abiotic stress tolerance, recent advancements attained in the fields of marker- assisted breeding and genetic transformation (based on gene expression) techniques has made it

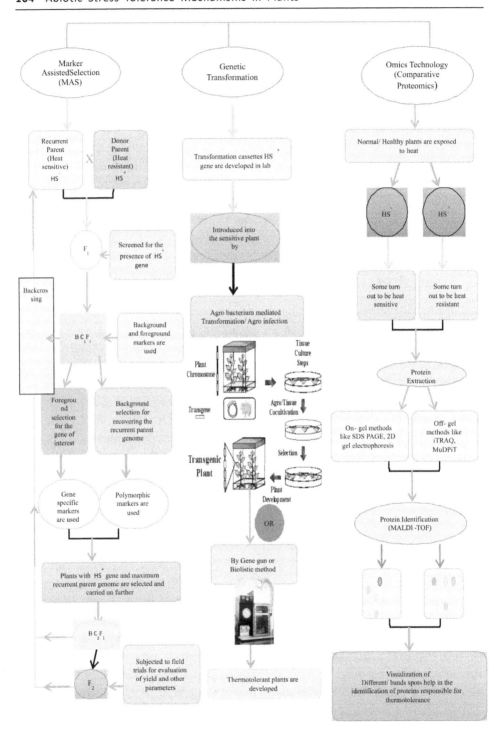

Fig. 4: Biotechnological Approaches Towards Heat Stress Tolerance

easier to develop stress tolerance in plants; with much higher success rates as compared to the conventional approaches (Foolad, 2005; Apse and Blumwald, 2002; Bohnert *et al.*, 2006). It has been found that by altering (up-regulating or down- regulating) the expression of multiple genes, enzymes as well as various biomolecules involved in different metabolic pathways that trigger thermotolerance the aim of the development of stress tolerant plant species can be easily achieved (Zhang *et al.*, 2001; Rontein *et al.*, 2002; Vinocur and Altman, 2005). Such remarkable contributions of various biotechnological approaches like genetic engineering coordinated with plant physiology has forced the researchers to accept this biotechnological era as a "second wave" (Yadav *et al.*, 2013) for the development of abiotic stress tolerant varieties in order to ensure food security in near future (Kasuga *et al.*, 1999).

4.1. Marker Assisted Selection and QTL Mapping for Heat Stress Tolerance

Conventional breeding strategies used to be efficient stress buster approaches but the complexities like the amount of time involved, requirement of sophisticated strategies for screening of tolerant varieties as well as dissecting the traits responsible for tolerance and understanding the complex mechanisms of inheritance involved in them limited their use in the development of stress tolerance (Bita and Gerats, 2013). The advent of molecular markers and marker assisted selection has proved a boon to the improvement of abiotic stress resistance in plants. Marker assisted selection (MAS) is based on the identification and availability of the genes/ markers/ QTLs linked to the stress tolerance ability of plants. In recent past, the enormous amount of research as well as the reduction in the sequencing cost/ base has made it possible to identify, map and characterize the QTLs associated with stress tolerance in crops (Foolad, 2005; Duran *et al.*, 2010). With the easy availability of the DNA sequence information and marker generation technologies, QTL mapping has been adopted as method of choice for the characterization of chromosomal locations carrying candidate genes for thermotolerance (Argyris *et al.*, 2011; Zhang *et al.*, 2012). Beneficiary aspects of identification and mapping of QTLs is, the ability to carry out gene pyramiding for stress tolerant genes and identification and understanding the stress tolerance mechanisms for other abiotic stresses (including heat) as well (Hirayama and Shinozaki, 2010; Roy *et al.*, 2011). QTL identification and MAS are closely related; as soon as the QTLs for heat tolerance and the markers linked to them are identified and isolated, they can be incorporated into the elite cultivars using MAS. Being a quantitative trait, the QTLs for heat tolerance are much susceptible to epistatic interactions (both gene× gene and gene× environment), leading to a reduction in the breeding efficiency

(Tester and Langridge, 2010; Lopes and Reynolds, 2010; Thomson *et al.*, 2010; Collins *et al.*, 2008). A number of QTL mapping studies have been carried out in various crops like rice, wheat, maize, *Arabidopsis* mutants, lettuce etc. and a number of loci attributing to heat tolerance have been identified (Ye *et al.*, 2012; Paliwal *et al.*,2012; Bai,2011, Hong *et al.*, 2003; Argyris *et al.*, 2008). Currently, the studies for the identification of QTLs have been performed using different yield parameters/ traits like thousand grain weight (TGW), yield (YLD), canopy temperature depression (CTD), Grain filling duration (GFD) and the traits related to senescence (Pinto *et al.*, 2010;Vijayalakshmi *et al.*, 2010).

4.2. Genetic Engineering and Transgenic Approaches for Thermo tolerance

These approaches play an important role in the development of heat stress tolerance in plants. With the help of the standardized transgenic protocols for important food crops, it becomes easy to transfer the thermotolerance from the resistant to the susceptible varieties. One of the main requirements for the application of transgenics to crop improvement is the isolation and characterization of the genes responsible for the desired trait of interest. With the help of these approaches, researches have efficiently exploited the constitutive expression of various genes, proteins (especially HSPs) and enzymes Zhang *et al.*, 2001; Rontein *et al.*, 2002 involved in the heat stress tolerance pathways to boost the thermotolerance in the crop plants. A number of efficient heat stress tolerant transgenic crops have been developed by enhancing the expression of heat shock.

The development of transgenic heat stress tolerant *Arabidopsis* involved the constitutive expression of the *AtHSF1*, a transcription factor triggering the synthesis of HSPs. The over expression of this factor boosted the thermotolerance in heat sensitive *Arabidopsis* (Lee *et al.*, 1995). Many other crops like carrot, tomato, tobacco, rice have been subjected to genetic transformation for the constitutive expression of the heat shock proteins like *Hsp17.7, MT-sHSP,* and *Hsp101* respectively (Malik *et al.*, 1999; Liu and Shono., 1999; Shanmiya *et al.*, 2004; Lee *et al.*, 2000). The transformation of *BADH* gene and Rubisco activase gene for the over expression of GB osmolytes and reversible decarboxylation of Rubisco respectively has lead to improved heat tolerance in crops (Yang *et al.*, 2005; Sharkey *et al.*, 2001). Since the immediate effect of heat stress is the alteration of the membrane fluidity, hence genetic manipulation approaches have been used for the silencing of the chloroplast omega-3 fatty acid desaturase gene in tobacco leading to stability of plasmalemma during heat stress (Mukrami *et al.*, 2000). Apart from these, the alternative ways in which transgenic approaches can act antagonistically to heat stress include over expression of the trans- acting factors like DREB2A, bZIP28 and WRKY, modification of the proteins involved in maintenance of osmotic balance, ROS detoxification, photosynthetic reactions and protein biosynthesis. Table 2 enlists the important genetically transformed crops in recent times. Inspite of all these success stories, the inadequate knowledge about the vital factors associated with stress tolerance remains as a bottle neck for the use of such approaches in crop improvement by developing HT- stress tolerant transgenicplants.

4.3. "Omics" Approaches

The recently evolved omics approaches like genomics, transcriptomics, micromics, proteomics and metabolomics brought in with it new ways and approaches for a better understanding of the mechanisms and pathways stimulating plant stress responses (Aprile *et al.*, 2004; Hasanuzzaman *et al.*, 2013). The role of micro-RNA and the proteins formed from them play an important role in osmoprotection and heat stress tolerance. Since the altered expression of the resistant target genes on micro RNA (miRNA) helped in the improved expression of target genes by inhibiting the post translational gene silencing, hence a better insight to the RNA based maintenance of the cellular homeostasis, tolerance and plasticity of plants during high temperature will prove helpful for the development of stress tolerance in plants (Chinnusamy *et al.*, 2009). Unlike transcriptomics, the knowledge of heat stress proteomics will help us to understand various biochemical metabolic pathways (metabolomics) active during HT- stress (Wienkoop *et al.*, 2008; Caldana *et al.*, 2011). The combined approach of all these omics approaches is directed towards the identification, isolation and exploitation of the gene cascade attributing to thermotolerance in plants (Vinocur and Altman, 2005). The determination of the chromatin structure, transcription binding patterns and the structure of the promoter regions when followed by transcriptome analysis and metabolite profiling, makes it easy to document the results generated due to altered transcription by up or down regulating the gene expression (Hirai *et al.*, 2005). Although, there are a number of different ways/ approaches that can be exploited for the development of heat stress tolerance, yet there exists a direct or indirect connection between them.

5. Conclusion and Future Perspectives

The increased rate of greenhouse gases like carbon dioxide and ozone responsible for global warming is accelerating the elevation in the atmospheric temperature. This rise in temperature above ambient dramatically decreases the crop growth, development and yield and the major issue of concern among researchers is to

Table 2: List of plants subjected to genetic transformation, role of the transgenes in HT- stress tolerance

S.No	Plant species (Transgenic plants)	Transformed genes/ Transgenes	Source of the transgenes	Role of Transgenes in heat stress tolerance	References
1.	Maize (Z. mays)	Hsp100, Hsp101	Arabidopsis thaliana	Produces heat shock proteins (HSPs) for HT tolerance	Queitsch et al., 2000; Katiyar-Agarwal et al., 2003
2.	Rice (O. sativa)	Hsp100, Hsp101	Arabidopsis thaliana	Induces synthesis of HSPs	Queitsch et al., 2000; Katiyar-Agarwal et al., 2003
3.	Arabidopsis (A. thaliana)	Hsp70	Trichoderma harzianum	Synthesis of heat shock proteins for thermotolerance	Montero-Barrientos et al., 2010
4.	Tobacco (N. tabacum)	Fad 7	Nicotiana tabacumand Oryza sativa	Desaturation of trienoic and hexa-decatrienoic fatty acids, increasing the level of unsaturated fatty acids for HT- tolerance	Sohn and Back, 2007; Murakami et al., 2000
5.	Carrot (Daucus carota)	Hsp17.7	Daucus carota	Induces sHsp synthesis	Malik et al., 1999; Murakami et al., 2004
6.	Tobacco (N. tabacum)	TLHS1	–	Synthesis of sHSP (Class I)	Park and Hong, 2002
7.	Arabidopsis (A. thaliana)	AtHSF1	–	Induces the fusion of HSF1 and GUS(β-glucuronidase) for increasing the production of HSPs with limited number of HSFs	Li et al., 1995

[Table Contd.

Contd. Table]

S.No	Plant species (Transgenic plants)	Transformed genes/ Transgenes	Source of the transgenes	Role of Transgenes in heat stress tolerance	References
8.	Arabidopsis (A. thaliana)	gusA	–	Triggers the synthesis of β-glucuronidase and helps the HSPs in the formation of active trimer	Li et al., 1995
9.	Tobacco (N. tabacum)	MT-sHSP	L. esculentum	Functions as an in vitro molecular chaperone	Liu and Shono, 1999; Sanmiya et al., 2004
10. Provides HT- tolerance	Tobacco (N. tabacum)	Dnak1 Ono et al., 2001	Aphanothece		
11.	Tobacco (N. tabacum)	BADH (betain aldehydede-hydrogenase)	Spinacia oleracea	Leads to the Over production of GB osmolyte which will provide enhanced heat tolerance	Yang et al., 2005
12.	Arabidopsis (A. thaliana)	Cod A (choline oxidaseA)	Arabidopsis globiformis	Accelerates Glycine betaine production for thermo-tolerance During the seedling stage	Alia et al., 1998
13.	Tobacco (N. tabacum)	ANP1/NPK1	–	Induces the production of H_2O_2 responsive MAPK kinase kinase(MAPKKK) for HT-tolerance	Kovtun et al., 2000
14.	Arabidopsis (A. thaliana)	Ascorbate peroxidase APX1 and HvAPX1	Pisum sativum and H. vulgare respectively	Confers HT- stress tolerance by H_2O_2 detoxification	Shi et al., 2001

find the possible ways and approaches for combating this heat induced terribleloss to the crops. Hence, the avoidance and the tolerance mechanism in plants against heat stress need to be understood efficiently. Though the heat stress stimulates andactivate the synthesis of antioxidants, osmoprotectants, heat stress proteins (HSPs), different metabolic and signaling cascades, yet understanding the nature and the expression of genes and the signaling cascade involved will be useful for uncovering new molecular and genetic engineering tools for the development of heat stress tolerant varieties (Rodriguez *et al.,* 2005; Zrobek-sokolnik, 2012). Along with the manipulation of the conventional field practices, subjecting heat sensitive plants to genetic manipulation for the production of thermotolerance inducing compounds are of potential importance for the development of HT- stress tolerant crops, which will ensure food security in near future.

6. References

1. Wahid, A.; Gelani, S.; Ashraf, M.; Foolad, M. R. Heat tolerance in plants: An overview. *Environmental and Experimental Botany.* **2007**, 61, 199–223.

2. Abdul-baki, A. A.; Stommel, J. R. Pollen viability and fruit-set of tomato genotypes under optimum-temperature and high-temperature regimes. *HortScience.* **1995**, 30, 115-117.

3. Ahamed, K. U.; Nahar, K.; Fujita, M.; Hasanuzzaman, M. Variation in plant growth, tiller dynamics and yield components of wheat (*Triticum aestivum* L.) due to high temperature stress. *Advances in Agriculture & Botanics.* **2010**, 2, 213-224.

4. Ahuja, I.; de Vos, R.C.H.; Bones, A.M.; Hall, R.D. Plant molecular stress responses face climatechange. *Trends Plant Sci.* **2010**, 15, 664–674.

5. Alia, H.H.; Sakamoto, A.; Murata, N. Enhancement of the tolerance of *Arabidopsis* to high temperatures by genetic engineering of the synthesis of glycinebetaine. *Plant J.* **1998**, 16, 155–161.

6. Anon, S.; Fernandez, J.A.; Franco, J.A.; Torrecillas, A.; Alarc on, J. J.; Sanchez-Blanco, M.J.;. Effects of water stress and night temperature preconditioning on water relations and morphological and anatomical changes of *Lotus creticus* plants. *Sci. Hortic.* **2004**, 101, 333–342.

7. Aprile, A.; Mastrangelo, A.M.; De Leonardis, A.M.; Galiba, G.; Roncaglia, E.; Ferrari, F.; De Bellis, L.; Turchi, L.; Giuliano, G.; Cattivelli, L. Transcriptional profiling in response to terminal drought stress reveals differential responses along the wheat genome. *BMC Genom.* **2009**, 10, 205.

8. Apse, M. P.; Blumwald, E. Engineering salt tolerance in plants. *Curr. Opin. Biotech.* **2002**, 13, 146–150.

9. Argyris, J.; Dahal, P.; Hayashi, E.; Still, D. W.; Bradford, K. J. Genetic variation for lettuce seed thermoinhibition is associated with temperature-sensitive expression of abscisic acid, gibberellin, and ethylene biosynthesis, metabolism, and response genes. *Plant Physiol.* **2008**, 148, 926–947.

10. Argyris, J.; Truco, M. J.; Ochoa, O.; McHale, L.; Dahal, P.; Van Deynze, A. A gene encoding an abscisic acid biosynthetic enzyme (LsNCED4) collocates with the high temperature germination locus Htg6. 1 in lettuce (*Lactuca* sp.). *Theor. Appl. Genet.* **2011**, 122, 95–108.

11. Ashraf, M.; Hafeez, M. Thermotolerance of pearl millet and maize at early growth stages: Growth and nutrient relations. *Biol. Plant.* **2004**, 48, 81–86.

12. Bai, J. Genetic Variation of Heat Tolerance and Correlation with Other Agronomic Traits in Maize (*Zea mays* L.) Recombinant Inbred Line Population. **2011**, 13572.

13. Barnabas, B.; Jager, K.; Feher, A. The effect of drought and heat stress on reproductive processes in cereals. *Plant Cell Environ.* **2008**, 31, 11–38.

14. Battisti, D. S.; Naylor, R. L. Historical warnings of future food insecurity with unprecedented seasonal heat. *Science.* **2009**,323, 240–244.

15. Varshney, R. K.; Bansal, K. C.; Aggarwal, P. K.; Datta, S. K.; Craufurd, P. Q. Agricultural biotechnology for crop improvement in a variable climate: hope or hype? *Trends Plant Sci.* **2011**, 16, 363–371.

16. Bohnert, H.J.; Gong, Q.; Li, P.; Ma, S. Unraveling abiotic stress tolerance mechanisms—getting genomics going. *Curr. Opin. Plant Biol.* **2006**, 9, 180–188.

17. Bukhov, N. G.; Mohanty, P. Elevated temperature stress effects on photosystems: characterization and evaluation of the nature of heat induced impairments. In: Singhal GS, Renger G, Sopory SK, Irrgang K-D, Govingjee (eds) Concepts in photobiology: photosynthesis and photomorphogenesis. Narosa Publishing House, New Delhi. **1999**, pp 617–648.

18. Caldana, C.; Degenkolbe, T.; Cuadros-Inostroza, A.; Klie, S.; Sulpice, R.; Leisse, A.; Steinhauser, D.; Fernie, A.R.; Willmitzer, L.; Hannah, M.A. High-density kinetic analysis of the metabolomic and transcriptomic response of *Arabidopsis* to eight environmental conditions. *Plant J.* **2011**, 67, 869–884.

19. Cao, Y.Y.; Duan, H.; Yang, L.N.; Wang, Z.Q.; Zhou, S.C.; Yang, J.C. Effect of heat stress during meiosis on grain yield of rice cultivars differing in heat tolerance and its physiological mechanism. *Acta Agron. Sin.* **2008**, 34, 2134–2142.

20. Carpentier, R. Effect of high-temperature stress on the photosynthetic apparatus. In: Pessarakli M (eds.) Handbook of plant and crop stress. Marcel Dekker Inc, New York. **1999**, pp 337– 348.

21. Chaitanya, K.V.; Sundar, D.; Reddy, A.R. Mulberry leaf metabolism under high temperature stress. *Biol. Plant.* **2001**, 44, 379–384.

22. Chinnusamy, V.; Zhu, J., Zhou, T., Zhu, J.K. Small RNAs: Big Role in Abiotic Stress Tolerance of Plants. In *Advances in Molecular Breeding toward Drought and Salt Tolerant Crops*; Jenks, M.A., Hasegawa, P.M., Jain, S.M., Eds.; Springer: Dordrecht, The Netherland. **2007**, pp. 223–260.

23. Christensen, J. H.; Christensen, O. B. A summary of the PRUDENCE model projections of changes in European climate by the end of this century. *Clim. Change.* **2007**, 81, 7–30.

24. Collins, N. C.; Tardieu, F.; Tuberosa, R. Quantitative trait loci and crop performance under abiotic stress: where do we stand? *Plant physiol.* **2008**, 147, 469–486.

25. Commuri, P. D.; Jones, R. J. High temperatures during endosperm cell division in maize: A genotypic comparison under in vitro and field conditions. *Crop Science.* **2001**, 41, 1122-1130.

26. Crafts-Brandner, S.J.; Salvucci, M.E. Sensitivity of photosynthesis in a C4 plant, maize, to heat stress. *Plant Physiol.* **2002**, 129, 1773–1780.

27. Craita, E. B.; Gerats, T. Plant tolerance to high temperature in a changing environment: scientific fundamentals and production of heat stress-tolerant crops. *Frontiers in plant science.* 2013,4, 273.

28. Djanaguiraman, M.; Prasad, P.V.V. Seppanen, M. Selenium protects sorghum leaves from oxidative damage under high temperature stress by enhancing antioxidant defense system. *Plant Physiol. Biochem.* **2010**, 48, 999–1007.

29. Djanaguiraman, M.; Prasad, P.V.V.; Al-Khatib, K. Ethylene perception inhibitor 1-MCP decreases oxidative damage of leaves through enhanced antioxidant defense mechanisms in soybean plants grown under high temperature stress. *Environ. Exp. Bot.* **2011**, 71, 215- 223.

30. Djanaguiraman, M.; Sheeba, J.A.; Devi, D.D.; Bangarusamy, U. Cotton leaf senescence can be delayed by nitrophenolate spray through enhanced antioxidant defense system. *J. Agron. Crop Sci.* **2009**, 195, 213–224.

31. Duque, A.S.; de Almeida, A.M.; da Silva, A.B.; da Silva, J.M.; Farinha, A.P.; Santos, D.; Fevereiro, P.; de Sousa Araujo, S. Abiotic stress responses in plants: Unraveling the complexity of genes and networks to survive. In *Abiotic Stress—Plant Responses and Applications in Agriculture*; Vahdati, K.; Leslie, C.; Eds.; InTech: Rijeka, Croatia. **2013**, pp. 3–23.

32. Duran, C.; Eales, D.; Marshall, D.; Imelfort, M.; Stiller, J.; Berkman, P. J. Future tools for association mapping in crop plants. *Genome.* **2013**, 53, 1017–1023.

33. Ebrahim, M.K.;Zingsheim, O.; El-Shourbagy, M.N.; Moore, P.H.; Komor, E. Growth and sugar storage in sugarcane grown at temperature below and above optimum. *J. Plant Physiol.* **1998**, 153, 593–602.

34. Edreira, J.I.R.; Otegui, M.E. Heat stress in temperate and tropical maize hybrids: Differences in crop growth, biomass partitioning and reserves use. *Field Crops Res.* **2012**, 130, 87–98.

35. Feller, U.; Crafts-Brandner, S.J.; Salvucci, M.E. Moderately high temperatures inhibit ribulose-1, 5-biphosphate carboxylase/oxygenase (Rubisco) activase-mediated activation of Rubisco. *Plant Physiol.* **1998**, 116, 539–546.

36. Fischer, R.; Edmeades, G.O. Breeding and cereal yield progress. *Crop Sci.* **2010**, 50, S-85–S-98.

37. Fitter, A.H.; Hay, R.K.M. *Environmental Physiology of Plants*, 3rd ed.; Academic Press: London, UK. **2002**.

38. Foolad, M.R. Breeding for abiotic stress tolerances in tomato. In *Abiotic Stresses: Plant Resistance through Breeding and Molecular Approaches*; Ashraf, M., Harris, P.J.C., Eds.; The Haworth Press Inc.: New York, NY, USA. **2005**, pp. 613–684.

39. Frova, C.; Sari-Gorla, M. Quantitative trait loci (QTLs) for pollen thermotolerance detected in maize. *Mol. Gen. Genomics.* **1994**, 245, 424–430.

40. Gan, Y.; Angadi, S.V.; Cutforth, H.; Potts, D.; Angadi, V.V.; Mcdonald, C.L. Canola and mustard response to short periods of temperature and water stress at different developmental stages. *Canadian Journal of Plant Science.* **2004**, 84, 697-704.

41. Greer, D.H. Weedon, M.M. Modelling photosynthetic responses to temperature of grapevine (*Vitis vinifera* cv. Semillon) leaves on vines grown in a hot climate. *Plant Cell Environ.* **2012**, 35, 1050–1064.

42. Gross, Y.; Kigel, J. Differential sensitivity to high temperature of stages in the reproductive development of common bean (*Phaseolus vulgaris* L.). *Field Crops Research.* **1994**, 36, 201-212.

43. Grover, A.; Mittal, D.; Negi, M.; Lavania, D. Generating high temperature tolerant transgenic plants: Achievements and challenges. *Plant Sci.* **2013**, 205–206, 38–47.

44. Guilioni, L.; W´ery, J.; Lecoeur, J. High temperature and water deficit may reduce seed number in field pea purely by decreasing plant growth rate. *Funct. Plant Biol.* **2003**, 30, 1151–1164.

45. Guilioni, L.; Wery, J.; Tardieu, F. Heat stress-induced abortion of buds and flowers in pea: is sensitivity linked to organ age or to relations between reproductive organs? *Ann. Bot.* **1997**, 80, 159–168.

46. Gunawardhana, M.D.M.; de Silva, C.S. Impact of temperature and water stress on growth yield and related biochemical parameters of okra. *Trop. Agric. Res.* **2011**, 23, 77–83.

47. Haldimann, P, & Feller, U. Inhibition of photosynthesis by high temperature in oak (*Quercus pubescens* L.) leaves grown under natural conditions closely correlates with a reversible heat-dependent reduction of the activation state of ribulose-1, 5-bisphosphate carboxylase/oxygenase. *Plant, Cell & Environment.* **2004**, 27, 1169-1183.

48. Hall, A.E. The mitigation of heat stress. **2011**.

49. Hall, A.E. Breeding for heat tolerance. *Plant Breed. Rev.* **1992**, 10, 129– 168.

50. Hasanuzzaman, M.; Gill, S.S.; Fujita, M. Physiological role of nitric oxide in plants grown under adverse environmental conditions. In *Plant Acclimation to Environmental Stress*; Tuteja, N., Gill, S.S., Eds.; Springer: New York, NY, USA. **2013**, pp. 269–322.

51. Hasanuzzaman, M.; Hossain, M.A.; da Silva, J.A.T.; Fujita, M. Plant Responses and Tolerance to Abiotic Oxidative Stress: Antioxidant Defenses is a Key Factor. In *Crop Stress and Its Management: Perspectives and Strategies*; Bandi, V., Shanker, A.K., Shanker, C., Mandapaka, M., Eds.; Springer: Berlin, Germany. **2012**, pp. 261–316.

52. Hasanuzzaman, M.; Hossain, M.A.; Fujita, M. Exogenous selenium pretreatment protects rapeseed seedlings from cadmium-induced oxidative stress by upregulating the antioxidant defense and methylglyoxal detoxification systems. *Biol. Trace Elem. Res.* **2012**, 149, 248–261.

53. Hasanuzzaman, M.; Hossain, M.A.; Fujita, M. Physiological and biochemical mechanisms of nitric oxide induced abiotic stress tolerance in plants. *Am. J. Plant Physiol.* **2010**, 5, 295–324.

54. Hasanuzzaman, M.; Hossain, M.A.; Fujita, M. Selenium in higher plants: Physiological role, antioxidant metabolism and abiotic stress tolerance. *J. Plant Sci.* **2010**, 5, 354–375.

55. Hasanuzzaman, M.; Nahar, K.; Alam, M.M.; Fujita, M. Exogenous nitric oxide alleviates high temperature induced oxidative stress in wheat (*Triticum aestivum*) seedlings by modulating the antioxidant defense and glyoxalase system. *Aust. J. Crop Sci.* **2012**, 6, 1314–1323.

56. Hasanuzzaman, M.; Nahar, K.; Fujita, M. Extreme Temperatures, Oxidative Stress and Antioxidant Defense in Plants. In *Abiotic Stress—Plant Responses and Applications in Agriculture*; Vahdati, K., Leslie, C., Eds.; InTech: Rijeka, Croatia. **2013**, pp. 169–205.

57. Hasanuzzaman, M.; Nahar, K.; Fujita, M. Plant response to salt stress and role of exogenous protectants to mitigate salt-induced damages. In *Ecophysiology and Responses of Plants under Salt Stress*; Ahmad, P., Azooz, M.M., Prasad, M.N.V., Eds.; Springer: New York, NY, USA. **2013**, pp. 25–87.

58. Hasanuzzaman, M.; Nahar, K.; Fujita, M.; Ahmad, P.; Chandna, R.; Prasad, M.N.V.; Ozturk, M. Enhancing Plant Productivity under Salt Stress—Relevance of Poly-Omics. In *Salt Stress in Plants: Omics, Signaling and Responses*; Ahmad, P., Azooz, M.M., Prasad, M.N.V., Eds.; Springer: Berlin, Germany. **2013**, pp. 113–156.

59. Hatfield, J.L.; Boote, K.J.; Kimball, B.A.; Ziska, L.H.; Izaurralde, R.C.; Ort, D.; Thomson, A.; Wolfe, D. Climate impacts on agriculture: Implications for crop production. *Agron. J.* **2011**, 103, 351–370.

60. Hirai, M.Y.; Klein, M.; Fujikawa, Y.; Yano, M.; Goodenowe, D.B.; Yamazaki, Y.; Kanaya, S.; Nakamura, Y.; Kitayama, M.; Suzuki, H.; *et al.* Elucidation of gene-to-gene and metabolite-to-gene networks in *Arabidopsis* by integration of metabolomics and transcriptomics. *J. Biol. Chem.* **2005**, 280, 25590–25595.

61. Hirayama, T.; Shinozaki, K. Research on plant abiotic stress responses in the post-genome era: past, present and future. *Plant J.* **2010**, 61, 1041–1052.

62. Hong, S.W.; Lee, U.; Vierling, E. *Arabidopsis* hot mutants define multiple functions required for acclimation to high temperatures. *Plant Physiol.* **2003**, 132, 757–767.

63. Hong, S.W.; Vierling, E. Mutants of *Arabidopsis thaliana* defective in the acquisition of tolerance to high temperature stress. *Proc. Natl. Acad. Sci. U.S.A.* **2000**, 97, 4392–4397.

64. Horvath, I.; Glatz, A.; Varvasovszki, V.; Torok, Z.; Pali, T.; Balogh, G. Membrane physical state controls the signaling mechanism of the heat shock response in Synechocystis PCC 6803: identification of hsp17 as a fluidity gene. *Proc Natl Acad Sci USA.* **1998**, 95:3513–3518.

65. Howarth, C.J. Genetic improvements of tolerance to high temperature. In Ashraf, M., Harris, P.J.C. (Eds.), Abiotic Stresses: Plant Resistance Through Breeding and Molecular Approaches. Howarth Press Inc., New York. **2005.**

66. Hurkman, W.J.; Vensel, W.H.; Tanaka, C.K.; Whitehand, L.; Altenbach, S.B. Effect of high temperature on albumin and globulin accumulation in the endosperm proteome of the developing wheat grain. *J. Cereal Sci.* **2009**, 49, 12–23.

67. IPCC. Intergovernmental panel on climate change report: Synthesis: Summary for policy makers. IPCC, World Meteorological Organization, Switzerland. **2007**.

68. Ismail, A.M.; Hall, A.E. Reproductive-stage heat tolerance, leaf membrane thermostability and plant morphology in cowpea. *Crop Sci.* **1999**, 39, 1762–1768.

69. Johkan, M.; Oda, M.; Maruo, T.; Shinohara, Y. Crop production and global warming In: Casalegno S (ed) Global warming impacts- Case studies on the economy, human health, and on urban and natural environments. *Rijeka: InTech.* **2011**, 139-152.

70. Kalra, N.; Chakraborty, D.; Sharma, A.; Rai, H.K.; Jolly, M.; Chander, S.; Kumar, P.R.; Bhadraray, S.; Barman, D.; Mittal, R.B.; *et al.* Effect of increasing temperature on yield of some winter crops in northwest India. *Curr. Sci.* **2008**, 94, 82–88.

71. Karim, M.A.; Fracheboud, Y.; Stamp, P. Heat tolerance of maize with reference of some physiological characteristics. *Ann. Bangladesh Agri.* **1997**, 7, 27–33.

72. Kasuga, M.; Liu, Q.; Miura, S.; Yamaguchi-Shinozaki, K.; Shinozaki, K. Improving plant drought, salt and freezing tolerance by gene transfer of a single stress-inducible transcription factor. *Nat Biotechnol.* **1999**, 17, 287–291.

73. Katiyar-Agarwal, S.; Agarwal, M.; Grover, A. Heat tolerant basmati rice engineered by Over-expression of hsp101. *Plant Mol. Biol.* **2003**, 51, 677–686.

74. Kinet, J.M.; Peet, M.M. Tomato. In: Wien, H.C. (Ed.), The Physiology of Vegetable Crops. CAB International, Wallingford, UK. **1997**, pp. 207– 258.

75. Kitano, M.; Saitoh, K.; Kuroda, T. Effect of high temperature on flowering and pod set in soybean. Scientific Report of the Faculty of Agriculture, Okayama University. **2006**, 95, 49-55.

76. Koini, M.A.; Alvey, L.; Allen, T.; Tilley, C.A.; Harberd, N.P.; Whitelam, G.C.; Franklin, K.A. High temperature-mediated adaptations in plant architecture require the bHLH transcription factor PIF4. *Curr. Biol.* **2009**, 19, 408–413.

77. Kosova, K.; Vítamvas, P.; Prašil, I.T.; Renaut, J. Plant proteome changes under abiotic stress-Contribution of proteomics studies to understanding plant stress response. *J. Proteom.* **2011**, 74, 1301–1322.

78. Kovtun, Y.; Chiu, W.L.; Tena, G.; Sheen, J. Functional analysis of oxidative stress-activated mitogen-activated protein kinase cascade in plants. *Proc. Natl. Acad. Sci. USA.* **2000**, 97, 2940–2945.

79. Kozai, N.; Beppu, K.; Mochioka, R.; Boonprakob, U.; Subhadrabandhu, S.; Kataoka, I. Adverse effects of high temperature on the development of reproductive organs in 'Hakuho' peach trees. *Journal of Horticultural Science and Biotechnology*. **2004**, 79, 533-537.

80. Kumar, S.; Kaur, R.; Kaur, N.; Bhandhari, K.; Kaushal, N.; Gupta, K.; Bains, T.S.; Nayyar, H. Heat-stress induced inhibition in growth and chlorosis in mungbean (*Phaseolus aureus* Roxb.) is partly mitigated by ascorbic acid application and is related to reduction in oxidative stress. *Acta Physiol. Plant.* **2011**, 33, 2091–2101.

81. Larcher, W. *Physiological Plant Ecology*; Springer: Berlin, Germany. **1995**.

82. Lee, B.H.; Won, S.H.; Lee, H.S.; Miyao, M.; Chung, W.I.; Kim, I.J.; Jo, J. Expression of the chloroplast-localized small heat shock protein by oxidative stress in rice. *Gene.* **2000**, 245, 283–290.

83. Lee, J.H.; Hubel, A.; Schoffl, F. Derepression of the activity of genetically engineered heat shock factor causes constitutive synthesis of heat shock proteins and increased thermotolerance in transgenic *Arabidopsis*. *Plant J.* **1995**, 8, 603–612.

84. Lin, C.; Li, C.; Lin, S.; Yang, F.; Huang, J.; Liu, Y.; Lur, H. Influence of high temperature during grain filling on the accumulation of storage proteins and grain quality in rice (*Oryza sativa* L.). *Journal of Agriculture and Food Chemistry.* **2010**, 58, 10545-10552.

85. Liu, J.; Shono, M. Characterization of mitochondria-located small heat shock protein from tomato (*Lycopersicon esculentum*). *Plant Cell Physiol.* **1999**, 40, 1297–1304.

86. Liu, N.; Ko, S.; Yeh, K.C.; Charng, Y. Isolation and characterization of tomato Hsa32 encoding a novel heat-shock protein. *Plant Sci.* **2006**, 170, 976–985.

87. Lobell, D.B.; Asner, G.P. Climate and management contributions to recent trends in U.S. agricultural yields. *Science.* **2003**, 299.

88. Lobell, D.B.; Field, C.B. Global scale climate–Crop yield relationships and the impacts of recent warming. *Environ. Res. Lett.* **2007**.

89. Lobell, D.B.; Schlenker, W.; Costa-Roberts, J. Climate trends and global crop production since 1980. *Science.* **2011**, 333, 616–620.

90. Lopes, M.S.; Reynolds, M. P. Partitioning of assimilates to deeper roots is associated with cooler canopies and increased yield under drought in wheat. *Funct. Plant Biol.* **2010**, 37, 147–156.

91. Los, D.A.; Murata, N. Membrane fluidity and its roles in the perception of environmental signals. *Biochim Biophys Acta,* **2004**, 1666:142–157.

92. Maestri, E.; Klueva, N.; Perrotta, C.; Gulli, M.; Nguyen, H.T.; Marmiroli, N. Molecular genetics of heat tolerance and heat shock proteins in cereals. *Plant Mol. Biol.* **2002**, 48, 667–681.

93. Maheswari, M.; Yadav, S.K.; Shanker, A.K.; Kumar, M.A.; Venkateswarlu, B. Overview of plant stresses: Mechanisms, adaptations and research pursuit. In *Crop Stress and Its Management: Perspectives and Strategies*; Venkateswarlu, B., Shanker, A.K., Shanker, C., Maheswari, M., Eds.; Springer: Dordrecht, The Netherlands. **2012**, pp. 1–18.

94. Malik, M.K.; Slovin, J.P.; Hwang, C.H.; Zimmerman, J.L. Modified expression of a carrot small heat shock protein gene, *Hsp17.7*, results in increased or decreased thermotolerance. *Plant J.* **1999**, 20, 89–99.

95. Marchand, F.L.; Mertens, S.; Kockelbergh, F.; Beyens, L.; Nijs, I. Performance of high arctic tundra plants improved during but deteriorated after exposure to a simulated extreme temperature event. *Glob. Change Biol.* **2005**, 11, 2078–2089.

96. Mazorra, L.M.; Nunez, M.; Echerarria, E.; Coll, F.; Sanchez-blanco, M.J. Influence of brassinosteroids and antioxidant enzymes activity in tomato under different temperatures. *Plant Biology.* **2002**, 45, 593-596.

97. McClung, C.R.; Davis, S.J. Ambient thermometers in plants: From physiological outputs towards mechanisms of thermal sensing. *Curr. Biol.* **2010**, 20, 1086–1092.

98. Mittler, R. Abiotic stress, the field environment and stress combination. *Trends in Plant Science.*

99. Mittler, R.; Blumwald, E. Genetic engineering for modern agriculture: Challenges and perspectives. *Ann. Rev. Plant Biol.* **2010**, 61, 443–462.

100. Mohammed, A. R.; Tarpley, L. Effects of high night temperature and spikelet position on yield-related parameters of rice (*Oryza sativa* L.) plants. *European Journal of Agronomy.* **2010**, 33, 117-123.

101. Momcilovic, I.; Ristic, Z. Expression of chloroplast protein synthesis elongation factor, EF-Tu, in two lines of maize with contrasting tolerance to heat stress during early stages of plant development. *J. Plant Physiol.* **2007**,164, 90–99.

102. Monjardino, P.; Smith, A.G.; Jones, R.J. Heat stress effects on protein accumulation of maize endosperm. *Crop Science.* **2005**, 45, 1203-1210.

103. Montero-Barrientos, M.; Hermosa, R.; Cardoza, R.E.; Gutierrez, S.; Nicolas, C.; Monte, E. Transgenic expression of the *Trichoderma harzianum hsp70* gene increases *Arabidopsis* resistance to heat and other abiotic stresses. *J. Plant Physiol.* **2010**, 167, 659–665.

104. Morales, D.; Rodriguez, P.; Dell'amico, J.; Nicolas, E.; Torrecillas, A.; Sanchez- Blanco, M.J. High-temperature preconditioning and thermal shock imposition affects water relations, gas exchange and root hydraulic conductivity in tomato. *Biol. Plant.* **2003**, 47, 203–208.

105. Moreno, A.A.; Orellana, A. The physiological role of the unfolded protein response in plants. *Biol. Res.* **2011**,44, 75–80.

106. Morita, S.; Yonermaru, J.; Takahashi, J. Grain growth and endosperm cell size under high night temperature in rice (*Oryza sativa* L.). *Annals of Botany.* **2005**, 95, 695-701.

107. Murakami, Y.; Tsuyama, M.; Kobayashi, Y.; Kodama, H.; Iba, K. Trienoic fatty acids and plant tolerance of high temperature. *Science.* **2000**, 287, 476–479.

108. Nakamoto, H.; Hiyama, T. Heat-shock proteins and temperature stress In: Pessarakli, M. (Ed.), Handbook of Plant and Crop Stress. Marcel Dekker, New York. **1999**, pp. 399–416.

109. Nash, D.; Miyao, M.; Murata, N. Heat inactivation of oxygen evolution in photosystem II particles and its acceleration by chloride depletion and exogenous manganese. *Biochim Biophys Acta.* **1985**, 807:127–133.

110. Nelson, G. C. *Climate Change: Impact on Agriculture and Costs of Adaptation.* International Food Policy Research Institute, Washington. **2009**.

111. Nollen, E.A.A.; Morimoto, R.I. Chaperoning signaling pathways: molecular chaperones as stress-sensing 'heat shock' proteins. *J. Cell Sci.* **2002**, 115, 2809–2816.

112. Omae, H.; Kumar, A.; Shono, M. Adaptation to high temperature and water deficit in the common bean (*Phaseolus vulgaris* L.) during the reproductive period. *J. Bot.* **2012**.

113. Ono, K.; Hibino, T.; Kohinata, T.; Suzuki, S.; Tanaka, Y.; Nakamura, T.; Takabe, T.; Takabe, T. Overexpression of DnaK from a halotolerant cyanobacterium *Aphanothece halophytica* enhances the high-temperature tolerance of tobacco during germination and early growth. *Plant Sci.* **2001**, 160, 455–461.

114. Ortiz, R.; Braun, H. J.; Crossa, J.; Crouch, J. H.; Davenport, G.; Dixon, J. Wheat genetic resources enhancement by the International Maize and Wheat Improvement Center (CIMMYT). *Genet. Resour. Crop Evol.* **2002**, 55, 1095– 1140.

115. Pagamas, P.; Nawata, E. Sensitive stages of fruit and seed development of chili pepper (*Capsicum annuum* L. var. Shishito) exposed to high-temperature stress. *Sci. Hort.* **2008**, 117, 21–25.

116. Paliwal, R.; Roder,M. S.; Kumar,U.; Srivastava, J.; Joshi, A. K. QTL mapping of terminal heat tolerance in hexaploid wheat (*T. aestivumL.*). *Theor. Appl. Genet.* **2002**, 125, 561-575.

117. Park, S.M.; Hong, C.B. Class I small heat shock protein gives thermotolerance in tobacco. *J. Plant Physiol.* **2002**, 159, 25–30.

118. Patel, D.; Franklin, K.A. Temperature-regulation of plant architecture. *Plant Signal. Behav.* **2009**, 4, 577–579.

119. Paulsen, G. M. High temperature responses of crop plants. In "Physiology and Determination of Crop Yield" (K. J. Boote, J. M. Bennett, T. R. Sinclair, and G. M. Paulsen, Eds.), pp. 365–389. American Society of Agronomy, Madison, WI. **1994**.

120. Peet, M.M.; Willits, D.H. The effect of night temperature on greenhouse grown tomato yields in warm climate. *Agric. Forest Meteorol.* **1998**, 92, 191–202.

121. Perez, D.E. BOBBER1 is a noncanonical Arabidopsis small heat shock protein required for both development and thermotolerance. *Plant Physiol.***2009**, 151, 241–252.

122. Perrotta, C.; Treglia, A.S.; Mita, G.; Giangrande, E.; Rampino, P.; Ronga, G.; Spano, G.; Marmiroli, N. Analysis of mRNAs from ripening wheat seeds: the effect of high temperature. *J. Cereal Sci.* **1998**, 27, 127–132.

123. Pinto, R.S.; Reynolds, M. P.; Mathews, K. L.; McIntyre, C. L.; Olivares-Villegas, J. J.; Chapman, S.C. Heat and drought adaptive QTL in a wheat population designed to minimize confounding agronomic effects. *Theor. Appl. Genet.* **2010**,121, 1001–1021.

124. Porch, T.G.; Jahn, M. Effects of high temperature stress on microsporogenesis in heat sensitive and heat-tolerant genotypes of *Phaseolus vulgaris*. *Plant, Cell & Environment.* **2001**, 24, 723-731.

125. Porter, J. R.; Gawith, M. Temperature and the growth and development of wheat: A review. *European Journal of Agronomy.* **1999**, 10, 23-36.

126. Porter, J.R. Rising temperatures are likely to reduce crop yields. *Nature.* **2005**, 436, 174.

127. Prasad, P.V.V.; Pisipati, S.R.; Momcilovic, I.; Ristic, Z. Independent and combined effects of high temperature and drought stress during grain filling on plant yield and chloroplast EF-Tu Expression in spring wheat. *Journal of Agronomy Crop Science.* **2011**, 197, 430-441.

128. Prasad, P. V. V, Boote, K. J, Allen Jr ,L. H. Adverse high temperature effects on pollen viability, seed-set, seed yield and harvest index of grain-sorghum [*Sorghum bicolor* (L.) Moench] Extreme Temperature Responses,

Oxidative Stress and Antioxidant Defense in Plants are more severe at elevated carbon dioxide due to higher tissue temperatures. *Agricultural and Forest Meteorology.* **2006**, 139, 237-251.

129. Queitsch, C.; Hong, S.W.; Vierling, E.; Lindquest, S. Heat shock protein 101 plays a crucial role in thermotolerance in *Arabidopsis. Plant Cell.* **2002**, 12, 479–492.

130. Radin, J.W.; Lu, Z.; Percy, R.G.; Zeiger, E. Genetic variability for stomatal conductance in Pima cotton and its relation to improvements of heat adaptation. *Proc. Natl. Acad. Sci. USA.* **1994**, 91, 7217–7221.

131. Rahman, M.A.; Chikushi, J.; Yoshida, S.; Karim, A.J.M.S. Growth and yield components of wheat genotypes exposed to high temperature stress under control environment. Bangladesh *Journal of Agricultural Research.* **2009**, 34, 361-372.

132. Rahman, M.M. Response of wheat genotypes to late-seeding heat stress. MS Thesis. Department of Crop Botany. Bangabandhu Sheikh Mujibur Rahman Agricultural University, Gazipur, Bangladesh. **2004.**

133. Rainey, K.; Griffiths, P. Evaluation of *Phaseolus acutifolius* A. Gray plant introductions under high temperatures in a controlled environment. *Genet. Resour. Crop Evol.* **2005**, 52, 117–120.

134. Yadav, R.C.; Solanke, A.U.K.; Kumar, P.; Pattanayak, D.; Yadav, N.R.; Kumar, P.A. Genetic Engineering for Tolerance to Climate Change-Related Traits. C. Kole (ed.), Genomics and Breeding for Climate-Resilient Crops, Vol. 1, Springer-Verlag Berlin Heidelberg. **2013.**

135. Rennenberg, H.; F. Loreto, A.; Polle, F.; Brilli, S.; Fares, R.S.; Gessler, A. Physiological response of forest trees to heat and drought. *Plant Biol.* **2006**, 8, 556–571.

136. Rodriguez, M.; Canales, E.; Borràs-Hidalgo, O. Molecular aspects of abiotic stress in plants. *Biotechnol. Appl.* **2005**, 22, 1–10.

137. Rontein, D.; Basset, G.; Hanson, A.D. Metabolic engineering of osmoprotectant accumulation in plants. *Metab. Eng.* **2002**, 4, 49–56.

138. Roy, S.J.; Tucker, E.J., Tester, M. Genetic analysis of abiotic stress tolerance in crops. *Curr. Opin. Plant Biol.* **2011**, 14, 232–239.

139. Ruelland, E.; Zachowski, A. How plants sense temperature. *Environ. Exp. Bot.* **2010**, 69, 225–232.

140. Saitoh, H. *Ecological and Physiology of Vegetable*; Nousangyoson Bunka Kyoukai: Tokyo, Japan. **2008**.

141. Sakata, T.; Higashitani, A. Male sterility accompanied with abnormal anther development in plants–genes and environmental stresses with special

reference to high temperature injury. *Int. J. Plant Dev.Biol.* **2008**, 2, 42–51.

142. Salvucci, M.E.; Crafts-Brandner, S.J.; Relationship between the heat tolerance of photosynthesis and the thermal stability of rubisco activase in plants from contrasting thermal environments. *Plant Physiol.* **2004a**, 134, 1460–1470.

143. Sanmiya, K.; Suzuki, K.; Egawa, Y.; Shono, M. Mitochondrial small heat shock protein enhances thermotolerance in tobacco plants. *FEBS Lett.* **2004**, *557*, 265–268.

144. Sarkar, N. Rice sHsp genes: genomic organization and expression profiling under stress and development. *BMC Genomics.* **2009,** 10, 393.

145. Sato, S.; Kamiyama, M.; Iwata, T.; Makita, N.; Furukawa, H.; Ikeda, H. Moderate increase of mean daily temperature adversely affects fruit set of *Lycopersicon esculentum* by disrupting specific physiological processes in male reproductive development. *Ann. Bot.* **2006**, 97, 731–738.

146. Sato, S.; Kamiyama, M.; Iwata, T.; Makita, N.; Furukawa, H.; Ikeda, H. Moderate increase of mean daily temperature adversely affects fruit set of *Lycopersicon esculentum* by disrupting specific physiological processes in male reproductive development. *Ann. Bot.* **2006**, 97, 731–738.

147. Savin, R.; Stone, P.J.; Nicolas, M.E.; Wardlaw, I.F. Effects of heat stress and moderately high temperature on grain growth and malting quality of barley. *Aust. J. Agric. Res.* **1997**, 48, 615–624.

148. Sayed, O.H. Adaptational responses of *Zygophyllum qatarense* Hadidi to stress conditions in a desert environment. *J. Arid Environ.* **1996**, 32, 445–452.

149. Schoffl, F.; Prandl, R.; Reindl, A. Molecular responses to heat stress. In *Molecular Responses to Cold, Drought, Heat and Salt Stress in Higher Plants*; Shinozaki, K., Yamaguchi-Shinozaki, K., Eds.; R.G. Landes Co.: Austin, TX, USA. **1999**, pp. 81–98.

150. Schuster, W.S.; Monson, R.K. An examination of the advantages of C3-C4 intermediate photosynthesis in warm environments. *Plant, Cell & Environment.* **1990**, 13, 903-912.

151. Sharkey, T.D. Effects of moderate heat stress on photosynthesis: importance of thylakoid reactions, rubisco deactivation, reactive oxygen species, and thermotolerance provided by isoprene. *Plant Cell Environ.* **2005**, 28, 269–277.

152. Sharkey, T.D.; Badger, M.R.; Von-Caemmerer, S.; Andrews, T.J. Increased heat sensitivity of photosynthesis in tobacco plants with reduced Rubisco activase. *Photosyn. Res.* **2001**, 67, 147–156.

153. Shi, W.M.; Muramoto, Y.; Ueda, A.; Takabe, T. Cloning of peroxisomal ascorbate peroxidase gene from barley and enhanced thermotolerance by overexpressing in *Arabidopsis thaliana*. *Gene* **2001**, 273, 23–27.

154. Simoes-Araujo, J.L.; Rumjanek, N.G.; Margis-Pinheiro, M. Small heat shock proteins genes are differentially expressed in distinct varieties of common bean. *Braz. J. Plant Physiol.* **2003**, 15, 33–41.

155. Sohn, S.O.; Back, K. Transgenic rice tolerant to high temperature with elevated contents of dienoic fatty acids. *Biol. Plant.* **2007**, 51, 340–342.

156. Srivastava, S.; Pathak, A.D.; Gupta, P.S.; Shrivastava, A.K.; Srivastava, A.K. Hydrogen peroxide-scavenging enzymes impart tolerance to high temperature induced oxidative stress in sugarcane. *J. Environ. Biol.* **2012**, 33, 657–661.

157. Sumesh, K.V.; Sharma-Natu, P.; Ghildiyal, M.C. Starch synthase activity and heat shock protein in relation to thermal tolerance of developing wheat grains. *Biol. Plant.* **2008**, 52, 749–753.

158. Sun, A.; Yi, S.; Yang, J.; Zhao, C.; Liu, J. Identification and characterization of a heat-inducible ftsH gene from tomato (*Lycopersicon esculentum* Mill.). *Plant Sci.* **2006**, 170, 551–562.

159. Sung, D.Y.; Kaplan, F.; Lee, K.J.; Guy, C.L. Acquired tolerance to temperature extremes. *Trends Plant Sci.* **2003**, 8, 179–187.

160. Suwa, R.; Hakata, H.; Hara, H.; El-Shemy, H.A.; Adu-Gyamfi, J.J.; Nguyen, N.T.; Kanai, S.;Lightfoot, D.A.; Mohapatra, P.K.; Fujita, K. High temperature effects on photosynthetic partitioning and sugar metabolism during ear expansion in maize (*Zea mays* L.) genotypes. *Plant Physiol. Biochem.* **2010**, 48, 124–130.

161. Suzuki, N.; Koussevitzky, S.; Mittler, R.; Miller, G. ROS and redox signalling in the response of plants to abiotic stress. *Plant Cell Environ.* **2012**, 35, 259–270.

162. Suzuki, N.; Miller, G.; Morales, J.; Shulaev, V.; Torres, M.A.; Mittler, R. Respiratory burst oxidases: The engines of ROS signaling. *Curr. Opin. Plant Biol.* **2011**, 14, 691- 699.

163. Tan, W.; Meng, Q.W.; Brestic, M.; Olsovska, K.; Yang, X. Photosynthesis is improved by exogenous calcium in heat-stressed tobacco plants. *J. Plant Physiol.* **2011**, 168, 2063–2071.

164. Tester, M.; Langridge, P. Breeding technologies to increase crop production in a changing world. *Science.* **2010**, 327, 818–822.

165. Tewolde, H.; Fernandez, C.J.; Erickson, C.A. Wheat cultivars adapted to post-heading high temperature stress. *J. Agron. Crop Sci.* **2006**,192, 111–120.

166. Thomson, M.J.; de Ocampo, M.; Egdane, J.; Rahman, M.A.; Sajise, A.G.; Adorada, D.L. Characterizing the Saltol quantitative trait locus for salinity tolerance in rice. *Rice.* **2010**, 3, 148–160.

167. Tsukaguchi, T.; Kawamitsu, Y.; Takeda, H.; Suzuki, K.; Egawa, Y.Water status of flower buds and leaves as affected by high temperature in heat tolerant and heat-sensitive cultivars of snap bean (*Phaseolus vulgaris* L.). *Plant Prod. Sci.* **2003**, 6, 4–27.

168. Tubiello, F.N.; Soussana, J.F.; Howden, S.M. Crop and pasture response to climate change. *Proc. Nat. Acad. Sci. USA.* **2007**, 104, 19686–19690.

169. Vara Prasad, P.V.; Craufurd, P.Q.; Summerfield, R.J. Fruit number in relation to pollen production and viability in groundnut exposed to shortepisodes of heat stress. *Ann. Bot.* **1999**, 84, 381–386.

170. Vijayalakshmi, K.; Fritz, A. K.; Paulsen, G. M.; Bai, G.; Pandravada, S.; Gill, B. S. Modeling and mapping QTL for senescence related traits in winter wheat under high temperature. *Mol. Breed.* **2010**, 26, 163–175.

171. Vinocur, B.; Altman, A. Recent advances in engineering plant tolerance to abiotic stress: achievements and limitations. *Curr. Opin. Biotechnol.* **2005**, 16, 123–132.

172. Vollenweider, P.; Gunthardt-Goerg, M.S. Diagnosis of abiotic and biotic stress factors using the visible symptoms in foliage. *Environ. Pollut.* **2005**, 137, 455–465.

173. Vu, J.C.V.; Gesch, R.W.; Pennanen, A.H.; Allen, L.H.J.; Boote, K.J.; Bowes, G. Soybean photosynthesis, Rubisco and carbohydrate enzymes function at supra-optimal temperatures in elevated CO_2. *J. Plant Physiol.* **2001**, 158, 295–307.

174. Wahid, A. Physiological implications of metabolites biosynthesis in net assimilation and heat stress tolerance of sugarcane (*Saccharum officinarum*) sprouts. *Journal of Plant Research.* **2007**, 120, 219-228.

175. Wahid, A.; Close, T.J. Expression of dehydrins under heat stress and their relationship with water relations of sugarcane leaves. *Biol. Plant.* **2007**, 51, 104–109.

176. Wang, L.J.; Li, S.H. Salicylic acid-induced heat or cold tolerance in relation to Ca2+ homeostasis and antioxidant systems in young grape plants. *Plant Science.* **2006**, 170, 685-694.

177. Wang, W.; Vinocur, B.; Shoseyov, O.; Altman, A. Role of plant heat-shock proteins and molecular chaperones in the abiotic stress response. *Trends Plant Sci.* **2004**, 9, 244–252.

178. Wang, X.; Cai, J.; Liu, F.; Jin, M.; Yu, H.; Jiang, D.; Wollenweber, B.; Dai, T.; Cao, W. Pre-anthesis high temperature acclimation alleviates the negative effects of post-anthesis heat stress on stem stored carbohydrates remobilization and grain starch accumulation in wheat. *J. Cereal Sci.* **2012**, 55, 331–336.

179. Waraich, E.A.; Ahmad, R.; Halim, A.; Aziz, T. Alleviation of temperature stress by nutrient management in crop plants: A review. *J. Soil Sci. Plant Nutr.* **2012**, 12, 221–244.

180. Wardlaw, I.F.; Blumenthal, C.; Larroque, O.; Wrigley, C.W. Contrasting effects of chronic heat stress and heat shock on kernel weight and flour quality in wheat. *Funct. Plant Biol.* **2002**, 29, 25–34.

181. Warland, J.S.; McDonald, M.R.; McKeown, A.M. Annual yields of five crops in the family Brassicacae in southern Ontario in relation to weather and climate. *Can. J. Plant Sci.* **2006**, 86, 1209–1215.

182. Weaich, K.; Briston, K.L.; Cass, A. Modeling preemergent maize shoot growth. II. High temperature stress conditions. *Agric. J.* **1996**, 88, 398–403.

183. Wienkoop, S.; Morgenthal, K.; Wolschin, F.; Scholz, M.; Selbig, J.; Weckwerth, W. Integration of metabolomic and proteomic phenotypes: analysis of data covariance dissects starch and RFO metabolism from low and high temperature compensation response in *Arabidopsis thaliana. Mol. Cell Proteom.* **2008**, 7, 1725–1736.

184. Wise, R.R.; Olson, A.J.; Schrader, S.M.; Sharkey, T.D. Electron transport is the functional limitation of photosynthesis in field-grown Pima cotton plants at high temperature. *Plant Cell Environ.* **2004**, 27, 717–724.

185. Wollenweber, B.; Porter, J.R.; Schellberg, J. Lack of interaction between extreme high temperature events at vegetative and reproductive growth stages in wheat. *J. Agron. Crop Sci.* **2003**, 189, 142–150.

186. Yang, X.; Chen, X.; Ge, Q.; Li, B.; Tong, Y.; Zhang, A.; Li, Z.; Kuang, T.; Lu, C. Tolerance of photosynthesis to photo inhibition, high temperature and drought stress in flag leaves of wheat: A comparison between a hybridization line and its parents grown under field conditions. *Plant Sci.* **2006**, 171, 389–397.

187. Yang, X.; Liang, Z.; Lu, C. Genetic engineering of the biosynthesis of glycine betaine enhances photosynthesis against high temperature stress in transgenic tobacco plants. *Plant Physiol.* **2005**, 138, 2299–2309.

188. Ye, C.; Argayoso, M.A.; Redoña, E.D.; Sierra, S.N.; Laza, M.A.; Dilla, C.J. Mapping QTL for heat tolerance at flowering stage in rice using SNP markers. *Plant Breed.* **2012**, 131, 33–41.

189. Yin, Y.; Li, S.; Liao, W.; Lu, Q.; Wen, X.; Lu, C. Photosystem II photochemistry, photoinhibition, and the xanthophyll cycle in heat-stressed rice leaves. *J. Plant Physiol.* **2010**, 167, 959–966.

190. Young, L.W.; Wilen, R.W.; Bonham-Smith, P.C. High temperature stress of *Brassica napus* during flowering reduces micro-and megagametophyte fertility, induces fruit abortion, and disrupts seed production. *J. Exp. Bot.* **2004**, 55, 485–495.

191. Zakaria, S.; Matsuda, T.; Tajima, S.; Nitta, Y. Effect of high temperature at ripening stage on the reserve accumulation in seed in some rice cultivars. *Plant Production Science.* **2002**, 5, 160-168.

192. Zhang, G.L.; Chen, L.Y.; Zhang, S.T.; Zheng, H.; Liu, G.H. Effects of high temperature stress on microscopic and ultrastructural characteristics of mesophyll cells in flag leaves of rice. *Rice Science.* **2009**, 16, 65-71.

193. Zhang, H.X.; Hodson, J.N.; Williams, J.P.; Blumwald, E. Engineering salt-tolerant *Brassica* plants: characterization of yield and seed oil quality in transgenic plants with increased vacuolar sodium accumulation. *Proc. Natl. Acad. Sci. USA.* **2001**, 98, 12832–12836.

194. Zhang, J.H.; Huang,W.D.; Liu, Y.P.; Pan, Q.H. Effects of temperature acclimation pretreatment on the ultrastructure of mesophyll cells in young grape plants (*Vitis vinifera* L. cv. Jingxiu) under cross-temperature stresses. *J. Integr. Plant Biol.* **2005**, 47, 959–970.

195. Zhang, W.B.; Jiang, H.; Qiu, P.C.; Liu, C.Y.; Chen, F.L.; Xin, D.W. Genetic overlap of QTL associated with low temperature tolerance at germination and seedling stage using BILs in soybean. *Can. J. Plant Sci.* **2012**, 92, 1–8.

196. Zhang, X.; Cai, J.; Wollenweber, B.; Liu, F.; Dai, T.; Cao, W.; Jiang, D. Multiple heat and drought events affect grain yield and accumulations of high molecular weight glutenin subunits and glutenin macropolymers in wheat. *J. Cereal Sci.* **2013**, 57, 134–140.

197. Zhang, Y.; Mian, M.A.R.; Bouton, J.H. Recent molecular and genomic studies on stress tolerance of forage and turf grasses. *Crop Sci.* **2006**, 46, 497–511.

198. Zinn, K. E.; Tunc-Ozdemir, M.; Harper, J. F. Temperature stress and plant sexual reproduction: uncovering the weakest links. *J. Exp. Bot.* **2010**, 61, 1959–1968.

199. Zrobek-sokolnik, A. Temperature stress and responses of plants. In: Ahmad P, Prasad MNV (eds) Environmental adaptations and stress tolerance of plants in the era of climate change. New York: Springer. **2012**, 113-134.

Abiotic Stress Tolerance Mechanisms in Plants, Pages 127–142
Edited by: Gyanendra K. Rai, Ranjeet Ranjan Kumar and Sreshti Bagati
Copyright © 2018, Narendra Publishing House, Delhi, India

3

HEAT SHOCK PROTEINS: ROLE AND MECHANISM OF ACTION

Suneha Goswami[1], Kavita Dubey[1], Khushboo Singh[1], Gyanendra K. Rai[2] and Ranjeet Ranjan Kumar[1]

[1]*Division of Biochemistry, IARI, New Delhi*
[2]*School of Biotechnology,*
Sher-e- Kashmir University of Agricultural Sciences and Technology of Jammu-180009 (J&K)
E-mail: ranjeetranjaniari@gmail.com

Abstract

Plants as sessile organisms are exposed to persistently changing stress factors. The primary stresses such as drought, salinity, cold and hot temperatures and chemicals are interconnected in their effects on plants. These factors cause damage to the plant cell and lead to secondary stresses such as osmotic and oxidative stresses. Plants cannot avoid the exposure to these factors but adapt morphologically and physiologically by some other mechanisms. Almost all stresses induce the production of a group of proteins called heat-shock proteins (Hsps) or stress-induced proteins. The induction of transcription of these proteins is a common phenomenon in all living things. These proteins are grouped in plants into five classes according to their approximate molecular weight: (1) Hsp100, (2) Hsp90, (3) Hsp70, (4) Hsp60 and (5) small heat-shock proteins (sHsps). Higher plants have at least 20 sHsps and there might be 40 kinds of these sHsps in one plant species. It is believed that this diversification of these proteins reflects an adaptation to tolerate the heat stress. Transcription of heat-shock protein genes is controlled by regulatory proteins called heat stress transcription factors (Hsfs). Plants show at least 21 Hsfs with each one having its role in regulation, but they also cooperate in all phases of periodical heat stress responses (triggering, maintenance and recovery). There are more than 52 plant species (including crop ones) that have been genetically engineered for different traits such as yield, herbicide and insecticide resistance and some metabolic changes. In conclusion, major heat-shock proteins have some kind of related roles in solving the problem of misfolding and aggregation, as well as their role as chaperones.

Keywords: Chaperones, Heat-shock proteins, Transcription factors, Stress, Heat stress, Heat tolerance

1. Introduction

The constant flow of energy through all biological organisms provides the dynamic driving force for the maintenance of biological processes such as cellular biosynthesis and transport. The maintenance of steady-state results in a meta-stable condition called **homeostasis.**Any undesired modulation disrupting the homeostasis is known as biological stress. Plants as sessile organisms are exposed to persistently changing environmental stress factors. Biological stress in plants divided into two categories: abiotic and biotic stress. Abiotic stress is a physical stress (e.g. temperature, drought, chemical, light, salt etc.) that the environment may impose on the plant. Biotic stress is a biological insult, (e.g., insects, pests, pathogens etc.) to which a plant may be exposed during its lifetime.

Abiotic stresses especially heat, drought and salinity stresses are the major problems in agriculture. They significantly affect the growth of plants and productivity of crops. It is considered as the major cause of >50% reduction in average yield of major crops. Heat stress is turning out to be a major problem in cultivation of various crops like wheat. Of late, a drastic decrease in the total seed setting and yield has been observed in many wheat growing regions of India mainly due to the terminal heat stress. The problem of heat stress is likely to exacerbate with the global climate change adding to the exasperations of the stake- holders. Heat stress has been shown to influence photosynthesis, cellular and sub-cellular membrane components, seed setting, protein content and antioxidant enzyme activity; thereby significantly limits the crop productivity (Georgieva, 1999). Besides mitigating the heat stress, crop productivity under the stress may be enhanced by adaptation strategies.

Numerous heat responsive proteins have been identified from different crop species. However, the expression pattern of these genes and proteins under heat stress are still not clear. Different stress associated proteins have been identified from crops like rice, maize, *Arabidopsis* etc. and their characterization has also been carried out in response to different stresses.

2. Heat stress

The primary stresses such as high temperature, drought, salinity, cold and chemicals are interconnected in their effects on plants. These factors cause damage to the plant cell and lead to secondary stresses such as osmotic and oxidative stresses. Plants cannot avoid the exposure to these factors but adapt morphologically and physiologically by some other mechanisms.Heat stress as well as other stresses can trigger some mechanisms of defense such as the expression of stress associated chaperones, the heat shock proteins (HSPs),that was not expressed under "normal" conditions (Kumar *et al.*, 2016). Almost all stresses induce the production of a group of proteins called heat-shock proteins (HSPs) or Stress-induced proteins. Heat stress/shock response is a universal phenomenon and heat shock proteins (HSPs) form the most crucial defense system in all living systems at the cellular level (Katschinski, 2004). The cyto-protective effects of HSPs were attributed primarily to their chaperone activities, which minimize the proteotoxicity induced by the accumulation of unfolded or denatured proteins upon stress (Katschinski, 2004). HSP synthesis is tightly regulated by different members of heat shock transcription factors (Hsfs) at transcriptional level (Morimoto, 1998). Hsfs alone can function in the maintenance of cellular homeostasis that include regulation of cell cycle, cell proliferation, redox homeostasis, cell death mechanisms etc. (Katschinski, 2004; Sreedhar *et al.*, 2006).

3. Heat Shock Proteins

During evolution, plants have developed sophisticated mechanisms to sense the subtle changes of growth conditions, and trigger signal transduction cascades, which in turn activate stress responsive genes and ultimately lead to changes at the physiological and biochemical levels.Abiotic stress especially thermal stress adversely affects the functioning of cellular and metabolic pathways in plants. One of the main effects is on functioning of normal cellular proteins. Under thermal stress there is aggregation and misfolding of important cellular proteins occurred. Plants have developed different defense mechanisms to adapt with these adverse conditions. Under the course of defense mechanism at molecular level, transcription and translation of special set of proteins like Heat Shock Proteins (HSPs) occurs (Kotak *et al.*, 2007, Kumar *et al.*, 2016). Diversification in HSPs may reflect an adaptation to tolerate the heat stress. These molecular chaperones assist in protein refolding under stress conditions, protects plants against stress by re-establishing normal protein conformation and thus cellular homeostasis.

Under stressful condition, cell response triggered the production of heat shock proteins (HSP). They were named heat shock protein as first described in relation to heat shock, but are now also known to be expressed during other stresses like exposure to cold, UV light, during wound healing or tissue remodeling. Many HSPs also functions as chaperone by stabilizing new proteins or by helping the refolding of damaged proteins of the cell due to stress (Figure 1). This increase in the expression of HSPs are transcriptionally regulated and the dramatic upregulation of the heat shock proteins is a key heat shock response and is induced primarily by heat shock factors (Hsfs) that are located in the cytoplasm in an inactive state. These factors are considered as transcriptional activators for heat shock (Baniwal *et al.*, 2004; Hu *et al.*, 2009). HSPs are found in virtually all living organisms, from bacteria to plants and humans.

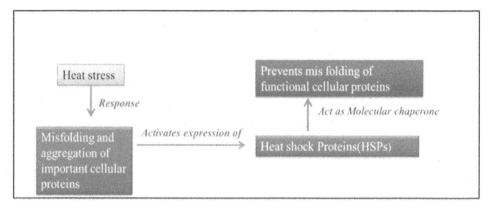

Fig. 1: Schematic representation of functional overview of HSP

3.1. Thermal Stability of HSPs

Incorrect protein folding into cells can cause several conformational disorders and in order to prevent such structural misfolding and to maintain homeostasis, cells have evolved an efficient protein quality control system (PQC) as an endogenous process. This PQC system needed molecular chaperones (including all HSP families) and their main function is to prevent inappropriate interactions, avoiding protein aggregation by assisting their correct folding and if protein correction is not possible, guiding them to cell degradation system. To maintain the thermal stability of proteins, the chaperone system changes from a folding to a storing function at heat shock temperatures. The temperature at which this change occurs depends on the presence of a thermosensor in at least one of the components of the chaperone systems. One of the most important chaperones is the Heat-shock protein 90 kDa (HSP90), which is responsible for the correct folding of a wide range of proteins. In the folding process, it is essential that HSP90 form complexes with co-chaperones, and providing a cooperative action during the maturation cycle of client proteins.

3.2. Classification of Heat Shock Proteins

Historically, the discovering of the heat-shock proteins was started, when the Italian Scientist F. Ritossa observed the expression of gene in the chromosomes puffing of *Drosophila melanogaster* after exposure to heat shock. Increase in protein synthesis was observed that occurred also by the use of other stress factors such as azide, salicylate and 2,4-dinitrophenol (Ritossa, 1962). After that report, these proteins were identified and named as heat-shock protein (HSP) (Tissieres*et al.*, 1974). Thereafter, various studies were started to find out the relationship of the synthesis of these proteins with the tolerance of stresses. On the other hand, Lin *et al.,* (1984) reported that the exposure of *Glycine max* seedlings to heat shock (from 28 to 45 °C) for 10 min (longer periods killed the seedlings) induce the synthesis of HSPs at the cost of other proteins synthesis.

Several types of heat shock proteins have been identified in almost all organisms (Bharti and Nover, 2002). HSPs are mainly characterized on the basis of the presence of a carboxylic terminal called heat-shock domain (Helm *et al.,* 1993). HSPs having molecular weights ranges from 10 to 200 kDa are characterized as chaperones where they participate in the induction of the signal during heat stress (Schoffl *et al.,* 1998). Heat-shock proteins of archaea have been classified on the basis of their approximate molecular weight as: (1) Heat shock protein of molecular weight 100 kDa: HSP100, (2) HSP90, (3) HSP70, (4) HSP60, and small heat-shock proteins (sHSPs) where the molecular weight ranges from 15 to 42 kDa (Trent, 1996). Schlesinger (1990) reported that in eukaryotic organisms, the principle heat-shock proteins of human beings do not differ from those of bacteria except for the presence of HSP33. Later, the HSPs of human beings were grouped into five families (Kregel, 2002) as in Table 1.

In plants, according to molecular weight,amino acid sequence homologies and functions, five classes of HSPs are characterized: (1) HSP 100, (2) HSP 90, (3) HSP 70, (4) HSP 60 (5) small heat-shock proteins (sHSPs) (Kotak *et al.,* 2007,Gupta *et al.,* 2010).

Table 1: Families of HSPs in human beings, their site, and suggested functions (Kregel, 2002)

HSP Families	Cellular location	Proposed functions
HSP27 (sHSP)	Cytosol, nucleus	Microfilament stabilization, antiapoptotic
HSP60	Mitochondria	Refolds proteins and prevent aggregation of denatured proteins, proapoptotic
HSP70		Antiapoptotic
HSP72(HSP70)	Cytosol, nucleus	Protein folding, cytoprotection
HSP73(HSP70)	Cytosol, nucleus	Molecular chaperones
HSP75(mHSP70)	Mitochondria	Molecular chaperones
HSP78(GRP78)	Endoplasmic reticulum	Cytoprotection, molecular chaperones
HSP90	Cytosol, endoplasmic reticulum, nucleus	Regulation of steroid hormone receptors, protein translocation
HSP110/104	Cytosol	Protein folding

The high molecular weight HSPs are characterized as molecular chaperone. Higher plants have at least 20 sHSPs and there might be 40 kinds of these sHSPs in one plant species.

The name of HSPs in bacteria differ from those in eukaryotic cells as given below but the nomenclature for sHSPs are same in both the organisms (Kotak *et al.*,2007).

Escherichia coli	Eukaryotic cells
ClpB	HSP100
HtpG	HSP90
Dnak	HSP70
GroEL	HSP60

3.3. Role of Different HSPs

Under thermal stress, the general role of HSPs is to act as molecular chaperones and regulating the protein folding,accumulation, localization and degradation of proteins in all plants and animal species (Hu *et al.*, 2009; Gupta *et al.*, 2010),indicated that HSPs protect the cells from injury and facilitate recovery and survival after a return to normal growth conditions. On the other hand, under non thermal stress, their function could be different: as it may protect the protein from damage and maintain the correct protein structure (Timperio *et al.*, 2008). As a chaperones, these proteins prevent the irreversible aggregation of other proteins and under heat stress, they participates in refolding of proteins (Tripp *et al.*, 2009). Each group of these HSPs has a unique mechanism and the role of each is as follows.

Class: HSP 100

This class of proteins is responsible for the reactivation of aggregated proteins (Parsell and Lindquist, 1993). They basically re-solubilize the non-functional protein aggregates and help to degrade irreversibly damaged polypeptides (Kim *et al.*, 2007). This class HSPs function is not restricted only to acclimation to high temperatures, but they also provides housekeeping functions, essential for chloroplast development (Lee *et al.*, 2006), and facilitating the normal situation of the organism after severe stress (Gurley, 2000).

Class: HSP 90

HSP90 can bind with HSP70 to form chaperone complexes and act as molecular chaperones, playing important role in signalling protein function and trafficking (Pratt and Toft, 2003; Kumar *et al.*, 2012), regulate the cellular signals such as the regulation of glucocorticoid receptor (GR) activity (Pratt *et al.*, 2004).

Cytoplasmic HSP 90 reacts with resistance protein (R), the signal receptor from the pathogen and participates in providing resistance from pathogens. Thus, HSP90 is considered as the essential component of innate-immune response and pathogenic resistance in rice (Thao *et al.,* 2007). Yamada *et al.,* (2007) reported that in *A. thaliana*, in the absence of heat stress, cytoplasmic HSP90 negatively inhibit the Hsf, but under heat stress this role is temporarily supressed, so that Hsf is active.

Class: HSP 70

The HSP 70 play role as a chaperone for newly translated proteins and prevents their accumulations as aggregates, helps in their proper folding, protein import and translocation, proteolytic degradation of unstable proteins by targeting the proteins to lysosomes or proteasomes (Su and Li, 2008). HSP 70 along with sHSPs play a crucial role in protecting plant cell from the detrimental effects of heat stress (Rouch *et al.,* 2004, Kumar *et al.,* 2016). HSP 70B present in the stroma of chloroplasts, also involved in photo-protection and repairing of photosystem II during and after the photoinhibition (Schroda *et al.,* 1999). A study on *A. thaliana* reported that HSP70 found in the stroma of chloroplast involved in the differentiation of germinating seeds (Su and Li, 2008). Structurally, HSP70 consists of a highly conserved N-terminal ATPase domain of 44 kDa and a C-terminal peptide-binding domain of 25 kDa. HSP70 family chaperones are considered to be the most highly conserved HSPs, with, 50% identical residues between the *Escherichia coli* homolog DnaK and the eukaryotic HSP70.

Class: HSP 60

A well-known chaperonin, responsible for assisting plastid proteins is Rubisco (Wang *et al.,* 2004). This class of HSPs participates in folding, aggregation and transport of many mitochondrial and chloroplast proteins (Lubben *et al.,* 1989). HSP60 prevent the aggregation of newly transcribed protein before their folding (Parsell and Lindquist, 1993). Functionally, plant chaperonins are limited and stromal chaperones (HSP 70 and HSP 60) are involved in attaining functional conformation of newly imported proteins to the chloroplast (Jackson-Constan *et al.,* 2001).

Class: HSP 40

HSP40 proteins regulate complex formation between polypeptides and HSP70 by different mechanisms. First, HSP40 interact with HSP70- polypeptide to stimulate its ATPase activity (Cyr *et al.,* 1992). Second, HSP40 proteins have polypeptide-binding domains (PPDs) that bind and deliver specific proteins to HSP70

(Cheetham and Caplan 1998). Third, within the same cellular compartment, specialized members of the HSP40 family are localized to different sites, which facilitate the interaction of different HSP70-HSP40 complex to bind unique proteins at that site (Shen *et al.,* 2002). This class of protein is also known as J-domain-containing protein (J-protein). It acts as a co-chaperone component of the HSP70 system, increasing HSP70 affinity for proteins (Kampinga and Craig, 2010). It has a conserved 70-amino acid J-domain that interacts with the nucleotide-binding domain (NBD) of HSP70 and participates in various virus-plant interactions. Similar to HSP70, the function of HSP40 in viral pathogenesis has been well established. For examples, the coat protein of *Potato virus Y* interacts with DnaJ-like protein (HSP40), which is important for cell-to-cell movement (Hofius *et al.,* 2007). The functions of HSP70 and HSP40 in plant immunity have been generally identified as chaperones in microbial pathogenesis, particularly, in viral movement. Several HSP70 and HSP40 were demonstrated as positive regulators in plant immunity. Overexpression or knock down of these HSPs enhance resistance and susceptibility to pathogen infections respectively, although the mechanisms remains unclear.

Class: sHSPs (Small HSPs)

The genes encode for small HSPs, there expression limited in the absence of environmental stress and occurs in some stages of growth and development of plants such as embryogenesis, germination, development of pollen grains, and fruit ripening (Sun *et al.,* 2002). Structurally these proteins have a common alpha-crystalline domain of 80–100 amino acid residues in the C-terminal region (Seo *et al.,* 2006, Kumar *et al.,* 2013). Functionally these proteins are responsible for the degradation of the proteins having unsuitable folding. The representative protein of this class of HSPs is the enzyme bound ubiquitin (molecular weight is 8.5 kDa) (Ferguson *et al.,* 1990). Unlike chaperones, these proteins have ATP independent activity (Miernyk, 1999). sHSPs can binds to partially folded or denatured proteins, preventing irreversible unfolding or wrong protein aggregation but they cannot refold the non-native proteins (Sun *et al.,* 2002). Nakamoto and Vigh (2007) concluded that under stress condition, small heat shock proteins play an important role in controlling the membrane quality and maintains membrane integrity.

4. HSPs/chaperones network

In the protective mechanism of HSPs/chaperones, many chaperones act in concert with the chaperone machinery network. During stress, several enzymes and structural proteins undergo detrimental structural and functional changes. Therefore, maintaining proteins in their functional conformations, preventing from aggregation of non-native proteins, refolding of denatured proteins to regain their functional conformation and removal of non-functional but potentially harmful polypeptides (arising from aggregation, misfolding or denaturation) are particularly important for cell survival under stress. Therefore, the different classes of HSPs/chaperones cooperate in cellular protection and play complementary and sometimes overlapping roles in the protection of proteins from stress. Small HSPs (sHSPs) bind to non-native proteins and prevent their aggregation, thus providing a reservoir of substrates for subsequent refolding by members of the HSP70/HSP100 chaperone families. The Chaperone/HSPs network under stress, how they regulate different proteins stability /degradation is presented in Figure 2.

The response of plants to heat shock resulted in changes in the level of enzymes, cellular membrane structure, photosynthesis activity, and protein metabolism (Singla *et al.*, 1997). It has been reported that high temperature changed the properties of membranes of nucleus, endoplasmic reticulum, mitochondria, and chloroplasts of rice plant *O. sativa* (Pareek *et al.*, 1998). Lipids in the thylakoid membranes of the chloroplast are very important to improve photosynthesis and hence stress tolerance.

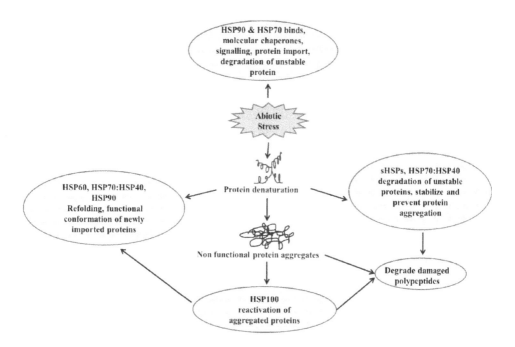

Fig. 2: The heat-shock proteins (HSPs) and chaperone network under abiotic stress

The transcription of these genes is controlled by regulatory proteins called heat shock transcription factors (Hsfs) located in the cytoplasm in an inactive state. So these factors are considered as transcriptional activators for heat shock

(Baniwal *et al.*, 2004; Hu *et al.*, 2009). Plants are characterized by a large number of transcriptional factors (Nover and Baniwal, 2006). These factors have been classified (Tripp *et al.*, 2009) into three classes according to the structural differences in their aggregation in triples, *i.e.* oligomerization domains as follows:

- Plant HsfA such as HsfA1 and HsfA2 in *L. esculentum*
- Plant HsfB such as HsfB1 in *L. esculentum*
- Plant HsfC

The synthesis of HSPs depends upon activity of special class of transcription factors called Heat Shock factors (Hsfs). Hsfs are modular transcription factors encoded by a large gene family in plants. Hsfs have three highly conserved features: the amino terminal DNA binding domain of approximately 100 amino acids (Harrison *et al.*, 1994) and a domain having three leucine zippers mediating multimerization (Wu *et al.*, 1994; Swamynathan, 1995) and an additional leucine zipper motif at the carboxy terminus. Hsfstrimerizes via the formation of a triple stranded α-helical coiled coil, involving the three conserved leucine zippers next to the DNA binding domain (Peteranderl and Nelson, 1993). Hsfs bind to heat shock elements (HSE) in a sequence specific and reversible manner, leading to the activation of transcription of heat shock proteins (Morimoto *et al.*, 1994; Goswami *et al.*, 2016).

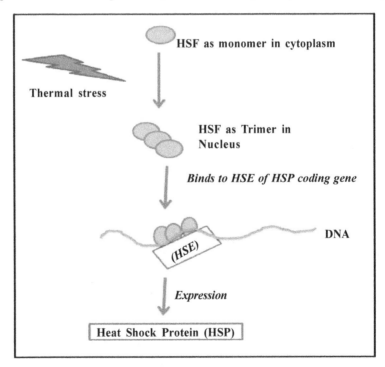

Fig. 3: Mechanism of action of Heat Shock Proteins

5. Genetically modified plants for heart stress tolerance

Several plant species (more than 52) have been genetically modified for different traits including crop plants like tomato, potato, soybean, maize, rice, cotton etc. Other non-crop transgenic plants were also developed for different abiotic stress tolerance in laboratory. High temperature stress is one of the major abiotic stresses for crop plants. The plant reaction to high temperature stress resulted in changes in cell membrane stability, photosynthesis activity, enzyme denatured, and protein synthesis (Goswami *et al.*, 2015). High temperature also changes the properties of membranes of mitochondria, chloroplast, endoplasmic reticulum and nucleus. Lipids in the thylakoid membrane of chloroplast is important for membrane stability and also for photosynthetic efficiency which may disturbed due to very high or very low temperature. By increasing the expression of glycerol 3- phosphate acyltransferase enzyme in tobacco plant, degree of lipid unsaturation were increased which make the plants cold tolerant. Increase in the degree of saturation of membrane lipids may leads to increase the heat tolerance of the plants. Other way to develop thermotolerance of plants, by changing the level of HSPs expression, Hsfs expression, increase in the synthesis of osmolytes in the cells, modify the endogenous genes of crop plants such as rubiscoactivase, oxygen evolving enhancer proteins, signaling molecules like calcium dependent protein kinase (CDPK), mitogen activated protein kinases (MAPK), genes involved in starch biosynthesis pathways etc. through site directed mutagenesis and make them thermotolerant. Some examples for the attempts taken for developing thermotolerant crop plants are given in Table 2.

Table 2: Transgenic attempts to enhance thermotolerance of plants

Phenotypes	Gene	Function	Plant
Heat stress tolerant	HsfA1	Transcription factor	Tomato
Heat stress tolerant	Hsf3	Transcription factor	Arabidopsis
Heat stress sensitive	HSP70	HSP	Arabidopsis
Heat stress tolerant	HSP17.7	HSP	Carrot
Heat stress tolerant	Hvapx1	Active oxygen species (AOS) metabolism	Barley
Heat stress tolerant	Fad7	Fatty acid desaturation	Tobacco

Source: Sample of a larger table of Sung *et al.,* (2003)

6. Conclusion

Although many attempts have been made to have genetically modified plants for stress tolerance especially crop plants. Mostly attempts were for one trait, while in nature the prevailing conditions were more than one stress, hence the stress combination should be dealt with as a new state of abiotic stress.

7. References

1. Baniwal, S.K.; Bharti, K.; Chan, K.Y.; Fauth, M.; Ganguli, A.; Kotak, S.; Mishra, S.K.; Nover, L.; Port, M.; Scharf, K.;Tripp, L.; Weber, C.; Zielinski, D.; von Koskull-Doring, P. Heat stress response in plants: a complex game with chaperones and more than 20 heat stress transcription factors. J. Biosci. **2004**, 29, 471–487.

2. Bharti, K.; Nover, L. Heat stress-induced signaling. In: Scheel, D., Wasternack, C. (Eds.), Plant Signal Transduction: Frontiers in Molecular Biology. Oxford University Press, Oxford, UK. **2002**, pp. 74– 115.

3. Cheetham, M.E.; Caplan, A.J. Structure function and evolution of DnaJ: conservation and adaptation of chaperone function. *Cell Stress Chaperones.* **1998**, 3, 28–36.

4. Cyr, D. M.; Lu, X.; Douglas, M. G. Regulation of HSP70 function by a eukaryotic DnaJ homolog. *J Biol Chem.* **1992**, 267, 20927–20931.

5. Ferguson, D.L.; Guikema, J.A.; Paulsen, G.M. Ubiquitin pool modulation and protein degradation in wheat roots during high temperature stress. *Plant Physiol.* **1990**, 92, 740–746.

6. Georgieva, K. Some mechanisms of damage and acclimation of the photosynthetic apparatus due to high temperature. *Bulg. J. Plant Physiol.* **1999**, 25, 89-99.

7. Goswami, S.; Kumar, R.R.; Dubey, K.; Singh, J.P.; Tiwari, S.; Kumar, A.; Smita, S.; Mishra, D.C.; Kumar, S.; Grover, M.; Padaria, J.C. SSH analysis of endosperm transcripts and characterization of heat stress regulated expressed sequence tags in bread wheat. *Frontiers in Plant Science.* **2016**, 7.

8. Goswami, S.; Kumar, R.R.; Sharma, S.K.; Kala, Y.K.; Singh, K.; Gupta, R.; Dhawan, G.; Rai, G.K.; Singh, G.P.; Pathak, H.; Rai, R.D. Calcium trigger protein kinases induced signal transduction for augmenting the thermotolerance of developing wheat grain under heat stress. J. *Plant Biochem. Biotechnol.* **2015**, 24, 441–452.

9. Gupta, S.C.; Sharma, A.; Mishra, M.; Mishra, R.; Chowdhuri, D.K. Heat shock proteins in toxicology: how close and how far? *Life Sci.* **2010**, 86, 377–384.

10. Gurley, W.B. HSP101: a key component for the acquisition of thermotolerance in plants. *Plant Cell.* **2000**, 12, 457–460.

11. Harrison, C. J.; Bohm, A.A.; Nelson, H.C.M. Crystal structure of the DNA binding domain of the heat shock transcription factor. *Science.* **1994**, 263, 224–227.

12. Helm, K.W.; Lafayete, P.R.; Nago, R.T.; Key, J.L.; Vierling, E. Localization of small heat shock proteins to the higher plant endomembrane system. *Mol. Cell. Biol.* **1993**, 13, 238–247.

13. Hofius, D.; Maier, A.T.; Dietrich, C.; Jungkunz, I.; Bornke, F.; Maiss, E.; Sonnewald, U. Capsid protein-mediat.ed recruitment of host DnaJ-like proteins is required for Potato virus Y infection in tobacco plants. *J Virol.* **2007**, 81, 11870–11880.

14. Hu, W.; Hu, G.; Han, B. Genome-wide survey and expression profiling of heat shock proteins and heat shock factors revealed overlapped and stress specific response under abiotic stresses in rice. *Plant Sci.* **2009**, 176, 583–590.

15. Jackson-Constan, D.; Akita, M.; Keegstra, K. Molecular chaperones involved in chloroplast protein import. *Biochim. Biophys. Acta.* **2001**, 1541, 102–113.

16. Kampinga, H.H.; Craig, E.A. The HSP70 chaperone machinery: J proteins as drivers of functional specificity.*Nature Reviews Molecular Cell Biology.* **2010**, 11, 579–592.

17. Katschinski, D.M. On heat and cells and proteins; *News Physiol. Sci.* **2004**, 19, 11–15.

18. Kim, H.J.; Hwang, N.R.; Lee, K.J. Heat shock responses for understanding diseases of protein denaturation. *Mol. Cells.* **2007**, 23, 123–131.

19. Kotak, S.; Larkindale, J.; Lee, U.; von Koskull-Doring, P.; Vierling, E.; Scharf, K.D. Complexity of the heat stress response in plants. *Curr. Opin. Plant Biol.* **2007**, 10, 310–316.

20. Kregel, K.C. Heat shock proteins: modifying factors in physiological stress responses and acquired thermotolerance. *J. Appl. Physiol.* **2002**, 92, 2177–2186.

21. Kumar R.R.; Sharma, S.K.; Goswami, S.; Singh, R.; Pathak, H.; Rai, R. D. Characterization of differentially expressed stress-associated proteins in starch granule development under heat stress in wheat (*Triticumaestivum* L.). *Indian J. Biochem. Biophys.* **2013**, 50:126-138.

22. Kumar, R.R.; Goswami, S.; Gupta, R. *et al.* The Stress of Suicide: Temporal and Spatial Expression of Putative Heat Shock Protein 70 Protect the Cells from Heat Injury in Wheat (*Triticumaestivum*). *J Plant Growth Regul.* **2016**, 35: 65–82.

23. Kumar, R.R.; Goswami, S.; Sharma, S.K.; Pathak, H.; Rai, G.K.; Rai, R.D. Genome wide identification of target heat shock protein 90 genes and their differential expression against heat stress in wheat. *International Journal of Biochemistry Research and Review.* **2012**, 2, 15-30.

24. Lee, U.; Rioflorido, I.; Hong, S.W.; Larkindale, J.; Waters, E.R.;Vierling, E. The Arabidopsis ClpB/ HSP100 family of proteins: chaperones for stress and chloroplast development. *Plant J.* **2006**, 49, 115–127.

25. Lin, C.Y.; Roberts, J.K.; Key, J.L. Acquisition of thermotolerance in soybean seedlings: synthesis and accumulation of heat shock proteins and their cellular localization. *Plant Physiol.* **1984**, 74, 152–160.

26. Lubben, T.H.; Donaldson, G.K.; Viitanen, P.V.; Gatenby, A.A. Several proteins imported into chloroplasts form stable complexes with the GroEL-related chloroplast molecular chaperone. *Plant Cell.* **1989**, 1, 1223–1230.

27. Miernyk, J.A. Protein folding in the plant cell. *Plant Physiol.* **1999**, 121, 695–703.

28. Mogk, A. et al. Small heat shock proteins, ClpB and the DnaK system form a functional triade in reversing protein aggregation. *Mol. Microbiol.* **2003**, 50, 585–595.

29. Morimoto, R.I. Regulation of heat shock transcriptional response: cross talk between a family of heat shock factors, molecular chaperones, and negative regulators; *Genes Dev.* **1998**, 12, 3788–3796.

30. Morimoto, R.I.; Jurivich, D.A.; Kroeger, P.E.; Mathur, S.K.; Murphy, S.P.; Nakai, A.; Sarge, K.; Abravaya, K.; Sistonen, L.T. Regulation of heat shock gene transcription by a family of heat shock factors; in the biology of heat shock proteins and molecular chaperones (eds) R I Morimoto, A Tissieres and C Georgopulos (New York: Cold Spring Harbor Laboratory Press). **1994**, pp 417–455.

31. Nakamoto, H.; Vigh, L. The small heat shock proteins and their clients. *Cell Mol. Life Sci.* **2007**, 64, 294–306.

32. Pareek, A.; Singla, S.L.; Grover, A. Plant HSP90 family with special reference to rice. *J. Biosci.* **1998**, 23, 361–367.

33. Parsell, P.A.; Lindquist, S. The function of heat-shock proteins in stress tolerance, degradation and reactivation of damaged proteins. *Annu. Rev. Genet.* **1993**, 27, 437–496.

34. Peteranderl, R.; Nelson, H.C.M. Trimerization of the heat shock transcription factor by a triple stranded alpha helical coiled coil; *Biochemistry.* **1993**, 31, 12272–12276.

35. Pratt, W.B.; Galigniana, M.D.; Harrell, J.M.; Deranco, D.B. Role of HSP90 and the HSP90-binding immunophilins in signalling protein movement. *Cell Signal.* **2004**, 16, 857–872.

36. Pratt, W.B.; Toft, D.O. Regulation of signaling protein function and trafficking by the HSP90/HSP70-based chaperone machinery. *Exp. Biol. Med.* **2003**, 228, 111–133.

37. Ritossa, F. A new puffing pattern induced by heat shock and DNP in Drosophila. *Experientia.* **1962**, 18, 571–573.

38. Rouch, J.M.; Bingham, S.E.; Sommerfeld, M.R. Protein expression during heat stress in thermo-intolerance and thermotolerance diatoms. *J. Exp. Mar. Biol. Ecol.* **2004**, 306, 231–243.

39. Schlesinger, M.J. Heat shock proteins. *J. Biol. Chem.* **1990**,265, 12111–12114.

40. Schoffl, F.; Prandl, R.; Reindl, A. Regulation of the heat shock response. *Plant Physiol.* **1998**, 117, 1135–1141.

41. Schroda, M.; Vallon, V.; Wollman, F.; Beck, C.F. A chloroplast-targeted heat shock protein 70 (HSP70) contributes to the photoprotection and repair of photosystem II during and after photoinhibition. *Plant Cell.***1999**, 11, 11165–11178.

42. Seo, J. S.; Park, T. J.; Lee, Y. M.; Park, H. G.; Yoon, Y. D.; Lee, J. S. Small heat shock protein 20 gene (HSP20) of the intertidal copepod Tigriopusjaponicus as a possible biomarker for exposure to endocrine disruptors. *Bulletin of Environmental Contamination & Toxicology.* **2006**, 76(4).

43. Shen, Y.; Meunier, L.; Hendershot, L.M. Identification and characterization of a novel endoplasmic reticulum (ER) DnaJ homologue which stimulates ATPase activity of BiP in vitro and is induced by ER stress. *J Biol Chem.* **2002**, 277, 15947–15956.

44. Singla, S.L.; Preek, A.; Grover, A. High temperature. In:Prasad, M.N.V. (Ed.), Plant Ecophysiology. John Wiley, New York. **1997**, pp. 101–127.

45. Sreedhar, A.S.; Deepu, O.; Abhishek, A.; Srinivas, U.K. Heat shock transcription factors: a comprehensive review; in *Stress response: a molecular biology approach* (eds) as (Research SignpostISBN). **2006**, 81, 308-0109-4).

46. Su, P.H.; Li, H.M. Arabidopsis stromal 70-kDa heat shock proteins are essential for plant development and important for thermotolerance of germinating seeds. *Plant Physiol.* **2008**, 146, 1231– 1241.

47. Sun, W.; Motangu, M.V.; Verbruggen, N. Small heat shock proteins and stress tolerance in plants. *Biochim. Biophys. Acta.* **2002**, 1577, 1–9.

48. Sung, D.Y.; Kaplan, F.; Lee, K.J.; Guy, C.L. Acquired tolerance to temperature extremes. *Trends Plant Sci.* **2003**, 8,179–187.

49. Swamynathan, S.K. Heat shock response in higher Eukaryotes: Cloning of heat shock transcription factor 1 (rHsf1) and mechanism of heat induction of albumin, *PH D.* Thesis, Jawaharlal Nehru University, New Delhi. **1995**.

50. Thao, N.P.; Chen, L.; Nakashima, A.;Hara, S.; Umemura, K.; Takahashi, A.; Shirasu, K.; Kawasaki, T.; Shimamoto, K. RAR1 and HSP90 form a complex with Rac/RopGTPase and function2 in innate-immune responses in rice. *Plant Cell.* **2007**, 19, 4035–4045.

51. Timperio, A.M.; Egidi, M.G.; Zolla, L. Proteomics applied on plant abiotic stresses: role of heat shock proteins (HSP). *J. Proteomics.* **2008**, 71, 391–411.

52. Tissieres, A.; Mitchell, H.K.; Tracy, U.M. Protein synthesis in salivary glands of D. Melanogaster. Relation to chromosome puffs.J. *Mol. Biol.* **1974**, 84, 389–398.

53. Trent, J.D. A review of acquired thermotolerance, heat-shock proteins and molecular chaperones in Archaea. *Fems Microbiol. Rev.* **1996**, 18, 249–258.

54. Tripp, J.; Mishra, S.K.; Scharf, K.D. Functional dissection of the cytosolic chaperone network in tomato mesophyll protoplasts. *Plant Cell Environ.* **2009**, 32, 123–133.

55. Wang, W.; Vinocur, B.; Shoseyov, O.; Altman, A. Role of plant heat-shock proteins and molecular chaperones in the abiotic stress response. *Trends Plant Sci.* **2004**, 9, 244–252.

56. Wu, C.; Clos, J.; Giorgi, G.; Haroun, R.I.; Kim, S.J.; Rabindran, S.K.; Westwood, J.T.; Wisniewski, J.; Yim, G. Structure and regulation of heat shock transcription factor; in the biology of heat shock proteins and molecular chaperones (eds) R I Morimoto, A Tissieres and C Georgopulos (New York: Cold Spring Harbor Laboratory Press). **1994**, pp 395–416.

57. Yamada, K.; Fukao, Y.; Hayashi, M.; Fukazawa, M.; Suzuki, I.; Nishimura, M. Cytosolic HSP90 Regulates the Heat Shock Response That Is Responsible for Heat Acclimation in *Arabidopsis thaliana.* *J. Biol. Chem.* **2007**, 282, 37794–37804.

Abiotic Stress Tolerance Mechanisms in Plants, Pages 143–174
Edited by: Gyanendra K. Rai, Ranjeet Ranjan Kumar and Sreshti Bagati
Copyright © 2018, Narendra Publishing House, Delhi, India

4

REACTIVE OXYGEN SPECIES GENERATION, ANTIOXIDANTS AND REGULATING GENES IN CROPS UNDER ABIOTIC STRESS CONDITIONS

Gyanendra K. Rai[1], Sreshti Bagati[1], Pradeep Kumar Rai[2],
Vibha Raj Shanti[1], DikshaBhadwal[1], R. R. Kumar[3], Monika Singh[4]
and Praveen Singh[5]

[1]*School of Biotechnology,
Sher-e- Kashmir University of Agricultural Sciences and Technology of Jammu-180009 (J&K)*
[2]*ACHR, Udheywala,
Sher-e- Kashmir University of Agricultural Sciences and Technology of Jammu-180009 (J&K)*
[3]*Division of Biochemistry, IARI, Pusa New Delhi*
[4]*GL Bajaj Institute of Technology and Management, Greater Noida (U.P.)*
[5]*Division of PBG, Sher-E- Kashmir University of Agricultural Sciences and Technology of Jammu*
E-mail: gkrai75@gmail.com

Abstract

Among the abiotic stresses such as drought, cold, salt and heat causes reduce the plant growth and crop yield worldwide. Reactive oxygen species (ROS) i.e. hydrogen peroxide (H_2O_2), superoxide ($O_2^{\bullet-}$), hydroxyl radical ($OH^{\bullet-}$) and singlet oxygen ($^1O_2$) are by products of metabolisms, and are specifically controlled by enzymatic and non- enzymatic antioxidant defense mechanisms. Under abiotic stress conditions, ROS are significantly accumulated, which reason oxidative damage and ultimately resulting in cell death. Recently, reported that, ROS have been also recognized as key players in the complex signaling network of plants stress responses. The involvement of reactive oxygen species (ROS) in signal transduction indicates that there must be synchronized function of network regulations to maintain the ROS at non-toxic levels,among ROS generation and ROS-scavenging pathways.Research evidence showed that ROS play vital roles in abiotic stress responses of crop plants for the activation of stress-response and defense pathways. More decisively, manipulating ROS levels offers an opportunity to improve plantstress tolerance towards abiotic stress conditions. This chapter presents an understanding to knowledge about homeostasis regulation of ROS in crop plants. In this chapter, we are summarizing the defense machinery i.e. enzymatic and non-enzymatic antioxidants, genes involved in abiotic stress tolerance of plants through ROS regulation. Comprehensive knowledge of ROS action and their regulation will enable to develop strategies to development of abiotic stress tolerance crop plants.

Keywords: Reactive oxygen species, Drought stress, Signaling

1. Introduction

Under favorable conditions, ROS is constantly being generated at basal levels. However, they are unable to cause damage, as they are being scavenged by different antioxidant mechanisms (Foyer and Noctor, 2005). The delicate balance between ROS generation and ROS scavenging is disturbed by the different types of stress factors like salinity, drought, extreme temperatures, heavy metals, pollution, high irradiance, pathogen infection, etc. (Figure 1). The survival of plants therefore, depends on many important factors like change in growth conditions, severity and

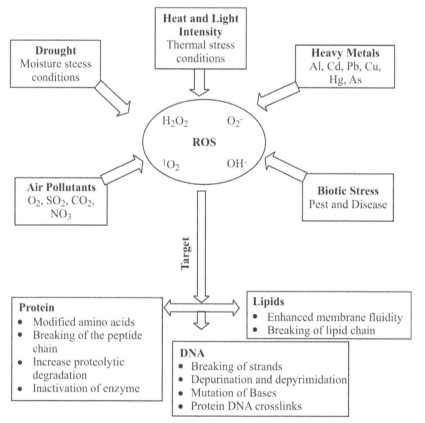

Fig. 1: Different reasons for the generation of Reactive Oxygen Species (ROS) and their impact

duration of stress conditions and the capacity of the plants to quickly adapt to changing energy equation (Miller *et al.,* 2010). Estimates show that only 1–2% of the O_2 consumption by plant tissues, leads to the formation of ROS.Among the abiotic stress drought is one of the most ecological stresses in agriculture (Boyer, 1982; Kumar *et al.,* 2013). Drought stress enhances the reactive oxygen species (ROS) in the cells, which can cause harsh oxidative damage to the crops, thus inhibiting growth, quality and yield. The balance among the production and scavenging of ROS is commonly known as redox homeostasis. However, when ROS production overwhelms the cellular scavenging capacity, thus unbalancing the cellular redox homeostasis, the result is a rapid and transient excess of ROS, known as oxidative stress (Mullineaux and baker, 2010; Sharma *et al.,* 2012). Plants have natural antioxidant defence mechanisms for scavenging/neutralising the excess ROS and prevent damages to cells. Therefore, this review will address oxidative stress in crops under drought stress conditions. An overview of the principal antioxidant gene involved in crops in ROS detoxification under drought stress condition, will be presented. Furthermore, signaling by reactive oxygen species (ROS) in crops enhancing the stress tolerance will also be covered.

2. Reactive Oxygen Species (ROS)

The generation of ROS in plants is triggered by different kinds of environmental stresses, such as high light, high or low temperature, salinity, drought, nutrient deficiency and pathogen attack (Figure 1). Among them drought stress promotes over production of reactive oxygen species (ROS), potentially causing oxidative damage to lipids, proteins and nucleic acids which ultimately results in cellular damage and death (Gill and Tuteja, 2010). Plants have evolved enzymatic and non-enzymatic antioxidant defence mechanism for scavenging and detoxifying ROS (Mittler, 2002). Reactive oxygen species (ROS), including hydrogen peroxide (H_2O_2), superoxide radical (O_2^-), hydroxyl radical (OH) and singlet oxygen ($^1O_2$) etc., resulting from excitation or incomplete reduction of molecular oxygen, are harmful by-products of basic cellular metabolism in aerobic organisms (Apel and Hirt, 2004; Miller *et. al.,* 2010). Under optimal growth conditions, intracellular ROS are mainly produced at a low level in organelles. However, ROS are dramatically accumulated during stress conditions. Under abiotic stress condition, limitation of CO_2 absorption, caused by stress-induced stomatal closure, favours photo respiratory production of H_2O_2 in the peroxisome and production of superoxide and H_2O_2 or singlet oxygen ($^1O_2$) by the reduced photosynthetic electron transport chain (Apel and Hirt, 2004; Noctor *et al.,* 2014). In addition to organelles, plasma membrane together with apoplast is the main site for ROS generation in response to endogenous signals and exogenous environmental stimuli. Several types of enzymes, such as NADPH oxidases, amine oxidases, polyamine oxidases, oxalate oxidases, and a large family of class III peroxidases, that localized at the cell surface or apoplast are contributed to production of apoplast ROS (Apel and Hirt, 2004; Cosio and Dunad, 2009; Gill and Tuteja, 2010). Besides the toxicity of ROS, ROS are also considered to be signaling molecules that regulate plant development, biotic and abiotic stress responses (Apel and Hirt, 2004; Mittler et al., 2004). Many excellent reviews have focused on ROS metabolism (Apel and Hirt, 2004; Noctor et al., 2014), ROS sensory and signaling networks (Miller et al., 2010; Suzuki et al., 2012; Baxter et al., 2014; Jun and Chan, 2015), as well as the cross-talk with other signaling molecules function in developmental and stress response processes (Suzuki et al., 2012; Noctor et al., 2014). Jun and Chan (2015) reviewed ROS regulation during abiotic stress responses in crop plants. Suzuki et al. (2012) reviewed enzymatic and non-enzymatic antioxidants and their roles in abiotic stress tolerance of crop plants.

2.1. Reactive oxygen species (ROS) generation sites in plant cells

The Reactive oxygen species (ROS) is generated under both normal and adverse conditions at different cell organs i.e. chloroplasts; mitochondria, peroxisomes, plasma membranes, endoplasmic reticulum and cell wall (Table 1). Chloroplasts and peroxisomes are the major sources of ROS production under the presence of light, while under dark conditions the mitochondrion is the important producer of Reactive oxygen species (Choudhary et al., 2013; Das and Ray Choudhary, 2014).

Table 1: Various members of Reactive Oxygen Species (ROS) and their generation sites

ROS	Generation Sites	Mode of action
Superoxide (O_2^-)	Membrane, Chloroplast, Mitochondria	React with double bond containing compounds such as (Fe-S) protein
Hydroxyl radical (OH^-)	Membrane, Chloroplast, Mitochondria	Extremely reactive with all biomolecules
Singlet oxygen ($^1O_2$)	Membrane, Chloroplast, Mitochondria, Peroxisomes, Endoplasmic reticulum	Oxidizes protein and forms OH^- via O^-
Hydrogen peroxide (H_2O_2)	Membrane, Chloroplast, Mitochondria	Oxidizes proteins, PUFAs and DNA

3. Antioxidant defense system in crop plants

Crop plant defense system contains of plentiful non-enzymatic and enzymatic antioxidative protection mechanisms that work together with ROS-generating pathway to sustain ROS homeostasis. Earlier studies revealed significant roles of antioxidant compounds in ROS homeostasis in crop plants.

3.1. Non-Enzymatic Antioxidants

Non-enzymatic antioxidants i.e. Glutathione (GSH), Ascorbic acid, carotenoids, tocopherols, and flavonoids are also decisive for Reactive oxygen species (ROS) homeostasis in crop plant (Gill and Tuteja, 2010, Kumar *et al.,*2013). Besides traditional non-enzymatic and enzymatic antioxidants, increasing evidences showed that sugars, including disaccharides, raffinose family oligosaccharides and fructans, have a dual role with respect to ROS (Couee *et al.,* 2006, Keunen *et al.,* 2013). Soluble sugars were directly linked with the production rates of ROS by regulation ROS producing metabolic pathways, such as mitochondrial respiration or photosynthesis. Conversely, they also feed NADPH-producing metabolism to participate in antioxidative processes (Couee *et al.,* 2006). Researchers have reported a two-fold increase in α-tocopherol in turf grass under the water stress (Ledford and Niyogi, 2005; Kiddle *et al.,*2003). α-tocopherols are lipophilic antioxidants synthesized by all plants and interacts with the polyunsaturated acyl groups of lipids, stabilize membranes, and scavenge and quench various reactive oxygen species (ROS) and lipid soluble byproducts of the oxidative stress (Fath *et al.,* 2002; Cvetkovska *et al.,* 2005). Singlet oxygen quenching by tocopherols is highly efficient, and it is estimated that a single α-tocopherol molecule can neutralize up to 120 singlet oxygen molecules *in vitro* before being degraded (Wu and Tang, 2004). Because of their chromanol ring structure, tocopherols are capable of donating a single electron to form the resonance - stabilized tocopheroxyl radical (Bechtold *et al.,* 2005). These also function as recyclable chain reaction terminators of polyunsaturated fatty acid (PUFA) radicals generated by lipid oxidation (Hare *et al.,*1998) and scavenge lipid peroxy radicals and yield a tocopheroxyl radical that can be recycled back to the corresponding α-tocopherol by reacting with ascorbate or other antioxidants (Igamberdiev and Hill, 2004). α-tocopherol level has been found to increase in photosynthetic plant tissue in response to various abiotic stresses (Noctor, 2006) and have been reported to scavenge and quench various ROS and lipid oxidation products, stabilize membranes, and modulate signal transduction (Shao and Chu, 2005; Noctor, 2006). Antioxidants including α-tocopherol and ascorbic acid have been reported to increase following Trizol treatment in tomato and thus may have a role in protecting membranes from oxidative damage contributing to chilling tolerance (Shao *et al.,*2007). Similarly, increase in the levels of antioxidants and antioxidant enzymes were observed in wheat in response to trizol (Shao *et al.,*2006). It has also been shown that water deficiency may result in an increase of tocopherol concentration in plant tissues (Wu *et al.,*2007; Shao *et al.,*2007). Some evidence implies that tocopherol content of soybean leaves increases with the decrease of rainfall (Shao *et al.,*2006): which is consistent with the reports (Shao *et al.,*2007) that subjecting spinach to water deficit increased the content of α-tocopherol in the leaves. Based on the studies of 10 different grass species under the water stress, the researchers found that drought stress led to an increase in α-tocopherol concentration from 1 to 3 fold in 9 out of 10 species (Shao *et al.,*2007; Pourcel *et al.,*2007) and it was pointed out that the species with a high stress are protected through tocopherol. Moreover, highly significant correlations have been observed between the stress tolerance and α-tocopherol concentration (the precursor of α-tocopherol; Spearman's rank correlation coefficient $r = 0.731$).

A fundamental role of AsA in the plant defense system is to protect metabolic processes against H_2O_2 and other toxic derivatives of oxygen. Acting essentially as a reductant and reacting with and scavenging many types of free radicals, AsA reacts non-enzymatically with superoxide, hydrogen peroxide, and singlet oxygen. As mentioned earlier, it can react indirectly by regenerating α-tocopherol or in the synthesis of zeaxanthin in the xanthophyll cycle. Ascorbic acid is synthesized in the mitochondria and is transported to other cell components through a proton – electrochemical gradient or through the facilitated diffusion. Further, it has also been implicated in the regulation of cell elongation (Yabuta *et al.,* 2004; Rio *et al.,* 2006). In the ascorbate-glutathione cycle, two molecules of AsA are utilized by APX to reduce H_2O_2 to water with concomitant generation of monode-hydroascorbate which is a radical with a short life time and can disproportionate into dehydroascorbate and ascorbic acid, the electron donor is usually NADPH and the reaction is catalyzed by monodehydro ascorbate reductase or ferredoxin in water-water cycle in the chloroplasts (Gapper and Dolan, 2006). In plant cells, the most important reducing substrate for H_2O_2 removal is ascorbic acid (Sagi and Fluhr, 2006; Foyer and Noctor, 2005; Rio *et al.,* 2006) its direct protective role has been demonstrated in rice also where partial protection against damage caused by a release from flooding conditions was provided by the pretreatment with ascorbic acid (Shcolnick and Keren, 2006). A continuous oxidative assault on plants during the drought stress leads to the appearance of an arsenal of enzymatic and non-enzymatic plant antioxidant defenses to counter the oxidative stress in plants (Shao *et al.,* 2007). Water stress has been shown to result in significant increase in AsA concentration in the turf grass. AsA showed a reduction under drought stress in maize and wheat, suggesting its vital involvement in deciding the oxidative response (Shao *et al.,* 2005). AsA and glutathione are also involved in the neutralization of secondary products of ROS reactions (Foyer and Noctor,

2003; Hare *et al.,* 1998). Therefore AsA influences many enzyme activities, and minimizes the damage caused by oxidative process in synergistic mode with other antioxidants (Foyer and Noctor, 2003; Foyer and Noctor, 2005). GSH acts as an antioxidant and is involved directly in the reduction of most active oxygen radicals generated due to stress; it has been found to help to withstand oxidative stress in transgenic lines of tobacco (Foyer and Noctor, 2005; Rio *et al.,*2006).

Carotenes form a key part of the plant antioxidant defense system but are very susceptible to oxidative destruction. β-carotene present in the chloroplasts of all green plants is exclusively bound to the core complexes of PSI and PSII (Havaux, 1998) and protection against damaging effects of ROS at this site is essential for chloroplast functioning. Here, β-carotene, in addition to its function as an accessory pigment, acts as an effective antioxidant and plays a unique role in protecting as well as sustaining photochemical processes (Havaux, 1998). A major protective role of β-carotene in photosynthetic tissue may be through direct quenching of triplet chlorophyll, which prevents the generation of singlet oxygen and protects from oxidative damage (Farooq *et al.,* 2009). Water stress, among other changes, has the ability to reduce the tissue concentrations of chlorophylls and carotenoids (Havaux, 1998; Kiani *et al.,* 2008), primarily through the production of ROS in the thylakoids (Niyogi, 1999; Reddy *et al.,* 2004). However, reports dealing with the strategies to improve the pigment contents under the water-stress are scarce.

3.2. Enzymatic Antioxidants

Plants have evolved an efficient enzymatic defense system to protect themselves against oxidative damage and fine modulation of low levels of ROS for signal transduction. ROS-scavenging enzymes of plants include superoxide dismutase (SOD), ascorbate peroxidase (APX), catalase (CAT), glutathione peroxidase (GPX), monodehydroascorbatereductase (MDHAR), dehydroascorbatereductase (DHAR), glutathione reductase (GR), glutathione *S*-transferase (GST), and peroxiredoxin (PRX). These antioxidant enzymes are located in different sites of plant cells and work together to detoxify ROS. SOD acts as the first line of defense and converts $O_2^{\bullet-}$ into H_2O_2. CAT, APX, and GPX then detoxify H_2O_2. In contrast to CAT, APX requires an ascorbic acid (AsA) and/or a glutathione (GSH) regenerating cycle involved MDHAR, DHAR, and GR. GPX, GST, and PRX reduce H_2O_2 and organic hydroperoxides through ascorbate-independent thiol-mediated pathways using GSH, thioredoxin (TRX) or glutaredoxin (GRX) as nucleophile (Dietz *et al.,* 2006; Meyer *et al.,*2012; Noctor *et al.,* 2014). Non-enzymatic antioxidants include GSH, AsA, carotenoids, tocopherols, and flavonoids are also crucial for ROS homeostasis in plant (Gill and Tuteja, 2010). Besides traditional enzymatic and non-enzymatic antioxidants, increasing evidences indicated that soluble sugars, including disaccharides, raffinose family oligo saccharides and fructans, have a dual role with respect to ROS (Couee *et al.,*2006; Keunen *et al.,*2013). Soluble sugars were directly linked with the production rates of ROS by regulation ROS producing metabolic pathways, such as mitochondrial respiration or photosynthesis. Conversely, they also feed NADPH-producing metabolism to participate in antioxidative processes (Couee *et al.,*2006).

In addition to the antioxidative system, avoiding ROS production by alleviating the effects of stresses on plant metabolism may also be important for keeping ROS homeostasis. Alternative oxidases (AOX) can prevent the excess generation of ROS in the electron transport chains of mitochondria (Maxwell *et al.,* 1999). By diverting electrons flowing through electron-transport chains, AOX can decrease the possibility of electron leaking to O_2 to generate $O_2^{\bullet-}$. Other mechanisms, such as leaf movement and curling, photosynthetic apparatus rearranging, may also represent an attempt to avoid the over-reduction of ROS by balancing the amount of energy absorbed by the plant with the availability of CO_2 (Mittler, 2002).

4. ROS regulation and abiotic stress tolerance genes in crops

4.1. Antioxidant genes

4.1.1. *Superoxide Dismutase (SOD)*

SOD (E.C.1.15.1.1) belongs to the family of metallo enzymes omnipresentinall aerobic organisms.Under environmental stresses, SOD forms the first line of defense against ROS-induced damages. The SOD catalyzes the removal of $O_2^{\bullet-}$ by dismutating it in to O_2 and H_2O_2. This removes the possibility of OH^- formation by the Haber-Weiss reaction. SODs are classified in to three isozymes based on the metal ion it binds, Mn-SOD (localized in mitochondria), Fe-SOD (localized in chloroplasts), and Cu/Zn-SOD (localized incytosol, peroxisomes, and chloroplasts) (Mittler, 2002). SOD has been found to be upregulated by abiotic stress conditions (Boguszewska *et al.,* 2010).

$$O_2^{\bullet-} + O_2^{\bullet-} + 2H^+ \rightarrow 2H_2O_2 + O_2$$

The rice (*japonica*) genome has eight genes that encode putative SODs, including two cytosolic copper-zinc SODs (*cCuZn-SOD1* and *cCuZn-SOD2*), one putative Cu Zn-SOD-like *(CuZn-SOD-L)*, one plastidic SOD (*pCuZn-SOD*), two iron SODs (*Fe-SOD2* and *Fe-SOD3*), and one manganese SOD (*Mn-SOD1*) (Nath*etal.,* 2014). Overexpression of *Mn-SOD1* showed less mitochondrial O_2^- under stress and reduced the stress induction of *OsAOX1a/b* in rice plants (Li *et al.,* 2013).

Table 2: List of all the enzymatic and non-enzymatic antioxidants along with their functions and cellular localization

Non-enzymatic Antioxidants	Function	Subcellular location
Ascorbic Acid (AA)	Detoxifies H_2O_2 via action of APX	Cytosol, Chloroplast, Mitochondria, Peroxisome, Vacuole and Apoplast
Reduced Glutathione (GSH)	Acts as a detoxifying co-substrate for enzymes like peroxidases, GR and GST	Cytosol, Chloroplast, Mitochondria, Peroxisome, Vacuole and Apoplast
α -Tocopherol	Guards against and detoxifies products of membrane LPO	Mostly in membranes
Carotenoids	Quenches excess energy from the photo systems, LHCs	Chloroplasts and other non-green plastids
Flavonoids	Direct scavengers of H_2O_2 and $^1O_2$ and $OH^•$	Vacuole
Proline	Efficient scavenger of $OH^•$ and $^1O_2$ and prevent damages due to LPO	Mitochondria, Cytosol, and Chloroplast

Enzymatic antioxidants	Enzyme code	Reaction catalyzed	Subcellular location
Superoxide dismutase(SOD)	1.15.1.1	$O^{•-}+ 2\,O_2^{•-}+ 2H^+ \rightarrow 2H_2O_2 + O_2$	Peroxisomes, Mitochondria, Cytosol and Chloroplast
Catalase(CAT)	1.11.1.6	$2H_2O_2 \rightarrow O_2+ 2H_2O$	Peroxisome and Mitochondria
Ascorbate peroxidase(APX)	1.11.1.1	$1\,H_2O_2+ AA \rightarrow 2\,H_2O + DHA$	Peroxisomes, Mitochondria, Cytosol and Chloroplast
Monodehydroascorbatereductase (MDHAR)	1.6.5.4	$2MDHA + NADH \rightarrow 2AA + NAD$	Mitochondria, Cytoplasm and Chloroplast
Dehydroascorbatereductase (DHAR)	1.8.5.1	$DHA + 2GSH \rightarrow AA + GSSG$	Mitochondria, Cytoplasm and Chloroplast
Glutathione reductase(GR)	1.6.4.2	$GSSG + NADPH \rightarrow 2GSH + NADP+$	Mitochondria, Cytoplasm and Chloroplast
Guaiacol peroxidase(GPX)	1.11.1.7	$H_2O_2 + DHA \rightarrow 2H_2O + GSSG$	Mitochondria, Cytoplasm, Chloroplast and ER

4.1.2. Catalase (CAT)

CAT(E.C.1.11.1.6) is a tetramericheme-containing enzyme responsible for catalysing the dismutation of H_2O_2 in to H_2O and O_2. It has high affinity for H_2O_2, but lesser specificity for organic peroxides (R-O-O-R). It has a very high turnover rate ($6x10^6$ molecules of H_2O_2 to H_2O and O_2 min^{-1}) and is unique amongst antioxidant enzymes in not requiring a reducing equivalent. Peroxisomes are the hot spots of H_2O_2 production due to β-oxidation of fatty acids, photo respiration, purine catabolism and oxidative stress (Mittler, 2002). However, recent reports suggest that CAT is also found in other sub cellular compartments such as the cytosol, chloroplast and the mitochondria, though significant CAT activity is yet to be seen (Mhamdi et al., 2010). Angiosperms have been reported to have three CAT genes. CAT1 is expressed in pollen sand seeds (localized in peroxisomes and cytosol), CAT2 pre dominantly expressed in photosynthetic tissues but also in roots and seeds (localized in peroxisomes and cytosol) and finally CAT3 is found to be expressed in leaves and vascular tissues (localized in the mitochondria). Stressful conditions demand greater energy generation and expenditure of the cell. This is fulfilled by increased catabolism which generates H_2O_2. CAT removes the H_2O_2 in an energy efficient way.

$$H_2O_2 \rightarrow H_2O + (1/2)\ O_2$$

4.1.3. Ascorbate peroxidase(APX)

APX (E.C.1.1.11.1) is an integral component of the Ascorbate- Glutathione (ASC-GSH) cycle. While CAT pre dominantly scavenges H_2O_2 in the peroxisomes, APX performs the same function in the cytosol and the chloroplast. The APX reduces H_2O_2 to H_2O and DHA, using Ascorbicacid (AA) as a reducing agent.

$$H_2O_2 + AA \rightarrow 2H_2O + DHA$$

The APX family comprises of five isoforms based on different amino acids and locations, viz., cytosolic, mitochondrial, peroxisomal, and chloroplastid (stromal and thylakoidal) (Sharma and Dubey, 2004). Since APX is widely distributed and has a better affinity for H_2O_2 than CAT, it is a more efficient scavenger of H_2O_2 at times of stress.

There are eight ascorbate peroxidase (APX) genes in rice, including two cytosolic APXs (OsAPX1 and OsAPX2), two chloroplastic APXs (OsAPX7 and OsAPX8), two mitochondrial APXs (OsAPX5 and OsAPX6) and two peroxisomal APXs (OsAPX3 and OsAPX4) (Teixeira et al., 2004, 2006). Cytosolic ascorbate peroxidase (APXs) i.e. OsAPX1 and OsAPX2, have vital roles in abiotic stress resistance in rice (Sato et al., 2011 Zhang et al., 2013).

Interestingly, rice mutants double silenced for cytosolic APXs (APX1/2s) exhibit significant changes in the redox status indicated by higher H_2O_2 levels and increased glutathione and ascorbate redox states, triggering alterations in the ROS signaling networks and making the mutants able to cope with abiotic stress similar to non-transformed plants (Bonifacio *et al.*, 2011).

4.1.4. *Glutathione Reductase(GR)*

GR (E.C.1.6.4.2) is a flavoproteinoxidoreductase which uses NADPH as a reductant to reduce GSSG to GSH. Reduced glutathione (GSH) is used up to regenerate AA from MDHA and DHA, and as a result is converted to its oxidized form (GSSG). GR, a crucial enzyme of ASC-GSH cycle catalyzes the formation of a disulphide bondin glutathione disulphide to maintain a high cellular GSH/GSSG ratio. It is predominantly found in chloroplasts with small amounts occurring in the mitochondria and cytosol. GSH is a low molecular weight compound which plays the role of a reductant to prevent thiol groups from getting oxidized, and react with detrimental ROS members like $1O_2$ and $OH^{\bullet}$.

$$GSSG + NADPH \rightarrow 2GSH + NADP$$

Some earlier studies of the ROS-scavenging enzymes, such as GST (Dixion and Edwards, 2010), TRX, and GRX (Meyer *et al.*, 2012), have evolved into large multigene families with varied functions that cope with a variety of adverse environmental conditions. Recent mutational and transgenetic plants analyses revealed special member of multigene enzyme family as a key player in ROS homeostasis regulation in crop plants. *OsTRXh1*, encodes h-type TRX in rice, regulates the redox state of the apoplast and participates in plant development and stress responses (Zhang *et al.*, 2011, Das and Ray choudhary, 2014). OsTRXh1 protein possesses reduction activity and secreted into the extracellular space. Overexpression of *OsTRXh1* produce less H_2O_2 under salt stress, reduce the expression of the salt-responsive genes, lead to a salt-sensitive phenotype in rice. In another study, Perez-Ruiz *et al.*(2006) reported that rice NADPH thioredoxinreductase (NTRC) utilizes NADPH to reduce the chloroplast 2-Cys PRX BAS1, thus protects chloroplast against oxidative damage by reducing H_2O_2.

The involvement of ROS in signal transduction implies that there must be coordinated function of regulation networks to maintain ROS at non-toxic levels in a delicate balancing act between ROS production and ROS-scavenging pathways, and to regulate ROS responses and subsequent downstream processes (Mittler *et al.*, 2004, You and Chan, 2015). Numerous studies from different plant species observed that the generation of ROS and activity of various antioxidant enzymes increased during abiotic stresses (Damanik *et al.*, 2010; Selote *et al.*, 2010; Tang *et al.*, 2010; Tarun and Ekmekci, 2011). There is an increasing body of literature concerning the mechanisms by which regulation of antioxidative protection system response to abiotic stresses in crops. Intrinsic to this regulation is ROS production and signalling that integrated with the action of hormone and small molecules.

4.1.5. Abscisic acid (ABA)

Abscisic acid (ABA) is the key regulator of abiotic stress resistance in plants, and regulates large number of stress-responsive genes by a complex regulatory network so as to confer tolerance to the environmental stresses (Raghwendra *et al.*, 2010; Cutler *et al.*, 2010). ABA-induced stress tolerance is partly linked with the activation of antioxidant defense systems, including enzymatic and non-enzymatic constituents, which protects plant cells against oxidative damage (Zang *et al.*, 2012a, 2014; Hung *et al.*, 2012). Water stress-induced ABA accumulation and exogenous ABA treatment triggers the increased generation of ROS, then leads to the activation of the antioxidant system in crops (Jiang and Zhang, 2002a, b; Ye *et al.*, 2011). Small molecules, such as Ca^{2+}, calmodulin (CaM), NO and ROS have been demonstrated to play vital roles in ABA-induced antioxidant defense (Jiang and Zhang, 2003; Hu *et al.*, 2007). In rice, a Ca^{2+}/ CaM-dependent protein kinase (CCaMK), OsDMI3, is necessary for ABA-induced increases in the expression and the activities of SOD and CAT. ABA-induced H_2O_2 production activates OsDMI3, and the activation of OsDMI3 also enhances H_2O_2 production by increasing the expression of NADPH oxidase genes (Shi *et al.*, 2012). Further study indicated that OsDMI3 functions upstream of OsMPK1, to regulate the activities of antioxidant enzymes and the production of H_2O_2 in rice (Shi *et al.*, 2014). Recent study provides evidence to show that rice histidine kinase OsHK3 functions upstream of OsDMI3 and OsMPK1, and is necessary for ABA-induced antioxidant defense (Wen *et al.*, 2015). Zhang *et al.* (2012a) reported that C2H2-type ZFP, ZFP182, is involved in ABA-induced antioxidant defense. Another C2H2-type ZFP, ZFP36, is also necessary for ABA-induced antioxidant defense (Zhang *et al.*, 2014). Moreover, ABA-induced H_2O_2 production and ABA-induced activation of OsMPKs promote the expression of *ZFP36*, and *ZFP36* also up-regulates the expression of NADPH oxidase and MAPK genes and the production of H_2O_2 in ABA signaling (Zhang *et al.*, 2014). In maize, ABA and H_2O_2 increased the expression and the activity of ZmMPK5, which is required for ABA-induced antioxidant defense. The activation of ZmMPK5 also enhances the H_2O_2 production by increasing the expression and the activity of NADPH oxidase, thus there is a positive feedback loop involving NADPH oxidase, H_2O_2, and ZmMPK5 in ABA signaling (Zhang *et al.*, 2006; Hu *et al.*, 2007; Lin *et al.*, 2009; Ding *et al.*, 2009). Subsequent experiments confirmed that ABA-induced H_2O_2 production mediates NO generation in maize leaves, which, in turn, activates MAPK and increases the expression and the activities of antioxidant enzymes in ABA signaling (Zhang *et al.*, 2007). Moreover, a maize CDPK gene, *ZmCPK11*, acts upstream of ZmMPK5, is essential for ABA-induced up-regulation of the expression and activities of SOD and APX, and the production of H_2O_2 in maize leaves (Ding *et al.*, 2013). Hu *et al.* (2007) found that Ca^{2+}-CaM is required for ABA-induced antioxidant defense and functions both upstream and downstream of H_2O_2 production in leaves of maize plants. Afterward, Ca^{2+}/CaM-dependent protein kinase, ZmCCaMK, was reported to be essential for ABA-induced antioxidant defense, and H_2O_2-induced NO production is involved in the activation of ZmCCaMK in ABA signaling (Ma *et al.*, 2012).

OsABA8ox3, encoding ABA 8-hydroxylase involved in ABA catabolism, is also a key gene regulating ABA accumulation and antioxidative stress capability under drought stress (Nguyen *et al.*, 2015). *OsABA8ox3* RNAi plants exhibited significant improvement in drought stress tolerance. Consistent with this, *OsABA8ox3* RNAi plants showed increased SOD and CAT activities and reduced MDA levels during dehydration treatment. In another study, over expression of the 9-*cis*-epoxy carotenoid dioxygenase gene from *Stylosanthesguianensis* (*SgNCED1*) in the transgenic tobacco increased ABA content and tolerance to drought and salt stresses (Zhang *et al.*, 2009). Moreover, enhanced abiotic stresses tolerance in transgenic plants is associated with ABA- induced production of H_2O_2 and NO, which, in turn, activate the expression and activities of ROS-scavenging enzymes (Zhang *et al.*, 2009).

Brassinosteroids are a group of steroid hormones and important for a broad spectrum of plant growth and development processes, as well as responses to biotic and abiotic stresses (Bajguz and Hayat, 2009; Divi and Krishna, 2009; Yang *et al.*, 2011; Zhu *et al.*, 2013a). Numerous studies have shown that BR can activate antioxidant defense systems to improve stress tolerance in crops (Ozdemir *et al.*, 2004; Xia *et al.*, 2009). Zhang *et al.* (2010) reported that ZmMPK5 is required for NADPH oxidase-dependent self-propagation of ROS in BR-induced antioxidant defense systems in maize. Further study founded that a 65 kDa microtubule-associated protein (MAP65), ZmMAP65-1a, directly phosphorylated by ZmMPK5, is required for BR-induced antioxidant defense (Zhu *et al.*, 2013b). Recently, Ca^{2+} and maize CCaMKgene, *ZmCCaMK*, was demonstrated to be required for BR-induced antioxidant defense (Yan *et al.*, 2015)

5. Transcription Factors

Transcriptional factors (TFs) are one of the important regulatory proteins involved in abiotic stress responses. They play essential roles downstream of stress signaling cascades, which could alter the expression of a subset of stress-responsive genes simultaneously and enhance tolerance to environmental stress in plants. Members of AP2/ERF (APETALA2/ethylene response factor), zinc finger, WRKY, bZIP (basic leucine zipper), and NAC (NAM, ATAF, and CUC) families have been characterized with roles in the regulation of plant abiotic stress responses (Yamaguchi-Shinozaki and Shinozaki, 2006; Ariel *et al.*, 2007; Ciftci-Yilmaz and Mittler, 2008; Fang *et al.*, 2008), and some of them have been demonstrated to be involved in ROS homeostasis regulation and abiotic stress resistance in crops.

Proteins containing zinc finger domain(s) were widely reported to be key players in the regulation of ROS-related defense genes in *Arabidopsis* and other species. For example, the expression of some zinc finger genes in *Arabidopsis*, *ZAT7*, *ZAT10* and *ZAT12*, is intensely up-regulated by oxidative stress in At APX1 knock out plants (Miller *et al.*, 2008). Subsequent experiments showed that these zinc finger proteins were involved in ROS regulation and multiple abiotic stresses tolerance (Davletova *et al.*, 2005; Mittler *et al.*,;Ciftci-Yilmaz *et al.*, 2007). The zinc finger proteins are divided in to several types, such as C_2H_2, C_2C_2, C_2HC, CCCH and C_3HC_4, based on the number and the location of characteristic residues (Ciftci- Yilmaz and Mittler, 2008). The signaling pathways participating in stomatal movement were well studied in the model plant *Arabidopsis*, but were largely unknown in crops. Huang *et al.* (2009) identified a drought and salt tolerance (*dst*) mutant, and the DST was cloned by the map-based cloning. DST encoded a C_2H_2-type zinc finger transcription factor that negatively regulated stomatal closure by direct regulation of genes related to H_2O_2 homeostasis, which identified a novel signaling pathway of DST-mediated H_2O_2-induced stomatal closure (Huang *et al.*, 2009). Loss of DST function increased the accumulation of H_2O_2 in guard cell, accordingly, resulted in increased stomatal closure and enhanced drought and salt tolerance in rice. Other two C_2H_2-type zinc finger proteins, ZFP36 and ZFP179, also play circle role in ROS homeostasis regulation and abiotic stress resistance in rice. *ZFP179* encodes a salt-responsive zinc finger protein with two C_2H_2-type zinc finger motifs (Sun *et al.*, 2010). The *ZFP179* transgenic rice plants increased ROS-scavenging ability and expression levels of stress-related genes, and exhibited significantly enhanced tolerance to salt and oxidative stress (Sun *et al.*, 2010). *ZFP36* is an ABA and H_2O_2-responsive C_2H_2-type zinc finger protein gene, and plays an important role in ABA- induced antioxidant defense and the tolerance of rice to drought and oxidative stresses (Zhang *et al.*, 2014). Moreover, ZFP36 is a major player in the regulation of the cross-talk involving NADPH oxidase, H_2O_2, and MAPK in ABA signaling (Zhang *et al.*, 2014). OsTZF1, a CCCH-tandem zinc finger protein, was identified as a negative regulator of leaf senescence in rice under stress conditions (Jan *et al.*, 2013). Meanwhile, OsTZF1confers tolerance to oxidative stress in rice by enhancing the expression of redox homeostasis genes and ROS-scavenging enzymes (Jan *et al.*, 2013). A cotton CCCH-type tandem zinc finger gene, *GhTZF1*, also serves as a key player in modulating drought stress resistance and subsequent leaf senescence by mediating ROS homeostasis (Zhou *et al.*, 2014b).

Members of AP2/ERF (APETALA2/ethylene response factor) transcription factor family, including DREB/CBF transcription factors, are especially important as they regulate genes involved in multiple abiotic stress responses (Mizoi *et al.*, 2012). During the initial phase of abiotic stresses, elevated ROS levels might act as a vital acclimation signal. But the key regulatory components of ROS-mediated a biotic stress response signaling are largely unknown. Rice salt-and H_2O_2-responsive ERF transcription factor, SERF1, has a critical role in regulating H_2O_2-mediated molecular signaling cascade during the initial response to salinity in rice (Schmidt *et al.*, 2013). SERF1regulates the expression of H_2O_2-responsive genes involved in salt stress responses in roots. SERF1 is also a phosphorylation target of a salt-responsive MAPK (MAPK5), and activation the expression of salt-responsive MAPK cascade genes (*MAPK5* and *MAPKKK6*), well established salt-responsive TF genes (*ZFP179* and *DREB2A*), and itself through direct interaction with the corresponding promoters in plants (Schmidt *et al.*, 2013). The SERF1 is essential for the propagation of the initial ROS signal to mediate salt tolerance. SUB1A, an ERF transcription factor found in limited rice accessions, limits ethylene production and gibberellin responsiveness during submergence, economizing carbohydrate reserves and significantly prolonging endurance (Fukao and Xiong, 2013). After floodwaters subside, submerged plants encounter reexposure to atmospheric oxygen, leading to post anoxic injury and severe leaf desiccation (Setter *et al.*, 2010; Fukao and Xiong, 2013). SUB1A also positively affects post submergence responses by restrained accumulation of ROS in aerial tissue during desubmergence (Fukao *et al.*, 2011). Consistently, SUB1A promptes the expression of ROS scavenging enzyme genes, resulting in enhanced tolerance to oxidative stress. On the other hand, SUB1A improves survival of rapid dehydration following desubmergence and water deficit during drought by increasing ABA responses, and activating stress- inducible gene expression (Fukao *et al.*, 2011). A jasmonate and ethylene-responsive ERF gene, JERF3, was isolated from tomato and involved in a ROS-mediated regulatory module in transcriptional networks that govern plant response to stress (Wu *et al.*, 2008). JERF3 modulates the expression of genes involved in osmotic and oxidative stresses responses by binding to the osmotic and oxidative-responsive related *cis*elements. The expression of these genes leads to reduce accumulation of ROS, resulting in enhanced abiotic stress tolerance in tobacco (Wu *et al.*, 2008).

The WRKY family proteins have one or two conserved WRKY domains comprising a highly conserved WRKYGQK hepta peptide at the N-terminus and a zinc-finger-like motif at the C-terminus (Eulgem *et al.*, 2000). The conserved

WRKYdomain plays important roles in various physiological processes by binding to the W-box in the promoter regions of target genes (Ulker and Somssich, 2004; Rushton *et al.*, 2010). Wang *et al.*(2015) reported a multiple stress-responsive WRKY gene, *GmWRKY27*, reduces ROS level and enhances salt and drought tolerance in transgenic soybean hairy roots. GmWRKY27 interacts with GmMYB174, which, in turn, acts in concert to reduce promoter activity and gene expression of *GmNAC29* (Wang *et al.*, 2015). Further experiments showed that GmNAC29 is a negative factor of stress tolerance for enhancing the ROS production under abiotic stress by directly activating the expression of the gene encoding ROS production enzyme. In another study, over expression of cotton WRKY gene, *GhWRKY17*, reduced transgenic tobacco plants tolerance to drought and salt stress. Subsequent experiments showed that GhWRKY17 involved in stress responses by regulating ABA signaling and cellular levels of ROS (Yan *et al.*, 2014). Sun *et al.*(2015) isolated a WRKY gene, *BdWRKY36*, from *B. distachyon*, and found it functions as a positive regulator of drought stress response by controlling ROS homeostasis and regulating transcription of stress related genes.

Members of other TF families also functioned in abiotic stress response through ROS regulation. ASR proteins are plant specific TFs and considered to be important regulators of plant response to various stresses. Wheat ASR gene, *TaASR1*, a positive regulator of plant tolerance to drought/osmotic stress, is involved in the modulation of ROS homeostasis by activating antioxidant defense system and transcription of stress responsive genes (Hu *et al.*, 2013). Soybean NACTF, GmNAC2, was identified as a negative regulator during abiotic stress, and participates in ROS signaling pathways through modulation of the expression of genes related to ROS scavenging (Jin *et al.*, 2013). Ramegowda *et al.*(2012) isolated a stress-responsive NAC gene, *EcNAC1*, from finger millet (*E. coracana*). Transgenic tobacco plants expressing *EcNAC1* increased ROS scavenging activity, up-regulated many stress responsive genes and exhibited tolerance to various abiotic stresses and MV-induced oxidativestress (Ramegowda *et al.*, 2012). Recently, a NAC transcription factor gene, *SNAC3*, functions as a positive regulator under high temperature and drought stress, was identified in rice (Fang *et al.*, 2015). *SNAC3* enhances the abiotic stresses tolerance by modulating H_2O_2 homeostasis state through controlling the expression of ROS- associated enzyme genes (Fang *et al.*, 2015).

In addition to TFs, transcriptional co-regulator as well as spliceosome component, *OsSKIPa*, a rice homolog of human Ski-interacting protein (SKIP), has been studied for effects on drought resistance (Hou *et al.*, 2009). *OsSKIPa*-overexpressing rice exhibited significantly enhanced drought stress tolerance at both the seedling and reproductive stages by increased ROS- scavenging ability and transcript levels of many stress-related genes (Hou *et al.*, 2009).

6. Similar to RCD One (SRO) Proteins

The SRO (Similar to RCD one) protein family was recently identified as a group of plant-specificproteins, and they are characterized by the plant-specific domain architecture which contains a poly (ADP-ribose) polymerase catalytic (PARP) and a C-terminal RCD1-SRO-TAF4 (RST) domain (Jaspers *et al.*, 2010). In addition to these two domains, some SRO proteins have an N-terminal WWE domain. Our limited knowledge of SRO proteins is mainly from the study in *Arabidopsis* mutant *rcd1* (*radical-induced cell death1*). *rcd1* exhibits pleiotropic phenotypes related to a wide range of exogenous stimulus responses and developmental processes, including sensitivity to apoplastic ROS and salt stress, resistance to chloroplastic ROS caused by methyl viologen (MV) and UV-B irradiation (Ahlfors *et al.*, 2004; Fujibe *et al.*, 2004; Katiyar-Agarwal *et al.*, 2006). RCD1 interacts with SOS1 and a large number of transcription factors which have been identified or predicted to be involved in both development and stress-related processes (Katiyar-Agarwal *et al.*, 2006; Jaspers *et al.*, 2009). Recent study demonstrated that RCD1 is possibly involved in signaling networks that regulate quantitative changes in gene expression in response to ROS (Brosche *et al.*, 2014). Recently, an SRO gene was also identified to be crucial for salinity stress resistance by modulating redox homeostasis in wheat (Liu *et al.*, 2014). Ta-*sro1*, the allele of the salinity-tolerant bread wheat cultivar Shanrong No.3, is derived from the wheat parent allele via point mutation. Unlike *Arabidopsis* SRO proteins, Ta-sro1 has PARP activity. Both the over expression of *Ta-sro1* in wheat and *Arabidopsis* promotes the accumulation of ROS by regulating ROS-associated enzyme. Ta- sro1 also enhances the activity of AsA-GSH cycle enzymes and GPX cycle enzymes, which regulate ROS content and cellular redox homeostasis (Liu *et al.*, 2014).

7. Calcium-Binding Proteins and Calcium Transporters

Calcium (Ca^{2+}) regulates numerous signaling pathways involved in growth, development and stress tolerance. The influx of Ca^{2+} into the cytosol is countered by pumping Ca^{2+} out from the cytosol to restore the basal cytosolic level, and this may be achieved either by P-type Ca^{2+}ATPases or antiporters. Huda *et al.*(2013) report the isolation and characterization of *OsACA6*, which encodes a member of the type IIBCa2+ATPase family from rice. Over expression of *OsACA6* confers tolerance to salinity and drought stresses in tobacco, which was correlated with reduced accumulation of ROS and enhanced the expression of stress responsive genes in plants (Huda *et al.*, 2013). In addition, over expression of *OsACA6* confers Cd^{2+} stress tolerance in transgenic lines by maintaining cellular ion homeostasis and modulating ROS scavenging pathway (Shukla *et al.*, 2014).

Annexins are calcium-dependent, phospholipid- binding proteins with suggested functions in response to environmental stresses and signaling during plant growth and development. OsANN1, a member of the annexin protein family in rice, has ATPase activity, the ability to bind Ca^{2+}, and the ability to bind phospholipids in a Ca^{2+}- dependent manner. OsANN1confers abiotic stress tolerance by modulating antioxidant accumulation and interacting with OsCDPK24 (Qiao et al., 2015).

8. Other Functional Proteins

Poly amines are low molecular weight aliphatic amines found in all living cells. Because of their cationic nature at physiological pH, PAs have strong binding capacity to negatively charged molecules (DNA, RNA, and protein), thus stabilizing their structure (Alcazar et al.,2010). The PAs biosynthetic pathway has been thoroughly investigated in many organisms, and arginine decarboxylase (ADC) plays a predominant role in the accumulation of PAs under stresses (Capell et al.,2004; Alcazar et al.,2010). Wang et al.(2011) isolated an arginine decarboxylase gene (*PtADC*) from *Poncirustrifoliata*. The transgenic tobacco and tomato plants elevated endogenous PAs level, accumulated less ROS and showed an improvement in drought tolerance. Jang et al.(2012) identified a highly oxidative stress-resistant T-DNA mutant line carried an insert ion in *OsLDC-like1* in rice. The mutant produced much higher levels of PAs compared to the wild type plants. Based on their results, the authors suggested that PAs mediate tolerance to abiotic stresses through their ability to decrease ROS generation and enhance ROS degradation. The12-oxo-phytodienoic acid reductases (OPRs) are classified into two sub groups, OPRI and OPRII. OPRII proteins are involved in jasmonic acid synthesis, while the function of OPRI is as yet unclear. Dong et al.(2013) characterized the functions of the wheat OPRI gene *TaOPR1*. Over expression of *TaOPR1* in wheat and *Arabidopsis* enhanced tolerance to salt stress by regulating of ROS and ABA signaling pathways (Dong et al.,2013). Helicases are ubiquitous enzymes that catalyse the unwinding of energetically stable duplex DNA or RNA secondary structures, and thereby play an important role in almost all DNA and/or RNA metabolic processes. OsSUV3, an NTP-dependent RNA/DNA helicase in rice, exhibits ATPase, RNA and DNA helicase activities (Tuteja et al.,2013). *OsSUV3* sense transgenic rice plants showed lesser lipid peroxidation and H2O2 production, alongwith higher activities of antioxidant enzymes, consequently resulting in increased tolerance to high salinity (Tuteja et al.,2013).

9. Conclusion and Future Perspectives

The discovery of the enzymatic activity of SOD 45years ago (Mc Cord and Fridovich, 1969) ushered in the field of ROS biology. During the last two decades, the main sources and places of ROS generation, and the key antioxidant molecules and enzymes that scavenge excess ROS have been produced in plant. However, current study about ROS homeostasis and signaling remains fragmental. Apoplastic ROS are rapidly produced in plants as a defense response to pathogen attack and abiotic stress. Whereas, in addition to NADPH oxidase, the function and regulation of other apoplastic ROS-associated enzymes, such as class III peroxidases, in stress responses signaling are largely unknown. On the other hand, 100s of genes that encode for ROS-metabolizing enzymes and regulators comprise ROS gene network in crop plants. Therefore, more than one enzymatic activity that produces or scavenges ROS exits in convinced cellular section. How these different enzymes are coordinated with in each compartment and between different compartments to adjust a particular ROS at an appropriate level during stresses is an important question needs to be addressed. There is increasing evidence suggesting the vital role of ROS signaling pathway in plant growth and stress tolerance. However, regulatory mechanisms at the biochemical level, the mechanisms of extra cellular ROS perception, transduction of ROS-derived signals, and especially the communication and interaction between different subcellular compartments in ROS signaling are still poorly understood. To build comprehensive regulation networks in ROS signaling and responses requires a combination of transcriptomics, proteomics and metabolomics approaches with analysis of mutant as well as protein–protein interactions. Plants need diverse responses and adjustments of multiple adaptation mechanisms to cope with the multiple stresses exist in nature. Comparison of transcription profiles of rice in response to multiple stresses suggested the central role of ROS homeostasis in different abiotic stresses (Mittal *et al.,*2012). Therefore, manipulating endogenous ROS levels provides us with an opportunity to improve common defense mechanisms against different stresses to ensure crop plants growth and survival under adverse growing condition. The functions of numerous stress responsive genes involved in ROS homeostasis regulation and abiotic stress resistance have been characterized in transgenic plants (Table1). As expected, transgenic crop plants harbored these genes enhanced tolerance to multiple abiotic stresses (Wu *et al.,*2008; Fukao *et al.,*2011; Lu *et al.,*2013; Campo *et al.,*2014). However, few studies have reported the abiotic stress tolerance of transgenic plant at the reproductive or flowering stage based on yield and/or setting rate, and very few of these tests were conducted under field conditions. Additionally, of the reported ROS-associated genes that involved in abiotic stresses just have been demonstrated its role in regulation of expression and/or activity of ROS-scavenging enzymes. Thus, network involving in function of these genes in ROS homeostasis to medicate abiotic stress resistance needs to be fully investigated, and the new components need to be integrated in to the signaling pathway. With a long-term goal to improve the abiotic stress resistance of crop plants by the utilizing of ROS regulation pathways, more and more key regulators need to be identified. It is also very important to clarify the mechanisms regulating ROS signaling pathways and their inter play during abiotic stresses. This can finally help to incorporate multiple necessary ROS- associated genes in to the genetic backgrounds of elite cultivars or hybrids to enhance their abiotic stress resistance under real agricultural field conditions.

10. References

1. Ahlfors, R.; Lang, S.; Overmyer, K.; Jaspers, P.; Brosche, M.; Tauriainen, A. et al.*Arabidopsis* radical-induced cell death belongs to the WWE protein-protein interaction domain protein family and modulates abscisic acid, ethylene, and methyl jasmonate responses. *Plant Cell.* **2004,** 16, 1925-1937.

2. Alcazar, R.; Altabella, T.; Marco, F.; Bortolotti, C.; Reymond, M.; Koncz, C. et al. Polyamines: molecules with regulatory functions in plant abiotic stress tolerance. *Planta.* **2010,** 231, 1237-1249.

3. Apel, K.; Hirt, H. Reactive oxygen species: metabolism, oxidative stress, and signal transduction. *Annu. Rev. Plant Biol.* **2004,** 55, 373-399.

4. Ariel, F. D.; Manavella, P. A.; Dezar, C. A.; Chan, R. L. The true story of the HD-Zip family. *Trends Plant Sci.* **2007,** 12, 419-426.

5. Bajguz, A.; Hayat, S. Effects of brassinosteroids on the plant responses to environmental stresses. *Plant Physiol. Biochem.* **2009,** 47, 1-8.

6. Baxter, A.; Mittler, R.; Suzuki, N.).ROS as key players in plant stress signalling. *Journal of Experimental Botany.* **2014,** 65, 1229-1240.

7. Bechtold, U.; Karpinski, S.; Mullineaux, P. M. The influence of the light environment and photosynthesis on oxidative signaling responses in plant-biotrophic pathogen interactions. *Plant Cell Environ.* **2005,** 28, 1046-1055.

8. Boguszewska, D.; Grudkowska, M.; Zagdañska, B. Drought-responsive antioxidant enzymes in potato (*SolanumtuberosumL.*). *Potato Research.* **2010,** 53, 373-382.

9. Bonifacio, A.; Martins, M. O.; Ribeiro, C.W.; Fontenele, A.V.; Carvalho, F. E.; Margis-Pinheiro, M.; *et. al.* Role of peroxidases in the compensation of cytosolic ascorbate peroxidase knock down in rice plants under abiotic stress. *Plant Cell Environ.* **2011,** 34, 1705-1722.

10. Boyer, J.S. Plant Productivity and Environment. *Science.* **1982,** 218, 443-448.

11. Brosche, M.; Blomster, T.; Salojarvi, J.; Cui, F.; Sipari, N.; Leppala, J. *et al.*Transcriptomics and functional genomics of ROS-induced cell death regulation by radical-induced cell death1. *PLoS Genet.* **2014,** 10, e1004112.

12. Campo, S.; Baldrich, P.; Messeguer, J.; Lalanne, E.; Coca, M.; San Segundo, B. Over expression of a calcium-dependent protein kinase confers salt and drought tolerance in rice by preventing membrane lipid peroxidation. *Plant Physiol.* **2014,** 165, 688-704.

13. Capell, T.; Bassie, L.; Christou, P. Modulation of the polyamine biosynthetic pathway in transgenic rice confers tolerance to drought stress. *Proc. Natl. Acad. Sci. U.S.A.* **2004,** 101, 9909-9914.

14. Choudhury, S.; Panda, P.; Sahoo, L.; Panda, S. K. Reactive oxygen species signaling in plants under abiotic stress. *Plant Signal. Behav.* **2013,** 8, e23681.

15. Ciftci-Yilmaz, S.; Mittler, R. The zinc finger network of plants. *Cell Mol. Life Sci.* **2008,** 65, 1150-1160.

16. Ciftci-Yilmaz, S.; Morsy, M. R.; Song, L.; Coutu, A.; Krizek, B. A.; Lewis, M.W.; et. al. The EAR-motif of the Cys2/His2-type zinc finger protein Zat7 plays a key role in the defense response of *Arabidopsis* to salinity stress. *Journal Biol. Chem.* **2007,** 282, 9260-9268.

17. Cosio, C.; Dunand, C. Specific functions of individual class III peroxidase genes. *Journal Experimental Botany.* **2009,** 60, 391–408.

18. Couee, I.; Sulmon, C.; Gouesbet, G.; ElAmrani, A. Involvement of soluble sugars in reactive oxygen species balance and responses to oxidative stress in plants. *Journal Experimental Botany.* **2006,** 57, 449-459.

19. Cutler, S. R.; Rodriguez, P. L.; Finkelstein, R. R.; Abrams, S. R. Abscisic acid: emergence of a core signalling network. *Annu. Rev. Plant Biol.* **2010,** 61, 651-679.

20. Cvetkovska, M.; Rampitsch, C.; Bykova, N.; Xing, T. Genomic analysis of MAP kinase cascades in Arabidopsis defense responses. *Plant Mol. Biol. Rep.* 2005, 23, 331-343.

21. Damanik, R. I.; Maziah, M.; Ismail, M. R.; Ahmad, S.; Zain, A. Responses of the antioxidative enzymes in Malaysian rice (*Oryza sativa* L.) cultivar sunder sub mergence condition. *Acta Physiol. Plant.* **2010,** 32, 739-747.

22. Das, K.; Roychoudhury, A. Reactive oxygen species (ROS) and response of antioxidants as ROS-scavengers during environmental stress in plants. *Frontier in Environ Science.* 2014, 2, 53.

23. Davletova, S.; Schlauch, K.; Coutu, J.; Mittler, R. The zinc-finger protein Zat12 plays a central role in reactive oxygen and abiotic stress signalling in *Arabidopsis. Plant Physiol.* **2005,** 139, 847-856.

24. Dietz, K. J.; Jacob, S.; Oelze, M. L.; Laxa, M.; Tognetti, V.; De Miranda, S. M. et al. The function of peroxiredoxins in plant organelle redox metabolism. *Journal of Experimental Botany.* **2006,** 57, 1697-1709.

25. Ding, H.; Zhang, A.; Wang, J.; Lu, R.; Zhang, H.; Zhang, J. *et al.* Identity of an ABA-activated 46kD amitogen-activated protein kinase from *Zea mays* leaves: partial purification, identification and characterization. *Planta.* **2009,** 230, 239-251.

26. Ding, Y.; Cao, J.; Ni, L.; Zhu, Y.; Zhang, A. Tan, M. *et al.* ZmCPK11 is involved in abscisic acid-induced antioxidant defence and functions upstream of ZmMPK5 in abscisic acid signalling in maize. *Journal of Experimental Botany.* **2013,** 64, 871–884.

27. Divi, U. K.; Krishna, P. Brassinosteroid: a biotechnological target for enhancing crop yield and stress tolerance. *Nature Biotechnology.* **2009,** 26, 131-136.

28. Dixon, D. P.; Edwards, R. Glutathione transferases. *Arabidopsis Book.* **2010,** 8, e0131.

29. Dong, W.; Wang, M.; Xu, F.; Quan, T.; Peng, K.; Xiao, L. *et al.* Wheat oxophytodienoatereductase gene TaOPR1confers salinity tolerance via enhancement of abscisic acid signalling and reactive oxygen species scavenging. *Plant Physiology.* **2013,** 161, 1217-1228.

30. Eulgem, T.; Rushton, P. J.; Robatzek, S.; Somssich, I. E. The WRKY super family of plant transcription factors. *Trends Plant Science.* **2000,** 5, 199-206.

31. Fang, Y.; Liao, K.; Du, H.; Xu, Y.; Song, H.; Li, X. *et al.* A stress- responsive NAC transcription factor SNAC3 confers heat and drought tolerance through modulation of reactive oxygen species in rice. *Journal of Experimental Botany.* **2015,** 66:6803.

32. Fang, Y.; You, J.; Xie, K.; Xie, W.; Xiong, L. Systematic sequence analysis and identification of tissue-specific or stress-responsive genes of NAC transcription factor family in rice. *Mol.Genet.Genomics.* **2008,** 280, 547-563.

33. Farooq, M.; Wahid, A.; Ito, O.; Lee, D. J.; Siddique, K. H. M. Advances in drought resistance of rice. *Crit. Rev. Plant Sci.* **2009,** 28:199-217.

34. Fath, A.; Bethke, P.; Beligni, V.; Jones, R. Active oxygen and cell death in cereal aleurone cells. *Journal of Experimental Botany.* **2002,** 53:1273-1282.

35. Foyer, C. H.; Noctor, G. Redox sensing and signaling associated with reactive oxygen in chloroplasts, peroxisomes and mitochondria. *Plant Physiol.* **2003,** 119:355- 364.

36. Foyer, C. H.; Noctor, G. Oxidant and antioxidant signaling in plants: a re-evaluation of the concept of 1750 oxidative stress in a physiological context. *Plant Cell Environ.* **2005,** 28, 1056-1071.

37. Foyer, C.H.; Noctor, G. Redox homeostasis and antioxidant signaling: a metabolic interface between stress perception and physiological responses. *Plant Cell.* **2005,** 17, 1866-1875.

38. Fujibe, T.; Saji, H.; Arakawa, K.; Yabe, N.; Takeuchi, Y.; Yamamoto, K. T. A methyl viologen-resistant mutant of *Arabidopsis*, which is allelic to ozone-sensitive rcd1, is tolerant to supplemental ultra violet-B irradiation. *Plant Physiol.* **2004,** 134, 275-285.

39. Fukao, T.; Xiong, L. Genetic mechanisms conferring adaptation to submergence and drought in rice: simple or complex? *Curr. Opin. Plant Biol.* **2013,** 16, 196-204.

40. Fukao, T.; Yeung, E.; Bailey-Serres, J. The submergence tolerance regulator SUB1A mediates cross talk between submergence and drought tolerance in rice. *Plant Cell.* **2011,** 23, 412-427.

41. Gapper, C.; Dolan, L. Control of plant development by reactive oxygen species. *Plant Physiology.* **2006,** 141, 341-345.

42. Gill, S. S.; Tuteja, N. Reactive oxygen species and antioxidant machinery in abiotic stress tolerance in crop plants. *Plant Physiol. Biochem.* **2010,** 48, 909-930.

43. Hare, P. D.; Cress, W. A.; Staden, J. V. Dissecting the roles of osmolyte accumulation during stress. *Plant Cell Environ.* **1998,** 21:535-553.

44. Havaux M (1998) Carotenoids as membrane stabilizers in chloroplasts. Trends Plant Sci. 3:147-151.

45. Hou, X.; Xie, K.; Yao, J.; Qi, Z.; Xiong, L. A homolog of human ski-interacting protein in rice positively regulates cell viability and stress tolerance. *Proc. Natl. Acad. Sci. U.S.A.* **2009,** 106, 6410-6415.

46. Hu, W.; Huang, C.; Deng, X.; Zhou, S.; Chen, L.; Li, Y. *et al.* TaASR1, a transcription factor gene in wheat, confers drought stress tolerance in transgenic tobacco. *Plant Cell Environ.* **2013,** 36, 1449-1464.

47. Hu, X.; Jiang, M.; Zhang, J.; Zhang, A.; Lin, F.; Tan, M. Calcium- calmodulin is required for abscisic acid induced antioxidant defense and functions both upstream and down stream of H_2O_2 production in leaves of maize (*Zea mays*) plants. *New Phytol.* **2007,** 173, 27-38.

48. Huang, J.; Sun, S.; Xu, D.; Lan, H.; Sun, H.; Wang, Z. *et al.* ATF IIIA- type zinc finger protein confers multiple abiotic stress tolerances in transgenic rice (*Oryza sativa* L.). *Plant Mol. Biol.* **2012,** 80, 337-350.

49. Huang, X. Y.; Chao, D. Y.; Gao, J. P.; Zhu, M. Z.; Shi, M.; Lin, H. X. A previously unknown zinc finger protein, DST, regulates drought and salt tolerance in rice via stomatal aperture control. *Genes Dev.* **2009,** 23, 1805-1817.

50. Huda, K.; M.; Banu, M.S.; Garg, B.; Tula, S.; Tuteja, R.; Tuteja, N. OsACA6, aP-type $IIBCa^{2+}$ ATPase promotes salinity and drought stress tolerance in

tobacco by ROS scavenging and enhancing the expression of stress-responsive genes. *Plant Journal.* **2013**, 76, 997-1015.

51. Igamberdiev, A. U.; Hill, R. D. Nitrate, NO and haemoglobin in plant adaptation to hypoxia: an alternative to classic fermentation pathways. *Journal of Experimental Botany.* **2004**, 55, 2473-2482.

52. Jan, A.; Maruyama, K.; Todaka, D.; Kidokoro, S.; Abo, M.; Yoshimura, E.; *et. al.* OsTZF1, a CCCH-tandem zinc finger protein, confers delayed senescence and stress tolerance in rice by regulating stress-related genes. *Plant Physiol.* **2013**, 161, 1202-1216.

53. Jang, S. J.; Wi, S. J.; Choi, Y. J.; An, G.; Park, K. Y. Increased polyamine biosynthesis enhances stress tolerance by preventing the accumulation of reactive oxygen species: T-DNA mutational analysis of *Oryza sativa* lysine decarboxylase-like protein1. *Mol. Cells.* **2012**, 34, 251-262.

54. Jaspers, P.; Blomster,T.; Brosche, M.; Salojarvi, J.; Ahlfors, R.; Vainonen, J. P.; *et. al.* Un equally redundant RCD1 and SRO1 mediate stress and developmental responses and interact with transcription factors. *Plant Journal.* **2009**, 60, 268-279.

55. Jaspers, P.; Overmyer, K.; Wrzaczek, M.; Vainonen, J. P.; Blomster, T.; Salojarvi, J.; *et. al.* The RST and PARP-like domain containing SRO protein family: analysis of protein structure, function and conservation in land plants. *BMC Genomics.* **2010**, 11, 170.

56. Jiang, M.; Zhang, J. Involvement of plasma-membrane NADPH oxidase in abscisic acid-and water stress-induced antioxidant defense in leaves of maize seedlings. *Planta.* **2002a**, 215, 1022-1030.

57. Jiang, M.; Zhang, J. Water stress-induced abscisic acid accumulation triggers the increased generation of reactive oxygen species and up-regulates the activities of antioxidant enzymes in maize leaves. *Journal of Experimental Botany.* **2002b**, 53, 2401-2410.

58. Jiang, M.; Zhang, J. Cross-talk between calcium and reactive oxygen species originated from NADPH oxidase in abscisic acid-induced antioxidant defence in leaves of maize seedlings. *Plant Cell Environ.* **2003**, 26, 929-939.

59. Jin, H. X.; Huang, F.; Cheng, H.; Song, H. N.; Yu, D. Y. Over expression of the Gm NAC2 gene, an NAC transcription factor, reduces abiotic stress tolerance in tobacco. *Plant Mol. Biol. Rep.* **2013**, 31, 435-442.

60. Jun, Y.;Zhulong Chan. ROS Regulation during Abiotic Stress Responses in Crop Plants.*Frontier in Plant Science.* **2015**, 6: 1092.

61. Katiyar-Agarwal, S.; Zhu, J.; Kim, K.; Agarwal, M.; Fu, X.; Huang, A. *et al.* The plasma membrane Na^+/H^+ antiporter SOS1 interacts with RCD1 and

functions in oxidative stress tolerance in *Arabidopsis*. *Proc. Natl. Acad. Sci. U.S.A.* **2006,** 103, 18816-18821.

62. Keunen, E.; Peshev, D.; Vangronsveld, J.; Van Den Ende, W.; Cuypers, A. Plant sugars are crucial players in the oxidative challenge during abiotic stress: extending the traditional concept. *Plant Cell Environ.* **2013,** 36, 1242-1255.

63. Kiani, S. P.; Maury, P.; Sarrafi, A.; Grieu, P. QTL analysis of chlorophyll fluorescence parameters in sunflower (*Helianthus annuus* L.) under well-watered and water stressed conditions. *Plant Science.* **2008,** 175:565-573.

64. Kiddle, G; Pastori, G. M.; Bernard, S. Effects of leaf ascorbate content on defense and photosynthesis gene expression in *Arabidopsis thaliana*. *Antioxidant Redox Signaling.* **2003,** 5(1), 23-32.

65. Kumar, R. R.; Goswami, S.; Singh, K.; Rai, Gyanendra K.; Rai, R. D. Modulation of redox signal transduction in plant system through induction of free radical /ROS scavenging redox-sensitive enzymes and metabolites. *Australian Journal of Crop Sciences.* **2013,** 7(11), 1744-1751.

66. Ledford, H. K.; Niyogi, K. K. Singlet oxygen and photo oxidative stress management in plants and algae. *Plant Cell Environ.* **2005,** 28, 1037-1045.

67. Li, C. R.; Liang, D. D.; Li, J.; Duan, Y. B.; Li, H.; Yang, Y. C. *et al.* Unravelling mitochondrial retrogradere gelation in the abiotic stress induction of rice ALTERNATIVE OXIDASE1 genes. *Plant Cell Environ.* **2013,** 36, 775-788.

68. Lin, F.; Ding, H.; Wang, J.; Zhang, H.; Zhang, A.; Zhang, Y. *et al.* Positive feedback regulation of maize NADPH oxidase by mitogen-activated protein kinase cascade in abscisic acid signalling. *Journal of Experimental Botany.* **2009,** 60, 3221-3238.

69. Liu, S.; Wang, M.; Wei, T.; Meng, C.; Xia, G. A wheat SIMILAR TO RCD-ONE gene enhances seedling growth and abiotic stress resistance by modulating redox homeostasis and maintaining genomic integrity. *Plant Cell.* **2014,** 26, 164-180.

70. Lu, W.; Chu, X.; Li, Y.; Wang, C.; Guo, X. Cotton GhMKK1 induces the tolerance of salt and drought stress, and mediates defence responses topathogen infection in transgenic *Nicotianabenthamiana*. *PLoSONE.* **2013,** 8:e68503.

71. Ma, F.; Lu, R.; Liu, H.; Shi, B.; Zhang, J.; Tan, M. *et al.* Nitric oxide-activated calcium/calmodulin-dependent protein kinase regulates the abscisic acid-induced antioxidant defence in maize. *Journal of Experimental Botany.* **2012,** 63, 4835-4847.

72. Maxwell, D. P.; Wang, Y.; Mcintosh, L. The alternative oxidase lowers mitochondrial reactive oxygen production in plant cells. *Proc. Natl. Acad. Sci. U.S.A.* **1999**, 96, 8271-8276.

73. Meyer, Y.; Belin, C.; Delorme-Hinoux, V.; Reichheld, J.P.; Riondet, C. Thioredoxin and glutaredoxin systems in plants: molecularmechanisms, crosstalks, and functional significance. *Antioxid.RedoxSignal.* **2012**, 17, 1124-1160.

74. Miller, G.; Suzuki, N.; Ciftci-Yilmaz, S.; and Mittler, R. Reactive oxygen species homeostasis and signalling during drought and salinity stresses. *Plant Cell Environ.* **2010**, 33, 453-467.

75. Miller, G.; Shulaev, V.; Mittler, R. Reactive oxygen signalling and abiotic stress. *Physiol. Plant.* **2008**, 133, 481-489.

76. Mittal, D.; Madhyastha, D. A.; Grover, A. Genome-wide transcriptional profiles during temperature and oxidative stress reveal coordinated expression patterns and overlapping regulons in rice. *PLoSONE.* **2012**, 7:e40899.

77. Mittler, R. Oxidative stress, antioxidants and stress tolerance. *Trends Plant Sci.* **2002**, 7, 405-410.

78. Mittler, R.; Vanderauwera, S.; Gollery, M.; Van Breusegem, F. Reactive oxygen gene network of plants. *Trends Plant Sci.* **2004**, 9, 490-498.

79. Mittler, R.; Kim, Y.; Song, L.; Coutu, J.; Coutu, A.; Ciftci-Yilmaz, S. *et al.* Gain-and loss-of-function mutations in Zat10 enhance the tolerance of plants to abiotic stress. *FEBS Lett.* **2006**, 580, 6537-6542.

80. Mizoi, J.; Shinozaki, K.; Yamaguchi-Shinozaki, K. AP2/ERF family transcription factors in plant abiotic stress responses. *Biochim. Biophys. Acta.* **2012**, 1819, 86-96.Mullineaux, P. M.; Baker, N. R. Oxidative stress: antagonistic signalling for acclimation or cell death? *Plant Physiol.***2010**,154: 521-525.

81. Nath, K.; Kumar, S.; Poudyal, R. S.; Yang, Y. N.; Timilsina, R.; Park, Y. S. *et al.* Developmental stage-dependent differential gene expression of superoxide dismutase isoenzymes and their localization and physical interaction network in rice (*Oryza sativa* L.). *Genes Genom.* **2014**, 36, 45-55.

82. Nguyen, H. T.; Cai, S. Jiang, G.; Ye, N.; Chu, Z.; Xu, X. *et al.* A key ABA catabolic gene, OsABA8ox3, is involved in drought stress resistance in rice. *PLoSONE.* **2015**, 10, e0116646.

83. Niyogi, K. K. Photo protection revisited: Genetic and molecular approaches. *Annu Rev Plant Physiol Plant Mol Biol.* **1999**, 50, 333-359.

84. Noctor, G. Metabolic signaling in defense and stress: the central roles of soluble redox couples. *Plant Cell Environ.* **2006**, 29, 409-425.

85. Noctor, G.; Mhamdi, A.; and Foyer, C. H. The roles of reactive oxygen metabolism in drought: notsocut and dried. *Plant Physiol.* **2014,** 164, 1636-1648.

86. Özdemir, F.; Bor, M.; Demiral, T.; Türkan, I. Effects of 24-epi brassinolideon seed germination, seedling growth, lipid peroxidation, proline content and antioxidative system of rice (*Oryza sativa* L.) under salinity stress. *Plant Growth Regul.* **2004,** 42, 203-211.

87. Perez-Ruiz, J. M.; Spinola, M. C.; Kirchsteiger, K.; Moreno, J.; Sahrawy, M.; Cejudo, F. J. Rice NTRC is a high-efficiency redox system for chloroplast protection against oxidative damage. *Plant Cell.* **2006,** 18, 2356-2368.

88. Pourcel, L.; Routaboul, J. M.; Cheynier, V. Flavonoid oxidation in plants: from biochemical properties to physiological functions. *Trends Plant Sci.* **2007,** 12(1):29-36.

89. Qiao, B.; Zhang, Q.; Liu, D.; Wang, H.; Yin, J.; Wang, R. *et al.* A calcium-binding protein, rice annexin OsANN1, enhances heat stress tolerance by modulating the production of H_2O_2. *Journal of Experimental Botany.* **2015,** 66, 5853-5866.

90. Raghavendra, A. S.; Gonugunta, V. K.; Christmann, A.; Grill, E. ABA perception and signalling. *Trends Plant Sci.* **2010,** 15, 395-401.

91. Ramegowda, V.; Senthil-Kumar, M.; Nataraja, K. N.; Reddy, M. K.; Mysore, K. S.; Udayakumar, M. Expression of a finger millet transcription factor, EcNAC1, in tobacco confers abiotic stress-tolerance. *PLoSONE.* **2012,** 7, e40397.

92. Reddy, V. S.; Day, I. S.; Thomas, T.; Reddy, A. S. KIC, a novel Ca^{2+} binding protein with one EF-hand motif, interacts with a microtubule motor protein and regulates trichome morphogenesis. *Plant Cell.* **2004,** 16, 185-200.

93. Rio, L. A. D.; Sandalio, L. M.; Corpas, F. J. Reactive oxygen species and reactive nitrogen species in peroxisomes. Production, scavenging and role in cell signaling. *Plant Physiology.* **2006,** 141, 330-335.

94. Rushton, P. J.; Somssich, I. E.; Ringler, P.; Shen, Q. J. WRKY transcription factors. *Trends Plant Sci.* **2010,** 15, 247-258.

95. Sagi, M.; Fluhr, R. Production of reactive oxygen species by plant NAPDH oxidases. *Plant Physiology.* **2006,** 141:336-340.

96. Sato,Y.; Masuta, Y.; Saito, K.; Murayama, S.; Ozawa, K. Enhanced chilling tolerance at the booting stage in rice by transgenic over expression of the ascorbate peroxidase gene, OsAPXa. *Plant Cell Rep.* **2011,** 30, 399-406.

97. Schmidt, R.; Mieulet, D.; Hubberten, H. M.; Obata, T.; Hoefgen, R.; Fernie, A. R.; *et. al.* Salt-responsive ERF1 regulates reactive oxygen species-dependent signalling during the initial response to salt stress in rice. *Plant Cell.* **2013,** 25, 2115-2131.

98. Selote, D. S.; Khanna-Chopra, R. Antioxidant response of wheat roots to drought acclimation. *Protoplasma.* **2010,** 245, 153-163.

99. Setter, T. L.; Bhekasut, P.; Greenway, H. Desiccation of leaves after de-submergence is one cause for intolerance to complete submergence of the rice cultivar IR42. *Functional Plant Biol.* **2010,** 37, 1096-1104.

100. Shao, H. B.; Chu, L.Y. Plant molecular biology in China: opportunities and challenges. *Plant Mol. Biol. Rep.* 2005, 23, 345-358.

101. Shao, H. B.; Chu, L. Y.; Zhao, C. X.; Guo, Q. J.; Liu, X. A.; Ribaut, J. M.) Plant gene regulatory network system under abiotic stress. *Acta Biol.* (Szeged). **2006,** 50(1-2), 1-9.

102. Shao, H. B.; Guo, Q. J.; Chu, L. Y.; Zhao, C. X.; Su, Z. L.; Hu, Y. C.; Cheng, J. F.) Understanding molecular mechanism of higher plant plasticity under abiotic stress. *Colloids Surf B.* **2007a,** 54(1), 37-45.

103. Shao, H. B.; Jiang, S. Y.; Li, F. M.; Chu, L. Y.; Zhao, C. X.; Shao, M. A.; Zhao, X. N.; Li, F. Some advances in plant stress physiology and their implications in the systems biology era. *Colloids Surf B.* **2007b,** 54(1):33-36.

104. Sharma, P.; Jha,A. B.; Dubey,R. S.; Pessarakli, M.; Reactive Oxygen Species, Oxidative Damage, and Antioxidative Defense Mechanism in Plants under Stressful Conditions. *Journal of Botany.* **2012,** 2012, 1-26

105. Sharma, P.; Dubey, R. S. Ascorbate peroxidase from rice seedlings: properties of enzyme isoforms, effects of stresses and protective roles of osmolytes. *Plant Sci.* **2004,** 167, 541-550.

106. Shcolnick, S.; Keren, N. Metal homeostasis in cyanobacteria and chloroplasts to balancing benefits and risks to the photosynthetic apparatus. *Plant Physiology.* **2006,** 141:805-810.

107. Shi, B.; Ni, L.; Liu, Y.; Zhang, A.; Tan, M.; Jiang, M. OsDMI3-mediated activation of OsMPK1 regulates the activities of antioxidant enzymes in abscisic acid signalling in rice. *Plant Cell Environ.* **2014,** 37, 341-352.

108. Shi, B.; Ni, L.; Zhang, A.; Cao, J.; Zhang, H.; Qin, T.; *et. al.* OsDMI3 is a novel component of abscisic acid signalling in the induction of antioxidant defense in leaves of rice. *Mol. Plant.* **2012,** 5, 1359-1374.

109. Shukla, D.; Huda, K. M.; Banu, M. S.; Gill, S. S.; Tuteja, R.; Tuteja, N. OsACA6, a P-type 2BCa^{2+} ATPase functions in cadmium stress tolerance

in tobacco by reducing the oxidative stress load. *Planta.* **2014,** 240, 809-824.

110. Sun, J.; Hu, W.; Zhou, R.; Wang, L.; Wang, X.; Wang, Q. *et. al.* The *Brachypodiumdistachyon*BdWRKY36 gene confers tolerance to drought stress in transgenic tobacco plants. *Plant Cell Rep.* **2015,** 34, 23–35.

111. Sun, S. J.; Guo, S. Q.; Yang, X.; Bao, Y. M.; Tang, H. J.; Sun, H. *et. al.* Functional analysis of a novel Cys2/His2-type zinc finger protein involved in salt tolerance in rice. *Journal of Experimental Botany.* **2010,** 61, 2807-2818.

112. Suzuki, N.; Koussevitzky, S.; Mittler, R.; Miller, G. ROS and redox signalling in the response of plants to abiotic stress. *Plant Cell Environ.* **2012,** 35, 259-270.

113. Tang, B.; Xu, S. Z.; Zou, X. L.; Zheng, Y. L.; Qiu, F. Z. Changes of antioxidative enzymes and lipid peroxidation in leaves and roots of water logging-tolerant and water logging-sensitive maize genotypes at seedling stage. *Agric. Sci. China.* **2010,** 9, 651-661.

114. Teixeira, F. K.; Menezes-Benavente, L.; Galvao, V. C.; Margis, R.; Margis-Pinheiro, M. Rice ascorbate peroxidase gene family encodes functionally diverse isoforms localized in different sub cellular compartments. *Planta.* **2006,** 224, 300-314.

115. Teixeira, F. K.; Menezes-Benavente, L.; Margis, R.; Margis- Pinheiro, M. Analysis of the molecular evolutionary history of the ascorbate peroxidase gene family: inferences from the rice genome. *Journal Mol. Evol.* **2004,** 59, 761–770.

116. Turan, O.; Ekmekci, Y. Activities of photosystem II and antioxidant enzymes in chick pea (*Cicerarietinum*L.) cultivars exposed to chilling temperatures. *Acta Physiol. Plant.* **2011,** 33, 67-78.

117. Tuteja, N.; Sahoo, R. K.; Garg, B.; Tuteja, R. OsSUV3 dual helicase functions in salinity stress tolerance by maintaining photosynthesis and antioxidant machinery in rice (*Oryzasativa*L.cv.IR64). *Plant Journal.* **2013,** 76, 115-127.

118. Ulker, B.; Somssich, I. E. WRKY transcription factors: from DNA binding towards biological function. *Curr.Opin.Plant Biol.* **2004,** 7, 491-498.

119. Wang, B. Q.; Zhang, Q. F.; Liu, J. H.; Li, G. H. Over expression of PtADC confers enhanced dehydration and drought tolerance in transgenic tobacco and tomato: effect on ROSelimination. *Biochem. Biophys. Res. Commun.* **2011**, 413, 10-16.

120. Wang, F.; Chen, H. W.; Li, Q. T.; Wei, W.; Li, W.; Zhang, W. K. *et. al.* GmWRKY27 interacts with GmMYB174 to reduce expression of GmNAC29 for stress tolerance in soybean plants. *Plant Journal.* **2015**, 83, 224-236.

121. Wen, F.; Qin, T.; Wang, Y.; Dong, W.; Zhang, A.; Tan, M.;*et. al.* OsHK3 is a crucial regulator of abscisic acid signalling involved in antioxidant defense in rice. *Journal Integr. Plant Biol.* **2015**, 57, 213-228.

122. Wong, H. L.; Pinontoan, R.; Hayashi, K.; Tabata, R.; Yaeno, T.; Hasegawa, K. *et. al.* Regulation of rice NADPH oxidase by binding of RacGTPase to its N-terminal extension. *Plant Cell.* **2007**, 19, 4022-4034.

123. Wu, G.; Wei, Z. K.; Shao, H. B. The mutual responses of higher plants to environment: physiological and microbiological aspects. *Colloids Surf B.* **2007**, 59:113-119.

124. Wu, Y. S.; Tang, K. X. MAP Kinase cascades responding to environmental stress in plants. *Acta Bot. Sin.***2004,** 46(2), 127-136.

125. Wu, L.; Zhang, Z.; Zhang, H.; Wang, X. C.; Huang, R. Transcriptional modulation of ethylene response factor protein JERF3 in the oxidative stress response enhances tolerance of tobacco seedlings to salt, drought, and freezing. *Plant Physiology.* **2008**, 148, 1953-1963.

126. Xia, X. J.; Wang, Y. J.; Zhou, Y. H.; Tao, Y.; Mao, W. H.; Shi, K.; *et. al.* Reactive oxygen species are involved in brassinosteroid-induced stress tolerance in cucumber. *Plant Physiology.* **2009**, 150, 801-814.

127. Yabuta, Y.; Maruta, T.; Yoshimura, K. Two distinct redox signaling pathways for cytosolic APX induction under photo-oxidative stress. *Plant Cell Physiol.* **2004,** 45(11), 1586-1594.

128. Yamaguchi-Shinozaki, K.; Shinozaki, K. Transcriptional regulatory networks in cellular responses and tolerance to dehydration and cold stresses. *Annu. Rev. Plant Biol.* **2006,** 57, 781-803.

129. Yan, H.; Jia, H.; Chen, X.; Hao, L.; An, H.; Guo, X. The cotton WRKY transcription factor GhWRKY17 functions in drought and salt stress in transgenic *Nicotianabenthamiana*through ABA signalling and the modulation of reactive oxygen species production. *Plant Cell Physiology.* **2014,** 55, 2060-2076.

130. Yan, J.; Guan, L.; Sun, Y.; Zhu, Y.; Liu, L.; Lu, R.; *et. al.* Calcium and ZmCCaMK are involved in brassinosteroid-induced antioxidant defense in maize leaves. *Plant Cell Physiology.* **2015,** 56, 883-896.

131. Yang, C. J.; Zhang, C.; Lu, Y. N.; Jin, J. Q.; Wang, X. L. The mechanisms of brassinosteroids action: from signal transduction to plant development. *Mol. Plant.* **2011,** 4, 588-600.

132. Ye, N.; Zhu, G.; Liu, Y.; Li, Y.; Zhang, J. ABA controls H_2O_2 accumulation through the induction of OsCATB in rice leaves under water stress. *Plant Cell Physiology.* **2011,** 52, 689-698.

133. Zhang, Z.; Zhang,Q.; Wu, J.; Zheng, X.; Zheng, S.; Sun, X. etal. Gene knock out study reveals that cytosolic ascorbate peroxidase2 (OsAPX2) plays a critical role in growth and reproduction in rice under drought, salt and cold stresses. *PLoSONE.* **2013,** 8, e57472.

134. Zhang, A.; Jiang, M.; Zhang, J.; Ding, H.; Xu, S.; Hu, X.; *et. al.* Nitric oxide induced by hydrogen peroxide mediates abscisic acid-induced activation of the mitogen-activated protein kinase cascade involved in antioxidant defense in maize leaves. *New Phytol.* **2007,** 175, 36-50.

135. Zhang, A.; Jiang, M.; Zhang, J.; Tan, M.; Hu, X. Mitogen-activated protein kinase is involved in abscisic acid-induced antioxidant defense and acts down stream of reactive oxygen species production in leaves of maize plants. *Plant Physiology.* **2006,** 141, 475-487.

136. Zhang, A.; Zhang, J.; Ye, N.; Cao, J.; Tan, M.; Jiang, M. ZmMPK5 is required for the NADPH oxidase-mediated self-propagation of apoplastic H_2O_2 in brassinosteroid-induced antioxidant defence in leaves of maize. *Journal of Experimental Botany.* **2010,** 61, 4399-4411.

137. Zhang, C. J.; Zhao, B. C.; Ge, W. N.; Zhang, Y. F.; Song, Y.; Sun, D. Y. *et al.* An apoplastic h-type thioredoxin is involved in the stress response through regulation of the apoplastic reactive oxygen species in rice. *Plant Physiology.* **2011,** 157, 1884-1899.

138. Zhang, H.; Liu, Y.; Wen, F.; Yao, D.; Wang, L.; Guo, J. *et al.* An ovel rice C_2H_2-type zinc finger protein, ZFP36, is a key player involved in abscisic acid-induced antioxidant defence and oxidative stress tolerance in rice. *Journal of Experimental Botany.* **2014,** 65, 5795-5809.

139. Zhang, H.; Ni, L.; Liu, Y.; Wang, Y.; Zhang, A.; Tan, M. *et al.* The C_2H_2-type zinc finger protein ZFP182 is involved in abscisic acid-induced antioxidant defense in rice. *Journal Integr.Plant Biol.* **2012a,** 54, 500-510.

140. Zhang,Y.; Tan, J.; Guo, Z.; Lu, S.; He, S.; Shu, W. *et al.* Increased abscisic acid levels in transgenic tobacco over-expressing9cis-epoxy carotenoid dioxygenase influence H_2O_2 and NO production and antioxidant defences. *Plant Cell Environ.* **2009,**32, 509-519.

141. Zhou, J.; Wang, J.; Li, X.; Xia, X. J.; Zhou, Y. H.; Shi, K. *et al.* H_2O_2 mediates the cross talk of brassino steroid and abscisic acid in tomato responses to heat and oxidative stresses. *Journal of Experimental Botany.* **2014a,** 65, 4371-4383.

142. Zhou, T.; Yang, X.; Wang, L.; Xu, J.; Zhang, X. GhTZF1 regulates drought stress responses and delays leaf senescence by inhibiting reactive oxygen species accumulation in transgenic *Arabidopsis*. *Plant Mol. Biol.* **2014b,** 85, 163-177.

143. Zhu, J. Y.; Sae-Seaw, J.; Wang, Z. Y. Brassino steroid signalling. *Development.* **2013a,**140, 1615-1620.

144. Zhu, Y.; Zuo, M.; Liang, Y.; Jiang, M.; Zhang, J.; Scheller, H.V. *et al.* MAP65-1 a positively regulates H_2O_2 amplification and enhances brassinosteroid-induced antioxidant defense in maize. *Journal of Experimental Botany.* **2013b,** 64, 3787-3802.

Abiotic Stress Tolerance Mechanisms in Plants, Pages 175–202
Edited by: Gyanendra K. Rai, Ranjeet Ranjan Kumar and Sreshti Bagati
Copyright © 2018, Narendra Publishing House, Delhi, India

5

ANTIOXIDANT DEFENSE SYSTEM IN PLANTS AGAINST ABIOTIC STRESS

**Mukesh K. Berwal[1], Ramesh Kumar[1], Krishna Prakash[2],
Gyanendra K. Rai[3] and K.B. Hebbar[2]**

[1]*Division of Crop Production, ICAR-Central Institute for Arid Horticulture, Bikaner*
[2]*Department of Crop Improvement, ICAR-CPCRI, Kasaragod-671124, Kerala*
[3]*School of Biotechnology,
Sher-e- Kashmir University of Agricultural Sciences and Technology of Jammu-180009 (J&K)*
E-mail: mukesh.kumar4@icar.gov.in, mkbiochem@gmail.com

Abstract

This chapter briefly introduces the formation of reactive oxygen species (ROS) as by-products of oxidation/reduction (redox) reactions, and the ways in which the antioxidant defense machinery is involved directly or indirectly in ROS scavenging. Major antioxidants, both enzymatic and non-enzymatic, that protect higher plant cells from oxidative stress damage are described. Biochemical and molecular features of the antioxidant enzymes superoxide dismutase (SOD), catalase (CAT), and ascorbate peroxidase (APX) are discussed because they play crucial roles in scavenging ROS in the different cell compartments and in response to stress conditions. Among the non-enzymatic defenses, particular attention is paid to ascorbic acid, glutathione, flavonoids, carotenoids, and tocopherols.

Keywords: Reactive oxygen species, ROS, antioxidant enzymes, antioxidant molecules, flavonoids.

1. Introduction

Plants are sessile forms of life and as a consequence they don´t have the capacity to "escape" from unfavorable environmental conditions. Therefore, plants often face the challenge to grow under stressful conditions such as water or light deficit or excess, low or high temperature, salinity, heavy metals, UV rays, insect and pests attack etc. These stresses wield adverse effects on plant growth and development by inducing many metabolic changes, such as the occurrence of an oxidative stress (Diaz-Vivancos *et al.*, 2008; Hernandez *et al.*, 2004a, b). As a principal cause of global crop failure, abiotic stresses decrease average yields for major crops by more than 50 % (Tuteja *et al.,* 2011). Abiotic stresses impact growth, development, and productivity, and significantly limit the global agricultural productivity mainly by impairing cellular physiology/biochemistry via elevating reactive oxygen species (ROS) generation. The production of ROS during these stresses results from pathways such as photorespiration, from the photosynthetic apparatus and from mitochondrial respiration. In addition, pathogens and wounding or environmental stresses (e.g. drought or osmotic stress) have been shown to trigger the active production of ROS by NADPH oxidases (Hammond-Kosack, and Jones 1996; Orozco-Cardenas and Ryan, 1999; Cazale, 1999, Pei *et al.,* 2000). The enhanced production of reactive oxygen species (ROS) during stress can pose a threat to cells but it is also thought that ROIs act as signals for the activation of stress-response and defense pathways (Desikin *et al.,* 2001; Knight and Knight, 2001). Thus, ROS can be viewed as cellular indicators of stress and as secondary messengers involved in the stress-response signal transduction pathway.

However, several anabolic and catabolic processes like photosynthesis and respiration occur as a part of common aerobic metabolism. It has been proved that ROS are generated in different cellular compartments as mitochondria, chloroplasts, peroxisomes, cytoplasm or in the extracellular space, known as apoplast by action of different enzymes (Table 1) (Gill and Tuteja, 2010; Sandalio *et al.,* 2013). In vegetative tissues, approximately 1-2% of total molecular oxygen consumption drives to the creation of ROS in normal conditions. This percentage increases when plants are subjected to stress conditions such as salinity, drought, cold stress or high temperatures. ROS are the species generated through the reduction of molecular oxygen (O_2), that includes some free radicals such as superoxide ($O_2^{\bullet-}$), hydroxyl radical ($OH^{\bullet}$), alkoxyl ($RO^{\bullet}$) and peroxyl ($ROO^{\bullet}$), and non-radical products like hydrogen peroxide (H_2O_2) and singlet oxygen ($^1O_2$), etc. (Halliwell and Gutteridge, 2007; Gill and Tuteja, 2010; Sandalio *et al.,* 2013). ROS generation is an unavoidable part and by-product in various metabolic processes, where 240 μM s^{-1} $O_2^{\bullet-}$, and 0.5 μM H_2O_2 can be observed in plants under optimal growth conditions. Further, abiotic stresses may significantly enhance the generation of varied ROS (and their reaction products) in plant cells, where stressed cells may exhibit accelerated ROS generation up to 720 μM s^{-1} $O2^{\bullet-}$ and 5–15 μM H_2O_2 (Mittler, 2011; Hasanuzzaman *et al.,* 2012).

$$O_2 + e^- \longrightarrow O2^{\bullet-} \qquad \text{Superoxide radical}$$

$$O_2^{\bullet-} + H_2O \longrightarrow HO_2^{\bullet} + OH^- \qquad \text{Hydroperoxyl radical}$$

$$HO_2^{\bullet} + e^- + H \longrightarrow H_2O_2 \qquad \text{Hydrogen Peroxide}$$

$$H_2O_2 + e^- \longrightarrow {}^{\bullet}OH + OH^- \qquad \text{Hydroxyl Radical}$$

Table 1: Production and scavenging of reactive oxygen species in plants (modified from Mittler, 2002).

Mechanisms	Organelles	Principal ROS	References
Production			
Photosynthesis ET and PS I or II	Chloroplast	O_2^-	Asada, 1999
Respiration Electron transport chain	Mitochondria	O_2^-	Dat et al., 2000, Maxwell et al., 1999
Glycolate oxidase	Peroxisome	H_2O_2	Corpas et al., 2001
Excited chlorophyll	Chloroplast	O_2^1	Asada and Takahashi, 1987
NADPH oxidase	Plasma membrane	O_2^-	Hammond and Jones, 1996
Fatty acid β-oxidation	Peroxisome	H_2O_2	Corpas et al., 2001
Oxalate oxidase	Apoplast	H_2O_2	Dat et al., 2000
Xanthine oxidase	Peroxisome	O_2^-	Corpas et al., 2001
Amine oxidase	Apoplast	H_2O_2	Allen and Fluhr, 1997
Peroxidase, Mn$^+$, and NADH	Cell wall	H_2O_2	Grant and Loake, 2000
Scavengers			
Enzymatic			
Superoxide dismutase	Chl, Cyt, Mit, Per, Apo	O_2^-	Bowler et al., 1992
Ascorbate peroxidase	Chl, Cyt, Mit, Per, Apo	H_2O_2	Asada and Takahashi, 1987
Catalase	Per	H_2O_2	Willekens et al., 1997
Glutathion peroxidase	Cyt	H_2O_2, ROOH	Dixon et al., 1998
Peroxidase	CW, Cyt, Vac	H_2O_2	Asada and Takahashi, 1987
Non-enzymatic			
Ascorbic acid	Chl, Cyt, Mit, Per, Apo	H_2O_2, O_2^-	Asada, 1999, Noctor and Foyer, 1998
Glutathione	Chl, Cyt, Mit, Per, Apo	H_2O_2	Asada, 1999, Noctor and Foyer, 1998
α-Tocopherol	Membranes	ROOH, O_2^1	Asada and Takahashi, 1987
Carotenoids	Chl	O_2^1	Asada and Takahashi, 1987

Abbreviations: Apo, apoplast; Chl, chloroplast; CW, cell wall; Cyt, cytosol; ET, electron transport; Mit, mitochondria; O_2^1, singlet oxygen; Per, peroxisome; PM, plasma membrane; PS, photosystem; Vac, vacuole.

Oxidative stress involves the increased production of some of the most damaging/reactive molecules in plants called reactive oxygen species (ROS) or reactive oxygen intermediates (ROIs), such as superoxide ($O_2^{\bullet-}$), hydrogen peroxide (H_2O_2), and hydroxyl radicals ($OH^{\bullet}$). ROS's are partially reduced forms of atmospheric oxygen, which are produced in vital processes such as photorespiration, photosynthesis and respiration (Mittler, 2002; Uchida *et al.*, 2002). To produce water in these processes, four electrons are required for perfect reduction of oxygen. But ROS typically results from the transference of one, two and three electrons, respectively, to O2 to form superoxide ($O_2^{\bullet-}$), peroxide hydrogen (H_2O_2) and hydroxyl radical ($HO^{\bullet}$) (Mittler, 2002). These species of oxygen are highly cytotoxic and can seriously react with vital biomolecules such as lipids, proteins, nucleic acids etc. causing lipid peroxidation, protein denaturation and DNA mutation, respectively (Breusegem *et al.,* 2001; Scandalios, 1993; Quiles and Lopez, 2004).

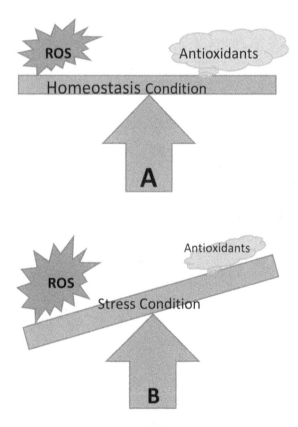

Fig. 1: The concept of homeostasis condition (A) and imbalance (B) between reactive oxygen species (ROS) and antioxidants.

Acclimatization of plants to changes in their environment requires a new state of cellular homeostasis achieved by a delicate balance between multiple pathways that reside in different cellular compartments. Under optimal growth conditions, ROS are mainly produced at a low level, being their rate of production drastically increased when plants are exposed to environmental stresses. Despite the fact that ROS has been considered as toxic molecules during stress conditions, recent studies have shown that ROS play a key role in plants as signal transduction molecules that regulate stress responses, as well as growth and development (Foyer and Noctor, 2005; Miller *et al.*, 2010; Barba-Espin *et al.*, 2011).

2. Antioxidant Defense System in Plants

ROS are continuously produced as byproducts during different metabolic processes in plants. The production and detoxification of ROS needs to be strictly controlled due to their highly reactive nature. ROS are produced in virtually all cell organelles but most notable in mitochondria, chloroplasts, apoplast, and peroxisomes, due to the occurrence of highly oxidizing metabolic processes. Plants have developed various antioxidant defense systems which are specifically localized to different sub cellular loci and are induced upon a stimulus. Diffusion of ROS among different cell compartments is very limited due to their highly reactive nature. This property requires ROS scavenging and detoxification to take place at or near to the location of their production. The ubiquitous occurrence of the sound antioxidant defense system is critical in order to prevent and survive oxidative stress. In plants, ROS production is kept under tight control by an efficient defense system includes avoidance like anatomical adaptations (leaf structure, epidermis etc.), C_4 or CAM metabolism, chlorophyll movements, photosystem (PS) and antenna modulations etc. (Mullineaux and Karpinski, 2002; Mittler, 2001)) and antioxidative system, which includes both enzymatic and non-enzymatic metabolites/scavengers that modulates intracellular ROS concentration under the cellular redox homeostasis. The antioxidant defense system is distributed in different cell organelles such as chloroplasts, mitochondria, peroxisomes, or apoplast. Plant antioxidant defense machinery, comprising antioxidant enzymes and nonenzymatic antioxidant components metabolize ROS and their reaction products in order to avert oxidative stress condition (Gill and Tuteja, 2010; Hasanuzzaman *et al.*, 2012). Among these non-enzymatic scavengers including low molecular weight compounds, ascorbic acid (AsA), glutathione (GSH), phenolic compounds, alkaloids, proline, and α-tocopherols. On the other hand, the enzymatic antioxidants like superoxide dismutase (SOD), catalase (CAT), peroxidase (POX), ascorbate peroxidase (APX), glutathione reductase (GR) play important role in creating tolerance against oxidative stress by scavenging reactive oxygen species. Under environmental stress conditions antioxidants function as redox buffers that interact with ROS and act as a metabolic interface that modulate the appropriate induction of acclimation/tolerance responses (Foyer and Noctor, 2005).

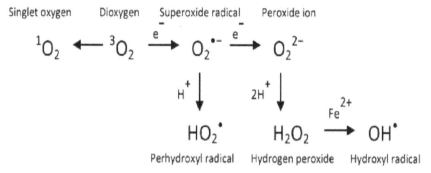

Fig. 2: **Reactive Oxygen Species Production in Plants (Adapted from Gill and Tuteja, 2010).**

2.1. Avoidance of ROS Production

Avoidance of ROS production could be as important as scavenging of ROS by antioxidant defense system. Because many abiotic stresses accompanied by an increasing rate of ROS production, avoiding or alleviating the effects of stresses such as drought or high light on plant metabolism will reduce the ROS production (Mittler, 2002). Adaptations in plants that might reduce ROI production during stress include: (1) anatomical adaptations such as leaf movement and curling, development of a refracting epidermis and hiding of stomata in specialized structures; (2) physiological adaptations such as C_4 and CAM metabolism; and (3) molecular mechanisms that rearrange the photosynthetic apparatus and its antennae in accordance with light quality and intensity or completely suppress photosynthesis (Mullineaux and Karpinski, 2002, Mittler, 2001). By balancing the amount of light energy absorbed by the plant with the availability of CO_2, these adaptations might represent an attempt to avoid the over-reduction of the photosynthetic apparatus and the transfer of electrons to O_2 rather than for CO_2 fixation. ROS production can also be reduced through the alternative channeling of electrons in the electron transport chains of the mitochondria and chloroplasts by a group of enzymes called alternative oxidases. Alternative oxidases (AOX) can divert electrons flowing through electron-transport chains and use them to reduce O_2 to water. Thus, they decrease ROS production by two mechanisms: they prevent electrons from reducing O_2 into O_2^- and reduce the overall level of O_2, the substrate for ROS production, in these organelles (Figure 3). Decreasing the amount of mitochondrial AOX increases the sensitivity of plants to oxidative stress (Maxwell *et al.,* 1999).

2.2. Non enzymatic ROS Scavengers

2.2.1. Ascorbic acid (Vitamin C)

Vitamin C (L-ascorbic acid) is most abundant, water-soluble antioxidant that predominantly acts to prevent or reduced the damage caused by ROS in Plants (Smirnoff 2005; Athar *et al.,* 2008). It is present in all plant parts, usually more in photosynthetic tissues, meristems and some fruits. Mature leaves with fully developed chloroplast and highest chlorophyll are rich in ascorbic acid concentration and remains in reduced form at normal physiological conditions (Smirnoff, 2000). It has been reported that about 30-40 % of ascorbic acid is present in chloroplast with higher stromal concentration (up to 50mM) (Foyer and Noctor, 2005). In plants, mitochondria are the main site for ascorbicacid metabolism, not only by synthesizing by L- galactono-γ-lactone dehydrogenase but also by regenerating it from its oxidized forms (Szarka *et al.,* 2007). The regeneration of ascorbic acid is extremely important because its fully oxidized form dehydroascorbic acid has a short half-life and would have been lost if not converted to reduced form back. It is considered as most powerful ROS scavenger due to its higher concentrations than other antioxidants in plants as well as its ability to donate electrons in a number of enzymatic as well as non-enzymatic reactions. Because of its high concentration, ascorbic acid serves as the major contributor to the cellular redox state and is important in maintaining photosynthetic function (Asada, 1994). Ascorbic acid is used to detoxify reactive oxygen species (ROS), for example, singlet oxygen ($^1O_2$), superoxideanion (O_2^-), hydroxyl radical and hydrogen peroxide in order to protect the photosynthetic apparatus and other cellular components from oxidative damage.

Ascorbic acid is also used by APX to convert H_2O_2 to water, and can directly scavenge other ROS that are produced during aerobic metabolic processes such as photosynthesis or respiration (Asada and Takahashi, 1987). Ascorbic acid is also important in determining the level of tolerance to many environmental stresses, including chilling, drought, salt, and exposure to heavy metals. For example, the vtc1 mutant (which has reduced level of ascorbic acid biosynthesis in cell) has reduced tolerance to 200 mMNaCl as determined by CO_2 assimilatory capacity and PSII function. Under salt stress, vtc1 plants had higher levels of H_2O_2 relative to wild-type plants despite having an elevated glutathione pool (Haung *et al.,* 2005).

2.2.2. Glutathione (GSH)

The tripeptide glutathione (γ-l-glutamyl-l-cysteinyl-glycine), the main non-protein thiol present in cells, is anabolized in the cytosol in two ATP dependent steps.

First, the formation of γ-glutamyl cysteine from glutamate and cysteine, mediated by the enzyme γ-glutamylcysteinesynthetase, followed by the formation of GSH by the activity of GSH synthetase, which uses γ-glutamyl cysteine and glycine as substrates (Meri *et al.*, 2007). Despite its exclusive synthesis in the cytosol, GSH is distributed in all cell compartments like endoplasmic reticulum (ER), nucleus, mitochondria, choloplast, vacuole, peroxisomes as well as in apoplast (Del Rio *et al.*, 2003; Ferreira *et al.*, 2002).The compartmentalization of GSH includes separate redox pools that are distinct from the cytoplasmic pool in terms of the balance between GSH and GSSG forms, their redox potential, and their control of cellular activities. GSH is one of the crucial metabolite in plants which is considered as most important intracellular ROS scavenger as well as it plays a central role in

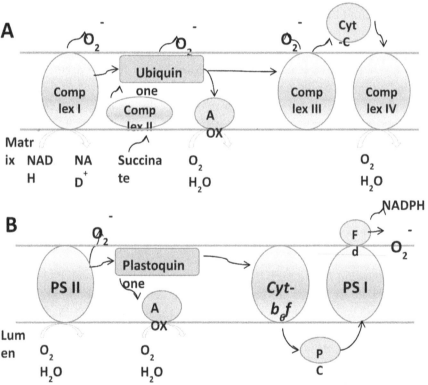

Fig. 3: Role of alternative oxidase (AOX) in reactive oxygen species (ROS) avoidance. In both the mitochondrial electron-transport chain (A) and the chloroplast electron-transport chain (b), AOX diverts electrons that can be used to reduce O_2 into O_2^- and uses these electrons to reduce O_2 to H_2O. In addition, AOX reduces the overall level of O_2, the substrate for ROS production, in the organelle. AOX is indicated in yellow and the different components of the electron-transport chain are indicated in brown, green or blue. Abbreviations: $Cytb_6f$, cytochrome b_6f; Cyt-C, cytochrome C ; Fd, ferredoxin; PC, plastocyanin; PSI, PSII, photosystems I and II. (Adapted from Mittler, 2002)

several physiological and developmental processes like signal transduction, sulphate transport regulation, detoxification of xenobiotics stress responsive gene expression, protection of thiol group (Gill and Tuteja, 2010; Alscher *et al.,* 2002, Sandalio *et al.,* 1988), cell differentiation, apoptosis, pathogen resistance, enzymatic regulations etc. (Scandalias, 1990).

The balance between GSH and GSSH play central role in maintaining the redox state (Foyer and Noctor, 2005) and GSH play critical role in maintaining the normal reduced state of the cell to counteract against the deleterious effect of ROS induced oxidative stress (Harinasut *et al.,* 2003). It is a potential scavenger of singlet oxygen radical ($^1O_2$) hydrogen peroxides (H_2O_2) and Hydroxyl Radicals (OH$^•$) (Noctor and Foyer, 1998, Gapinska *et al.,* 2008). A number of studies about the role of GSH in antioxidant defense battery of plant supports for its use as stress marker as its cellular concentration is correlated with the abiotic stress tolerance level of the plant. Pietrini *et al.,* (2003) reported the enhanced level of antioxidant in the leaves of *Phragmiteaustralis*Trinwas associated with higher accumulation of GSH which resulted in protection of photosynthetic enzymes against

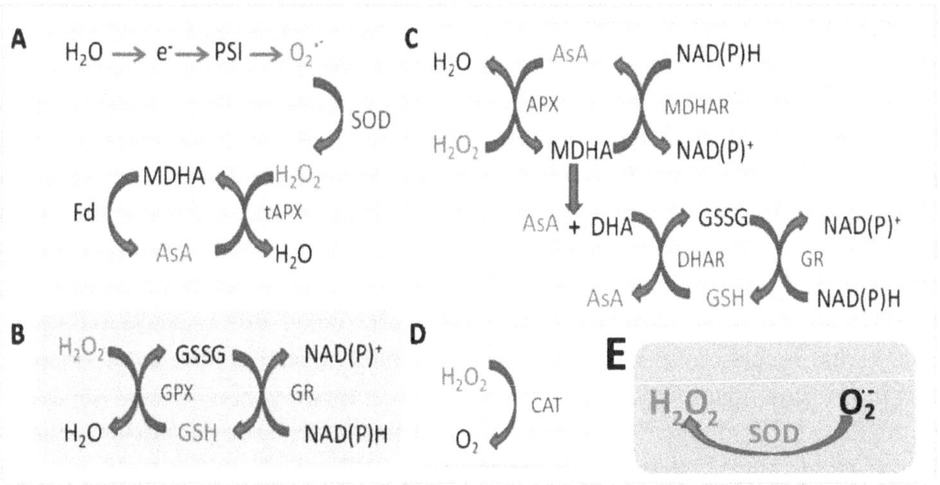

Fig. 4: Diverse ROS scavenging pathways in plants: A. water-water cycle (Mehler reaction); B. ascorbate–glutathione cycle; C. glutathione peroxidase cycle; D. Catalase (CAT); E. Superoxide dismutase (SOD). Superoxide dismutase (SOD) acts as the first line of defense converting $O_2^{•-}$ into H_2O_2, then ascorbate peroxidases (APX), glutathione peroxidases (GPX) and catalases eliminate H_2O_2. In contrast to CAT, both APX and GPX require ascorbate (AsA) or glutathione (GSH) regenerating cycles that use electrons from the photosynthesis (A) or NAD(P)H (B, C) as reducing power. Abbreviations: DHA - dehydroascorbate; DHAR - DHA reductase; Fd - ferredoxin; GR - glutathione reductase; GSSG - oxidized glutathione; MDA - monodehydroascorbate; MDAR - MDA reductase; PSI - photosystem I; tAPX - thylakoid-bound APX (Modified from Racchi, 2013).

thiophillic bursting of cadmium. Many other researchers also reported increased pool of GSH with increased cadmium concentration in *P. sativum* (Metwally *et al.*, 2005), *Sedum alfredii* (Rausch and Wachter, 2005), *Vignamungo* (Molina *et al.*, 2008). Sumithra *et al.*, (2006) observed higher antioxidant enzyme activity as well as GSH concentration in the leaves of Pusa Bold than CO4 cvs. of*Vignaradiata*which indicate that Pusa bold has efficient ROS scavenger capacity, therefore it can better withstand under the abiotic stress conditions than CO4. Xiang *et al.*, (2001) also reported that plants with lower GSH pool were highly sensitive towards even lower cadmium concentration than that of higher GSH pool plants.

2.2.3. α-*Tocopherols*

Tocopherols, (also known as vitamin E) is an important lipid soluble natural antioxidant, are considered as potential scavengers for ROS specifically lipid radicals (Holander-Czytko *et al.*, 2005). Natural vitamin E is known as α, β, γ, and ä according to the methyl or proton groups that are bound to their Benzene rings and the most common and biologically active form is alpha-tocopherol (Brigelius-Flohe *et al.*, 1999). ROS scavenging reactions of α-tocopherol take place via the α -tocopheroxyl radical as an intermediate, if a suitable ROS is present, a non-radical product can be formed by the coupling of the free radical with the α-tocopheroxyl radical. α-Tocopherol are considered as bio-membrane's antioxidants, specifically present in thylakoid membrane of chloroplast, where it protect the membrane stability by quenching ROS like singlet oxygen ($^1O_2$). A higher level occurrence of α-tocopherol was observed in many plant species like *Arabidopsos*, tobacco, *Brassica* etc. but is low in γ-tocopherol under different abiotic stress conditions. Tocopherol prevents the chain propagation step in lipid autooxidation and it has been quantified that one molecule of α-tocopherol is able to scavange up to 120 singlet oxygen ($^1O_2$) molecules by resonance energy transfer (Munn-Bosch *et al.*, 2005). It has been reported that under different oxidative stress conditions that triggers the expression of genes of tocopherol biosynthesis pathways as well as higher level of α-tocopherol, ascorbic acid and GSH in higher plants like tomato, Arabidopsis, tobacco etc. (Wu *et al.*,2007; Shao *et al.*, 2007; Bergmuller *et al.*, 2003; Giacomelli *et al.*, 2007). Srivastava *et al.*, (2005) observed an induction in α-tocopherol content under NaCl and copper stress. Trebst *et al.*, (2002) observed a quick disappearance of α-tocopherol and PS II activity under high light at blockage of α-tocopherol biosynthesis at the 4-hydroxyphenylpyruvate dioxygenase by the application of pyrazolynate herbicide, and they concluded that α-tocopherol is essential component to keep photosynthesis active.

2.2.4. *Proline*

It is well known that the amino acid proline is a widespread compatible osmo protector in plants, whose dramatic accumulation is observed under salt, drought and metal stress might be due to increase synthesis or decreased degradation. Therefore it is also considered as potent non-enzymatic ROS scavenger antioxidant which is used by plant, animals as well as by microbes to mitigate the adverse effect of ROS (Kuznetsov and Shevyakova, 1999, Chen and Dickman, 2005). Smirnoff (1989), tested the OH$^•$ scavenging capacity of proline along with Sorbitol, Manitol and Myo-inositol and found that proline as most effective OH$^•$ scavenger. Kant *et al.,* (2006) reported a high level of proline and its effective accumulation (> 200 mol/g fresh weight) in response to salt stress (600–700 mMNaCl) were found in *Th. halophila* plants. Increased accumulation of proline has been correlated with improved abiotic stress tolerance level in many plant species. Enhanced level of proline synthesis under drought or salt stress has been implicated as process to enhance cytoplasmic acidosis and maintain the ratio of $NADP^+/NADPH$ at redox state suitable for metabolism (Hare and Cress, 1997). Many other studies are supporting proline as an effective ROS quencher. It has been reported that proline protect the yeast cell from herbiside methyl voilogen and it was observed that the ability of proline to scavange ROS and ability to inhibit ROS mediated apoptosis can be crucial function in response to cellular stress (Chen and Dickman, 2005).

2.2.5. *Carotenoids*

Carotenoids are yellow, orange and red colored lipid soluble pigments that are uniquely synthesized in plants, algae, fungi, and bacteria which play a major role in the protection of plants against photooxidative processes through filtering of blue light (Sandmann, 2001). Carotenoids are 40-carbon isoprenoids with polyene chain that may contain as many as 15 double bonds. Plant carotenoids are encoded by nuclear genes and generally synthesized and located in plastids (Hirschberg, 2001). There are hundreds of carotenoids and most of them are efficient antioxidants; scavenge singlet molecular oxygen and peroxyl radicals. According to their structure most carotenoids exhibit absorption maximum at around 450 to 550 nm and play multiple functions in plant metabolism (Sieferman-Harms, 1987). Carotenoids acts as photoprotectants by quenching ROS before oxidative damage can occur, or by active non-photochemical quenching (NPQ), or heat dissipation of excess light energy (Demmig-Adams *et al.,* 1996; Frank andCogdell, 1996; Croce *et al.,* 1999; Niyogi, 1999).

Because of its importance in human health by preventing a range of diseases, the bioavailability of provitaminA carotenoids and their bio-efficacy have been extensively studied. Approximately 50 carotenoids are called "provitamin A" compounds, because the body can convert them into retinol, an active form of vitamin A (Tang, 2010).

2.3. Enzymatic ROS Scavengers

2.3.1. *Superoxide dismutase (SOD)*

Superoxide dismutase (SOD: EC 1.15.1.1) are ubiquitous metalloenzymes (Fridivich, 1975; Jackson *et. al.,* 1978) that constitute the first line of defense against reactive oxygen species (ROS) and one of the most effective components of the antioxidant defense system in plant cells against ROS toxicity. It constitutes one of the major enzymatic components of detoxification of superoxide radicals by catalyzing its dismutation to H_2O_2 (Hossain *et al.,* 2009).By removing $O_2^{\bullet-}$, SODs decrease the risk of $OH^{\bullet}$ formation via the metal catalyzed Haber-Weiss-type reaction because this reaction has a 10,000-fold faster rate than the spontaneous dismutation (Gill and Tuteja, 2010).This enzyme is unique that its activity determines the concentrations of $O_2^{\bullet-}$ and H_2O_2, the two Haber-Weiss reaction substrates, and it is therefore likely to be central in the antioxidant defense mechanism (Bowler *et al.,* 1992; Kandhari, 2004). The SOD system in higher plants exists in multiple isoforms that are developmentally regulated and highly reactive in response to exogenous stimuli.The significance of all the SODs has been confirmed in the direct or indirect efficient metabolism of different ROS and its reaction products in numerous studies (Alscher *et al.,* 2002; Gill and Tuteja, 2010; Kumar *et al.,* 2014).The multiple SOD isoforms are classified according to the active site metal into three major groups (types): Fe-SOD (iron cofactor), MnSOD (manganese cofactor) and Cu/ZnSOD (copper and zinc as cofactors; copper is the redox active catalytic metal). While in bacteria one more type of SOD called nickel SODs (Ni-SODs) has also been reported by many researchers with nickel as metal co-factor (Reddy and Venkaiah, 1982; Werner *et al.,* 1992; Alscher *et al.,* 2002; Pradedova *et al.,* 2009).

These multiple SOD isoforms are designated to specific cell compartments viz. Fe-SODs are located in plastids, Mn-SODs in mitochondrial matrix and peroxisomes and they have been also found in cell wall, while Cu/Zn-SODs occur in cytosol, peroxisomes, plastids, and possibly extracellular space (Alscher *et al.,* 2002; Kukavica *et al.,* 2009; Pilon *et al.,* 2011; Miller, 2012). All SOD are encoded by nuclear genes and targeted to their respective subcellular localization by a amino terminal guiding sequence. The activity of SOD isozymes can be detected by staining with nitrobluetetrazolium (NBT) and identification can be done on the basis to their sensitivity against KCN, H_2O_2 and $CHCl_3$:CH_3CH_2OH (Fridovich, 1975; Wang *et al.,* 2009; Kumar *et al.,* 2014).

Table 2: Major enzymes involved in ROS scavenging with their catalyzed reaction in plants

Enzyme	EC Number	Subcellular Localization	Reaction Catalyzed	References
Superoxide dismutase (SOD)	1.15.1.1	Cytosol (Cu/Zn-SOD) chloroplast (Cu/Zn-SOD, Fe-SOD) mitochondria and peroxisome (Mn-SOD)	$O_2^{-} + O_2^{-} + 2H+ = 2H_2O_2 + O_2$	Alscher et al., 2002; Del Rio, 2011, Bowler et al., 1992
Ascorbate peroxidase (APX)	1.11.1.11	Chloroplast mitochondria, cytosol	$H_2O_2 + AsA = H_2O + DHA$	Asada and Takahashi, 1987, Santos et al., 1996
Catalase	1.11.1.6	Peroxisomes	$H_2O_2 = H_2O + \frac{1}{2} O_2$	Willekenes et al., 1997, Mhamdi et al., 2010
Peroxidase	1.11.1.7	Cell wall, Cytosol, vacuole	$H_2O_2 + R(OH)_2 = 2H_2O + R(O)_2$	Asada and Takahashi, 1987
Glutathione reductase (GR)	1. 6.4.2	Cytosol, peroxisomes chloroplast, mitochondria	$GSSG + NAD(P)H = 2GSH + NAD(P)^{+}$	Kaur and Hu, 2009

It has been observed under numerous studies that the higher the SOD activity or higher number of isoforms, greater the potential to remove ROS. The up-regulation of SODs is implicated in combating over produced ROS due to biotic or abiotic stresses and has a crucial role in the survival of the plant under stressful environment. Kumar et al. (2011) studied the antioxidant isozyme variability in different genotypes of citrus and kumquat and concluded that kumquat had higher SOD and catalase activities as well as higher number of their isoforms consequently had greater potential to remove reactive oxygen species. It has been also observed in coconut that more drought tolerant genotypes had higher SOD activity as well as higher number of SOD isoforms than the susceptible genotypes (unpublished data). Significant increase in total leaf SOD activities as well as some extra SOD isoforms (in some studies) has been reported in many plant species under various types of abiotic stresses viz. drought, salt, heavy metals (Cu, Cd etc.) in a number of crops like *Arabidopsis*, mulberry (Harinasut et al., 2003), tomato (Gapinska et al., 2008), *Brassica juncea* (Mobin et al., 2007; Li et al., 2009)*Treaticumaestivum*

(Khan *et al.,* 2007), *Hordiumvulgare*(Guo *et al.,* 2004)*Vignamungo* citrus (Kumar *et al.,* 2011) etc. Increased transcript abundance for SOD is observed in response to abiotic and biotic stress to contrast oxidative stress exerts a significant role in stress tolerance. Transgenic plants over-expressing various SOD isoforms exhibit enhanced tolerance to oxidative stress and to various environmental stress factors; these results have been observed in many crops and model species including rice, potato, alfalfa, poplar, *Arabidopsis*, tobacco (Van Camp *et al.,* 1994). There have been many reports of development of stress tolerant plants with increased expression of different SODs viz. over expressed Mn-SOD in GM *Arabidopsis* (Wang *et al.,* 2004) and tomato (Wang *et al.,* 2007) exhibited higher tolerance to salt, Cu/Zn-SOD over expression in tobacco plant exhibited tolerance towards multiple stresses (Badawi *et al.,* 2004), furthermore, Lee *et al.,* (2007) reported that combined over expression of Cu/Zn-SOD and ascorbate peroxidase in GM *Festucaarundinacea* plant exhibited multiple tolerance against drought (Mthylvilogen), H_2O_2, Cu and Cd.

Table 3: Location of different SOD in plant cells

SOD Isozymes	Locations	Sensitivity
Mn-SOD	Mitochondria and peroxisomes	$CHCl_3:CH_3CH_2OH$
Cu/Zn-SOD	Chloroplast, cytosol and apoplast, peroxisomes	H_2O_2 and KCN
Fe-SOD	Chloroplast	H_2O_2, $CHCl_3:CH_3CH_2OH$

2.3.2. *Catalase*

Catalase (EC 1.11.1.6) is atetramericheme (Iron porphyrin) containing enzyme and is ubiquitous in the peroxisome, with the potential to directly dismutates the H_2O_2 into water and oxygen. Together with SOD and hydroxy peroxidases, catalases are part of a defense system for scavenging superoxide radicals (Mittler, 2002; Garg, 2009) and are important in the removal of H_2O_2 generated in peroxisomes as by product of fatty acid oxidation, photorespiration or purine catabolism. Catalase mediated reactions has one of the highest turnover rate, one molecule of catalase can covert approximately 6 million H_2O_2 molecules to water and oxygen per minute (Gill and Tuteja, 2010). Catalases, in most eukaryotes, including mammals and several fungi, monomeric CAT is encoded by a single gene, while in plant, CATs are encoded by a small gene family comprises of three to four isoenzyme genes in one species (Scandalios, 1997). The catalases isozymes are most thoroughly studied in many higher plants such as potato, tomato, cotton, maize, *Brassica*, castor barley and *Arabidopsis* and has been reported as low as

2 isoforms in barley (Azevedo *et al.*, 1998) and as many as 12 isoforms in *Brassica* (Frugoli *et al.*, 1996). On the basis of spatial and temporal expression pattern throughout the life cycle of the plant, catalases isoforms are categorized in three classes as CAT_1, CAT_2 and CAT_3. CAT1 and CAT_2 are localized in peroxisomes and acytosol whereas, CAT_3 in mitochondria. CAT_3s are mainly responsible for removal of H_2O_2 from glyoxysomes and are highly abundant is seed and young seedlings, while almost absent is later developmental stages (Willekens, 1995, Skadsen *et al.*, 1995). Catalase activity is a determining factor in protecting the photosynthetic cells against oxidative stress induced during abiotic stress conditions such as chilling, drought, salt and ozone (Willekens *et al.*, 1994; Dat *et al.*, 2000). Catalases along with APX, is a key enzyme in modulating the level of H_2O_2, which acts downstream of salicylic acid as a second messenger implicated in the signal transduction pathway that in plants leads to the development of systemic acquired resistance (SAR). Variable response of catalase has been reported in different species under oxidative stress for example a significant decline in catalase activity was observed in *Arabidopsis, Glycine max* and, *Capsicum annuum* (Cho and Seo, 2007; Balestrasse *et al.*, 2001; Leon *et al.*, 2002), while a significant increase in sunflower, wheat, *Brassica*, oats (Zhank and Kirkham, 1994; Khan *et al.*, 2007; Mobin and Khan, 2007; Hsu and Kao, 2007). Besides, the response of the antioxidant system is characterized by the degree and duration of stress. For example, in the two grass species activities of SOD, CAT and POD increased under soil drought, but when stress intensified, under severe stress the enzyme activities decreased, which correlated with the chlorophyll loss and destruction of membranes (Fu and Huang, 2001).

2.3.3. *Ascorbate peroxidase (APX)*

The main hydrogen peroxide-detoxification system in plant chloroplasts is the ascorbate-glutathione cycle, in which ascorbate peroxidase is a key enzyme (Asada, 1992). Ascorbate peroxidase is involved in scavenging H_2O_2 to water in water-water cycle and acsorbate-glutathione (ASH-GSH) cycle by utilizing ascorbate as electron donor, consequently, APX plays a role similar to that of CAT; however, the two enzymes exhibit distinctive features (Xu *et al.*, 2008). APX utilizes ascorbate as specific electron donor to reduce H_2O_2 to water with the parallel generation of monodehydroascorbate (MDHA), a univalent oxidant of ascorbate. DHA reductase utilizes glutathione (GSH) as an electron donor to regenerate ascorbate from its oxidized form. Thus, the ascorbate–glutathione cycle located in the stroma of the chloroplast represents the main pathway to prevent the accumulation of toxic levels of H_2O_2 in photosynthetic organisms. The importance of APX and ascorbate-glutathione cycle is not restricted to chloroplasts; it also plays a role in ROS scavenging in cytosol, mitochondria and peroxisomes (Asada, 1992, 1999; Mittler *et al.*, 2004; Noctor and Foyer, 1998; Shigeoka *et al.*, 2002). APX has higher affinity for H_2O_2 than catalase and peroxidase, consequently may play more important role in scavenging the ROS during oxidative stress. In response to environmental stress like salt, heavy metal, drought etc., increased APX activity in plants has been demonstrated along with other enzymes activities, such as CAT, SOD, and GSH reductase (Shigeoka *et al.*, 2002). A significant increased leaf APX activity has been recorded under heavy metal (Cadmium) stress in wheat (Khan *et al.*, 2007), *Brassica* (Mobin and Khan, 2007) and *Vignamungo* (Singh *et al.*, 2008). Sharma and dubey (2005) reported increased level of chloroplatic APX level while the activity decreased under heavy drought in rice. Under water stress a significant higher level of APX activity was recorded in *Phaseolus vulgaris* and *Phaseolusasperata*(Zlaev *et al.*, 2008; Yang *et al.*, 2008). It has also been demonstrated in higher plants that higher APX activity or over expression of chloroplastic APX is linked with abiotic stress tolerance (Yang *et al.*, 2009). Badavi *et al.* (2004) has reported enhanced plant tolerance against drought and slat stress under over expressed chloroplastic APX in tobacco plant. APX plays a cooperative role in protection of each organelle. High and low temperatures, high light, drought and salt stress, heavy metals and pathogen attack modulate APX gene transcription (Caverzan *et al.*, 2012). The over-expression in transgenic plant of different APX genes has defined the specific action of single genes in stress tolerance. Transgenic rice plants over-expressing a cytosolic APX1 gene (OsAPXa), exhibited enhanced cold tolerance at the booting stage (Sato *et al.*, 2011). The over-expression in tobacco of a tomato chloroplastictAPX gene was effective in minimizing photo-oxidative damage during high and low temperature stress (Sun *et al.*, 2010 a, b)

2.3.4. *Glutathione reductase*

Glutathione reductase (GR) is a flavo-protein oxidoreductase enzyme that catalyses the reduction of glutathione disulphide (GSSG) to the GSH and play critical role in maintained the ascorbate and glutathione pools in a reduced state which is the key to scavenge the ROS produced during oxidative stress (Figure 5). This enzyme uses NADPH as a reductant. GR is predominantly found in chloroplast. However, a small quantity of the enzyme is also present in mitochondria, cytosol, and peroxisomes (Foyer and Halliwell, 1976; Edwards *et al.*, 1990; Jimenez *et al.*, 1997). GR has been most extensively studied among antioxidant defense enzymes (Pooja *et al.*, 2008). The catalytic cycle of GR comprises of two phases; a reductive half-reaction phase during which FAD, (the prosthetic group of GR) is reduced by NADPH, and the oxidative phase, in which the resulting dithiol reacts with the glutathione disulphide and the last electron acceptor, GSSG, is reduced to two molecules of GSH. GR is the key enzyme of ROS detoxifying ascorbate-glutathione pathway (Halliwell-Asada pathway) functions at different cellular compartments viz. chloroplast, mitochondria, cytosol, peroxisomes, and apoplast. GR and GSH play an important role in determining the tolerance of the plant under various stresses (Rao *et al.,* 2008).

Increased level of GR activity has been reported in various plant species under different abiotic stresses. Transgenic plant studies proved that GR plays a prominent role in conferring resistance to oxidative stress caused by drought, ozone, heavy metals, high light, salinity, cold stress, etc. There has also been found an enhanced GR activity in *A. thaliana, Vignamungo, Triticumaestivum, Capsicumannum,* and *Brassicajuncea* following the cadmium treatments (Mobin and Khan, 2007; Singh *et al.,* 2008, Khan *et al.,* 2007). Sharma and Dubey (2005) reported an increased GR activity in *Oryzasativa* seedlings under drought stress. Foyer *et al.* (1995) reported 500-fold higher GR activity in chloroplast of poplar plant transformed with *E. coli*GRthan that of untransformed plants. They also reported two times higher levels of both glutathione and ascorbate in leaves as compared to control plants or plants that expressed the GR gene without the chloroplast-targeting sequence. Although these plants did not show increased protection from MV-induced inhibition of CO_2 assimilation, but they were found to be more resistant to photo inhibition caused by higher light intensity and chilling stress. In one fascinating study, Yoon *et al.*(2005) reported enhanced sensitivity to oxidative stress by transgenic tobacco plants with 30-70 % less GR activity. On the basis of numerous studied it has been suggested that GR plays an important role in regeneration of GSH for GSSH and consequently acts as protecting mechanism against oxidative stress also by maintaining the ascorbate pool (Ding *et al.,* 2009).

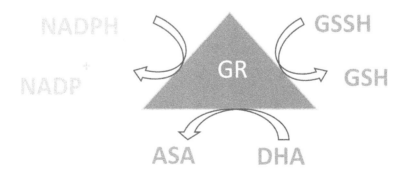

Fig. 5: Molecular mechanism of glutathione reductase (GR) in maintaining cellular redox of ascorbate (ASA) and glutathione (GSH), (Adapted from Yousuf*et al.*, 2012).

3. Conclusion and Future Perspectives

It is well established fact that various abiotic stresses lead to the increased generation of reactive oxygen species (ROS) in plants which are highly reactive and toxic, results in oxidative stress. Oxidative stress is a state in which ROS or free radicals, are generated inter or intra- cellularly, which produce their toxic effect on the cell. These ROS may adversely affect the membrane stability by producing lipid peroxidation, oxidative damage to nucleic acid and protein and make them non functional. However, the cells are outfitted with excellent antioxidant defense system. Antioxidant defense system directly or indirectly concerned with ROS scavenging in plants represents a powerful tool to neutralize oxidative stress at the cellular level. This antioxidant defense could be either non-enzymatic (ascorbate, glutathione, prolines, tocopherol and carotenoids) or enzymatic (superoxide dismutase, catalase, ascorbate peroxidase and glutathione reductase). This antioxidant defense system determined the capacity of crop plant to withstand under abiotic stress, in other words, strong antioxidant defense system leads to make the plant more tolerant against abiotic stress. Over expression of ROS scavenging enzymes (SOD, CAT, APX, GR) and higher level of cellular antioxidant molecules (ascorbate, glutathione, α-tocopherol, carotenoids), resulted in abiotic stress tolerance in various crop plants due to efficient ROS scavenging capacity. It is important to produce plant varieties with the ability to efficiently scavenge and/or control the level of cellular ROS for sustainability of agricultural production under the up-coming harsh environmental condition due to climate change.

4. References

1. Alscher, R.G.; Erturk, N.; Heath, L.S. Role of superoxide dismutases (SODs) in controlling oxidative stress in plants. *Journal of Experimental Botany*, **2002**, 53(372), 1331–1341.

2. Asada, K. Ascorbateperoxidase-Ahydrogenperoxide-scavenging enzyme in plants. *Physiologia Plantarum.* **1992**, 85, 235–241.

3. Asada, K.Mechanisms for scavenging reactive molecules generated in chloroplasts under light stress photo inhibition of photosynthesis: from molecular mechanisms to the field, In: Baker, N.R., Bowyer, J.R. (Eds.). BIOS Scientific Publishers, Oxford UK. **1994**, pp. 129-142.

4. Asada, K. The water cycle in chloroplasts: Scavenging of active oxygen's and dissipation of excess photons. *Annual Review of Plant Physiology and Plant Molecular Biology.***1999**,50, 601-639.

5. Asada, K., Takahashi,M.Production and scavenging of active oxygen in photosynthesis. In: Kyle, D.J. et al., (Eds.), Photo inhibition. Elsevier. **1987**, pp. 227-287.

6. Athar, H. R.; Khan, A.; Ashraf, M. Exogenously applied ascorbic acid alleviates salt-induced oxidative stress in wheat. *Environmental and Experimental Botany.* **2008**, *63*, 224-231.

7. Azevedo, R.A.; Alas, R.M.; Smith, R J.; Lea, P.A.Response of antioxidant enzymes to transfer from elevated carbon dioxide to air and ozone fumigation, in leaves and roots of wild-type and catalase-deficient mutant of barley. *Physiologia Plantarum.* **1998**,*104*, 280-292.

8. Badawi, G.H.; Yamauchi, Y.; Shimada, E.; Sasaki, R.; Kawano, N.; Tanaka, K.; Tanaka, K.Enhanced tolerance to salt stress and water deficit by over expressing superoxide dismutase in tobacco (*Nicotianatabacum*) chloroplasts. *Plant Science.* **2004**, *166*, 919-928.

9. Balestrasse, K.B.; Gardey, L.; Gallego, S.M.; Tomaro, M.L. Response of antioxidant defense system in soybean nodules and roots subjected to cadmium stress. *Australian Journal of Plant Physiology.* **2001**, *28*, 497-504.

10. Barba-Espin, G.Understanding the role of H_2O_2 during pea seed germination: a combined proteomic and hormone profiling approach. *Plant, Cell and Environment.* **2011**,*34*, 1907–1919.

11. Bergmuller, E.;Porfirova, S.;Dormann, P. Characterization of an *Arabidopsis* mutant deficient in g-tocopherol methyl transferase. *Plant Molecular Biology.* **2003**, *52*, 1181-1190.

12. Bowler, C.; Van, Montagu.; M.;Inze, D. Superoxide dismutase and stress tolerance. *Annual Review of Plant Physiology and Plant Molecular Biology.* **1992**, *43*,83-116.

13. Breusegem, F.V.;Vranova, E.;Dat, J.F.;Inze, D. The role of active oxygen species in plant signal transduction. *Plant Science.* **2001**, *161*, 405-414.

14. Brigelius-Flohe, R.; Traber, M.G. Vitamin E: Function and metabolism. *The FASEB Journal.* **1999**, *13*, 1145-1155.

15. Caverzan, A.;Passaia, G.;Barcellos, R.S.; Werner-Ribeiro, C.;Lazzarotto, F.; Margis-Pinheiro, M. Plant responses to stresses: Role of ascorbate peroxidase in the antioxidant protection. *Genetics and Molecular Biology.* **2012**, *35*, 1011-1019.

16. Cazale, A.C.; *etal.*MAP kinase activation by hypo-osmotic stress of tobacco cell suspensions: towards the oxidative burst response? *The Plant Journal.* **1999**, *19*, 297-307.

17. Chen, C.; Dickman, M.B. Proline suppresses apoptosis in the fungal pathogen *Colletotrichumtrifolii. PNAS.* **2005**, *102*, 3459-3464.

18. Cho, U.; Seo, N. Oxidative stress in *Arabidopsis thaliana* exposed to cadmium is due to hydrogen peroxide accumulation. *Plant Science.* **2005**, *168*, 113-120.

19. Corpas, F.J.; *et al.* Peroxisomes as a source of reactive oxygen species and nitric oxide signal molecules in plant cells. *Trends in Plant Science.* **2001**, 6, 145–150.

20. Croce. R.; Weiss, S.; Bassi, R. Carotenoid-binding sites of the major light-harvesting complex II of higher plants. *Journal of Biological chemistry.* **1999**, 274, 29613–23.

21. Dat, J.;V andenabeele, S.; Vranovα, E.; van Montagu, M.; Inzé, D.; van Breusegem, F. Dual action of the active oxygen species during plant stress responses. *Cellular and Molecular Life Sciences.* **2000**, 57, 779-795.

22. Del Rio, L.A. Peroxisomes as a cellular source of reactive nitrogen species signal molecules. *Archives of Biochemistry and Biophysics.* **2011**, 506 (1),1-11.

23. Del Rio, L.A.; Sandalio, L.M.; Altomare, D.A.; Zilinskas, B.A. Mitochondrial and peroxisomal manganese superoxide dismutase: differential expression during leaf senescence. *Journal of Experimental Botany.* **2003**, 54, 923-933.

24. Demmig-Adams, B.; Gilmore, A.M.; Adams III, W.W. *In vivo* function of carotenoids in higher plants. *The FASEB Journal.* **1996**, 10, 403-412.

25. Desikin, R.;etal. Regulation of the *Arabidopsis*transcriptosome by oxidative stress. *Plant Physiology.* **2001**, 127, 159-172.

26. Diaz-Vivancos, P. Alteration in the chloroplastic metabolism leads to ROS accumulation in pea plants in response to Plum pox virus. *Journal of Experimental Botany.* **2008**, 59, 2147-60.

27. Ding, S.; Lu, Q.; Zhang, Y.; Yang, Z.; Wen, X.; Zhang, L.; Lu C.Enhanced sensitivity to oxidative stress in transgenic tobacco plants with decreased glutathione reductase activity leads to a decrease in ascorbate pool and ascorbateredoxstate. *Plant Molecular Biology.* **2009**, 69, 577–592.

28. Dixon, D.P. etal. Glutathione-mediated detoxification systems in plants. *Current Opinion in Plant Biology.* **1998**, 1, 258–266.

29. Edwards, E.A.;Rawsthorne, S.; Mullineaux, P.M. Subcellular distribution of multiple forms of glutathione reductase in leaves of pea (*Pisumsativum* L.). *Planta.* **1990**, 180, 278-284.

30. Ferreira, R.R.; Fornazier, R.F.; Vitoria, A.P.; Lea, P.J.; Azevedo, R.A. Changes in antioxidant enzyme activities in soybean under cadmium stress. *Journal of Plant Nutrition.* **2002**, 25, 327-342.

31. Foyer, C.H.; Halliwell, B. The presence of glutathione and glutathione reductase in chloroplasts: a proposed role in ascorbic acid metabolism. *Planta.* **1976**, 133, 21-25.

32. Foyer, C.H.;Noctor, G. Redox homeostasis and antioxidant signalling: a metabolic interface between stress perception and physiological responses. *The Plant Cell.* **2005**, 17, 1866-1876.

33. Foyer, C.H.;Lelandais, M.;Galap, C.; Kunert, K.J. Effects of elevated cytosolic glutathione reductase activity on the cellular glutathione pool and photosynthesis in leaves under normal and stress conditions. *Plant Physiology.* **1991**, 97, 863-872.

34. Foyer, C.H.;Souriau, N.;Perret, S.;Lelandais, M.;Kunert, K.J.;Pruvost, C.;Jouanin, L. Over expression of glutathione reductase but not glutathione synthetase leads to increases in antioxidant capacity and resistance to photo inhibition in poplar trees. *Plant Physiology.* **1995**, 109, 1047-1057.

35. Frank, H.A.; Cogdell, R.J. Carotenoids in photosynthesis. *Photochemical and Photobiology.* **1996**, 63, 257-264.

36. Fridovich, I. Superoxide Dismutase. *Annual Review of Biochemistry.* **1975**, 44, 147-59.

37. Frugoli, J.A.;Zhong, H.H.; Nuccio, M.L.; McCourt, P.; McPeek, M.A.; Thomas, T.L.; McClung, C.R. Catalase is encoded by a multi gene family in *Arabidopsis thaliana* (L.). *Plant Physiology.* **1996**, 112, 327-336.

38. Fu, J.; Huang, B. Involvement of antioxidants and lipid peroxidation in the adaptation of two cool season grasses to localized drought stress. *Environmental and Experimental Botany.* **2001**, 45, 105-114.

39. Gapinska, M.;Sklodowska, M.;Gabara, B. Effect of short and long-term salinity on the activities of antioxidative enzymes and lipid peroxidation in tomato roots. *ActaPhysiologiae Plantarum.* **2008**, 30, 11-18.

40. Garg, N.; Manchanda, G. ROS generation in plants: boon or bane? *Plant Biosystems.* **2009**, 143, 8-96.

41. Giacomelli, L.; Masi, A.; Ripoll, D.R.; Lee, M.J.; van Wijk, K.J. *Arabidopsis thaliana* deficient in two chloroplast ascorbate peroxidases shows accelerated light induced necrosis when levels of cellular ascorbate are low. *Plant Molecular Biology.* **2007**, 65, 627-644.

42. Gill, S.S.;Tuteja, N. Reactive oxygen species and antioxidant machinery in abiotic stress tolerance in crop plants. *Plant Physiology and Biochemistry.* **2010**, 48, 909–930.

43. Grant, J.J.; Loake, G.J. Role of reactive oxygen intermediates and cognate redox signaling in disease resistance. *Plant Physiology.* **2000**, 124, 21-29.

44. Guo, T.; Zhang, G.; Zhou, M.; Wu, F.; Chen, J. Effects of aluminium and cadmium toxicity on growth and antioxidant enzyme activities of two barley genotypes with different Al resistance. *Plant and Soil.* **2004**, 258, 241-248.

45. Halliwell, B.; Gutteridge, J.M.C. Free radicals in biology and medicine. Oxford University Press, Oxford. **2007**.

46. Hammond-Kosack, K.E.; Jones, J.D.G. Resistance gene-dependent plant defense responses. *Plant Cell.* **1996**, 8, 1773-1791.

47. Hare, P.D.; Cress, W.A. Metabolic implications of stress-induced proline accumulation in plants. *Plant Growth Regulations.* **1997**, 21, 79-102.

48. Harinasut, P.;Poonsopa, D.;Roengmongkol, K.;Charoensataporn, R.Salinity effects on antioxidant enzymes in mulberry cultivar. *ScienceAsia.* **2003**, 29, 109-113.

49. Hasanuzzaman, M., Hossain, M.A.,daSilva, J.A.T., Fujita, M. Plant responses and tolerance to abiotic oxidative stress: antioxidant defense is a key factor.In: Bandi, V., Shanker, A.K., Shanker, C.,Mandapaka, M. (Eds.), Crop stress and its management: perspectives and strategies. Springer, Berlin. **2012**, pp. 261-316.

50. Hernandez, J.A. Oxidative stress induced by long-term plum pox virus infection in peach (*Prunuspersica* L. cv. GF305). *PhysiologiaPlantarum.* **2004a**, 122, 486-495.

51. Hernandez, J.A. Role of hydrogen peroxide and the redox state of ascorbate in the induction of antioxidant enzymes in pea leaves under excess light stress. *Functional Plant Biology.* **2004b**, 31, 359-368.

52. Hirschberg, J. Carotenoid biosynthesis in flowering plants. *Curr. Opin. Plant Biol.* **2001**, 4, 210-218.

53. Hossain, Z.; Lopez-Climent, M.F.; Arbona, V.; Perez-Clemente, R.M.; Gomez Cadenas, A. Modulation of the antioxidant system in citrus under waterlogging and subsequent drainage. *Journal of Plant Physiology.* **2009**, 166, 1391-1404.

54. Hsu, Y.T.; Kao, C.H. Cadmium toxicity is reduced by nitric oxide in rice leaves. *Plant Growth Regulations.* **2007**, 42, 227–238.

55. Jackson, C.A.; Moore, B.; Halliwell.; Foyer,C.H.; Hall, D.O. Subcellular localization and identification of superoxide dismutase in the leaves of higher plants. *European Journal of Biochemistry.* **1978**, 91, 339-344.

56. Jimenez, A.; Hernandez, J.A.; delRío, L.A.; Sevilla, F. Evidence for the presence of the ascorbate-glutathione cycle in mitochondria and peroxisomes of pea (*Pisumsativum* L.) leaves. *Plant Physiology.* **1997**, 114, 275-284.

57. Kandhari, P. Generic differences in antioxidant concentration in the fruit tissues of four major cultivars of apples. Master Thesis, University of Maryland, College Park. **2004**.

58. Kant, S.; Kant, P.; Raven, E.; Barak, S. Evidence that differential gene expression between the halophytes, *Thellungiella halophile*, and *Arabidopsis thaliana*, is responsible for higher levels of compatible osmolyteproline and higher control of Na$^+$ uptake in *T. halophile*. *Plant, Cell and Environment*. **2006**, 29, 1220-1234.

59. Kaur, N.; Hu, J.Dynamics of peroxisome abundance: a tale of division and proliferation. *Current Opinions in Plant Biology*. **2009**, 12(6), 781-788.

60. Khan, N.A.; Samiullah.; Singh, S.;Nazar, R.Activities of antioxidative enzymes, sulphur assimilation, photosynthetic activity and growth of wheat (*Triticumaestivum*) cultivars differing in yield potential under cadmium stress. *Journal of Agronomy and Crop Science*. **2007**, 193,435-444.

61. Knight, H.; Knight, M.R. Abiotic stress signaling pathways: specificity and cross-talk. *Trends in Plant Science*. **2001**, 6, 262–267.

62. Kukavica, B.;Mojovic, M.;Vucinic, Z.;Maksimovic, V.;Takahama, U.;Veljovic-Jovanovic, S. Generation of hydroxyl radical in isolated pea root cell wall, and the role of cell wall-bound peroxidase, Mn-SOD and phenolics in their production. *Plant and Cell Physiology*. **2009**, 50, 304-317.

63. Kumar, M.;Sugatha, P.; Hebbar, K.B. Superoxide dismutase isozymes and their heat stability in coconut (*Cocusnucifera* L.) leaves. *Annals of Biology*. **2014**, 30(4),593-597.

64. Kumar, N.; Ebel, R.C.;Roberts, P.D. Antioxidant isozyme variability in different genotypes of citrus and kumquat. *Journal of Crop Improvement*. **2011**, 25(1), 86-100.

65. Kuznetsov, V.V.; Shevyakova, N.I. Proline under Stress: Biological role, metabolism, and regulation. *Russian Journal of Plant Physiology*.**1999**, 46, 274–289.

66. Lee S.H.; Ahsan, N.; Lee, K.W.; Kim, D.H.; Lee, D.G.; Kwak, S.S.; Kwon, S.Y.; Kim, T.H.; Lee, B.H. Simultaneous over expression of both Cu/Zn superoxide dismutase and ascorbate peroxidase in transgenic tall fescue plants confers increased tolerance to a wide range of abiotic stresses. *Journal of Plant Physiology*. **2007**, 164, 1626–1638.

67. Leon, A.M.; Palma, J.M.;Corpas, F.J.; Gomez, M.; Romero-Puertas, M.C.; Chatterjee, D.;Mateos, R.M.; del Rio, L.A.; Sandalio, L.M. Antioxidant enzymes in cultivars of pepper plants with different sensitivity to candium. *Plant Physiology and Biochemistry*. **2002**, 40,813-820.

68. Li, Y.; Song, Y.; G. Shi, G.; Wang, J.; Hou, X.Response of antioxidant activity to excess copper in two cultivars of *Brassica campestris* ssp. chinensis Makino, *Acta Physiol. Plant*. **2009**, 31, 155-162.

69. Maxwell, D.P. The alternative oxidase lowers mitochondrial reactive oxygen production in plant cells. *Proceedings of the National Academy of Sciences, U. S. A.* **1999**, 96, 8271-8276.

70. Metwally, A.; Safronova, V.I.;Belimov, A.A.; Dietz, K.J. Genotypic variation of the response to cadmium toxicity in *Pisumsativum* L. *Journal of Experimental Botany.* **2005**, 56, 167-178.

71. Mhamdi, A.;Queval, G.;Chaouch, S.;Vanderauwera, S.; Van Breusegem, F.; Noctor, G. Catalase function in plants: a focus on *Arabidopsis* mutants as stress-mimic models. *Journal of Experimental Botany.* **2010**, 61(15), 4197-4220.

72. Miller, A.F. Superoxide dismutases: Ancient enzymes and new insights. *FEBS Letter.* **2012**, 586, 585-595.

73. Miller, G.;etal. Reactive oxygen species homeostasis and signaling during drought and salinity stresses. *Plant, Cell and Environment.* **2010**, 33, 453-497.

74. Mittler, R. Oxidative stress, antioxidants and stress tolerance. *Trends in Plant Science.* **2002**, 7, 405-410.

75. Mittler, R. Oxidative stress in plants. **2011**

76. Mittler, R. Living under a 'dormant' canopy: A molecular acclimation mechanism of the desert plant *Retamaraetam. Plant Journal.* **2001**, 25, 407–416.

77. Mobin, M.; Khan, N.A. Photosynthetic activity, pigment composition and antioxidative response of two mustard (*Brassica juncea*) cultivars differing in photosynthetic capacity subjected to cadmium stress. *Journal of Plant Physiology.* **2007**, 164, 601-610.

78. Molina, A.S.;Nievas, C.;Chaca, M.V.P.;Garibotto, F.; Gonzalez, U.; Marsa, S.M.; Luna, C.;Giménez, M.S.; Zirulnik, F. Cadmium-induced oxidative damage and antioxidative defense mechanisms in *Vignamungo* L. *Plant Growth Regulation.* **2008**, 56, 285-295.

79. Mullineaux, P.;Karpinski, S.Signal transduction in response to excess light: Getting out of the chloroplast. *Current Opinion in Plant Biology.* **2002**, 5, 43-48.

80. Munné-Bosch, S. The role of a-tocopherol in plant stress tolerance. *Journal of Plant Physiology.* **2005**, 162, 743-748.

81. Niyogi, K.K. Photoprotection revisited: Genetics and molecular approaches. *Annual Review of Plant Physiology and Plant Molecular Biology.* **1999**, 50, 333-359.

82. Noctor, G.; Foyer, C.H. A re-evaluation of the ATP: NADPH budget during C3 photosynthesis. A contribution from nitrate assimilation and its associated respiratory activity? *Journal of Experimental Botany.* **1998a**, 49, 1895–1908.

83. Noctor, G.; Foyer, C.H.Ascorbate glutathione: Keeping active oxygen under control. *Annual Review of Plant Physiology and Plant Molecular Biology.* **1998b**, 49, 249–279.

84. Orozco-Cardenas, M.; Ryan, C.A. Hydrogen peroxide is generated systemically in plant leaves by wounding and systemin via the octadecanoid pathway. *Proceedings of the National Academy of Sciences*, U. S. A. **1999**, 96,6553-6557.

85. Pei, Z.M.; *et al.* Calcium channels activated by hydrogen peroxide mediate abscisic acid signaling in guard cells. *Nature.* **2000**, 406, 731-734.

86. Pietrini, F.; Iannelli, M.A.;Pasqualini, S.; Massacci, A. Interaction of cadmium with glutathione and photosynthesis in developing leaves and chloroplasts of *Phragmitesaustralis*(Cav.) Trin. exSteudel. *Plant Physiology.* **2003**, 133, 829-837.

87. Pilon, M.;Ravet, K.; Tapken, W. The biogenesis and physiological function of chloroplast superoxide dismutases. *BiochimicaetBiophysicaActa.* **2011**, 1807, 989-998.

88. Pooja, B.M.;Vadez, V.; Sharma, K.K.Transgenic approaches for abiotic stress tolerance in plants: Retrospect and prospects. *Plant Cell Reports.* **2008**, 27, 411-424.

89. Pradedova, E.V.; Isheeva, O.D.; Salyaev, R.K.Superoxide dismutase of plant cell vacuoles. *Membrane and Cell Biology.* **2009**, 26(1), 21-30.

90. Quiles, M.J.; Lopez, N.I. Photoinhibition of photosystems I and II induced by exposure to high light intensity during oat plant grown effects on the chloroplastic NADH dehydrogenase complex. *Plant Science.* **2004**, 166, 815-823.

91. Racchi, M.L.Antioxidant defense on plants with attention to *Prunus* and *Citrus* spp. *Antioxidants.* **2013**,2, 340-369.

92. Rao, A.S.V.C.; Reddy, A.R. Glutathione reductase: a putative redox regulatory system in plant cells.In: Khan, N.A., Singh, S., Umar, S. (Eds.), Sulfur assimilation and abiotic stresses in plants. Springer, The Netherlands. **2008**, pp. 111-147.

93. Rausch, T.;Wachter, A. Sulfur metabolism: a versatile platform for launching defence operations, *Trends in Plant Science.* **2005**,10, 503-509.

94. Reddy, C.D.;Venkaiah, B.Isoenzymes of superoxide dismutase from mung bean (*Phaseolusaureus*) seedlings. *Current Science.* **1982**, 20,987-988.

95. Sandalio, L.M.; Rodriguez-Serrano, M.; Romero-Puertas, M.C.; delRíýo, L.A. Role of peroxisomes as a source of reactive oxygen species (ROS) signaling molecules. *Subcellular Biochemistry.* **2013**, 69, 231-255.

96. Sandalio, M.; Del Rio, L.A. Intra-organellar distribution of superoxide dismutase in plant peroxisomes (glyoxysomes and leaf peroxisomes). *Plant Physiology.* **1988**, 88, 1215-1218.

97. Sandmann, G. Genetic manipulation of carotenoid biosynthesis: strategies, problems and achievements. *Trends in Plant Science.* **2001**, 6, 14-17.

98. Santos, M.;Gousseau, H.; Lister, C.; Foyer, C.; Creissen, G.; Mullineaux, P. Cytosolic ascorbate peroxidase from *Arabidopsis thaliana* L. is encoded by a small multigene family. *Planta.* **1996**, 198(1), 64-69.

99. Sato, Y.;Masuta, Y.; Saito, K.; Murayama, S.; Ozawa, K. Enhanced chilling tolerance at the booting stage in rice by transgenic over-expression of the ascorbate peroxidase gene, OsAPXa. *Plant Cell Reports.* **2011**, 30,299-306.

100. Scandalios, J.G. Response of plant antioxidant defense genes to environmental stress. *Advances in Genetics.* **1990**, 28, 1–41.

101. Scandalios, J.G.Oxygen stress and superoxide dismutase. *Plant Physiology.* **1993**, 101, 712–726.

102. Shao, H.B.; Chu, L.Y.; Wu, G.; Zhang, J.H.; Lu, Z.H.; Hu, Y.C. Changes of some antioxidative physiological indices under soil water deficits among 10 wheat (*Triticumaestivum* L.) genotypes at tillering stage. *Colloids and Surfaces B: Biointerfaces.* **2007**, 54, 143-149.

103. Sharma, P.; Dubey, R.S.Modulation of nitrate reductase activity in rice seedlings under aluminium toxicity and water stress: Role of osmolytes as enzyme protectant. *Journal of Plant Physiology.* **2005**, 162, 854-864.

104. Shigeoka, S.; Ishikawa, T.;Tamoi, M.;Miyagawa, Y.; Takeda, T.;Yabuta, Y.; Yoshimura, K. Regulation and function of ascorbate peroxidase isoenzymes. *Journal of Experimental Botany.* **2002**, 53, 1305-1319.

105. Sieferman-Harms, D. The light harvesting function of carotenoids in photosynthetic membrane. *Plant Physiology.* **1987**, 69, 561-568.

106. Singh, S.; Khan, N.A.;Nazar, R.; Anjum, N.A. Photosynthetic traits and activities of antioxidant enzymes in blackgram (*Vignamungo* L. Hepper) under cadmium stress. *American Journal of Plant Physiology.* **2008**, 3, 25-32.

107. Skadsen, R.W.; Schulz-Lefert, P.; Herbt, J.M.Molecular cloning, characterization and expression analysis of two classes of catalase isozyme genes in barley. *Plant Molecular Biology.* **1995**, 29, 1005-1014.

108. Smirnoff, N. Ascorbic acid: metabolism and functions of a multifaceted molecule. *Current Opinion in Plant Biology.* **2000**, 3, 229-235.

109. Smirnoff, N. Ascorbate, tocopherol and carotenoids: metabolism, pathway engineering and functions.In: Smirnoff, N. (Ed.), Antioxidants and reactive oxygen species in plants. Blackwell Publishing Ltd., Oxford, UK. **2005**, pp. 53-86.

110. Smirnoff, N.;Cumbes, Q.J. Hydroxyl radical scavenging activity of compatible solutes. *Phytochemistry.* **1989**, 28, 1057-1060.

111. Srivastava, A.K.; Bhargava, P.L.; Rai, C. Salinity and copper-induced oxidative damage and changes in antioxidative defense system of *Anabaena doliolum. World Journal of Microbiology and Biotechnology.* **2005**, 22, 1291–1298.

112. Sumithra, K.;Jutur, P.P.; Carmel, B.D.; Reddy, A.R. Salinity-induced changes in two cultivars of *Vignaradiata*: responses of antioxidative and proline metabolism. *Plant Growth Regulation.* **2006**, 50, 11-22.

113. Sun, W.H.; Duan, M.; Shu, D.F.; Yang, S.; Meng, Q.W.Over-expression of tomato tAPX gene in tobacco improves tolerance to high or low temperature stress. *BiologiaPlantarum.* **2010a**, 54, 614-620.

114. Sun, W.H.; Duan, M.; Shu, D.F.; Yang, S.; Meng, Q.W. Over-expression of StAPX in tobacco improves seed germination and increases early seedling tolerance to salinity and osmotic stresses. *Plant Cell Reports.* **2010b**, 29, 917-926.

115. Szarka, A.;Horemans, N.; Kovacs, Z.;Grof, P.; Mayer, M.;Banhegyi, G.Dehydroascorbate reduction in plant mitochondria is coupled to the respiratory electron transfer chain. *PhysiologiaPlantarum.* **2007**, 129, 225-232.

116. Tang, G. Bioconversion of dietary Provitamin A carotenoids to Vitamin A in humans. *Am. J .Clin. Nutr.* **2010**, 91, 1468-1473.

117. Trebst, A.;Depka, B.;Holländer-Czytko, H. A specific role for tocopherol and of chemical singlet oxygen quenchers in the maintenance of photosystem II structure and function in *Chlamydomonasreinhardtii. FEBS Letter.* **2002**, 516, 156-160.

118. Tuteja, N.; Tiburcio, A.F.; Fortes, A.M.; Bartels, D. Plant abiotic stress: Introduction to PSB special issue. *Plant Signaling and Behavior.* **2011**, 6, 173-174.

119. Uchida, A.; Jagendorf, A.T.; Hibino, T.; Takabe, T. Effects of hydrogen peroxide and nitric oxide on both salt and heat stress tolerance in rice. *Plant Science.* **2002**, 163, 515-523.

120. Van Camp, W.;Willekens, H.; Bowler, C.; Van Montagu, M.;Inzé, D.;Reupold-Popp, P.;Sandermann, H. Jr.;Langebartels, C. Elevated levels of superoxide dismutase protect transgenic plants against ozone damage. *Nature Biotechnology.* **1994**, 12, 165-168.

121. Wang, J.;Xuequn, L.;Guanghui, Y.U.Identification of superoxide dismutase isoenzymes in tobacco pollen. *Frontiers of Biology in China.* **2009**, 4(4), 442-445.

122. Wang, Y.; Wisniewski, M.;Meilan R.; Uratsu, S.L.; Cui, M.G.;Dandekar, A.; Fuchigami, L. Ectopic expression of Mn-SOD in *Lycopersiconesculentum* leads to enhanced tolerance to salt and oxidative stress. *Journal of Applied Horticulture.* **2007,** 9, 3-8.

123. Wang, Y.; Ying, Y.; Chen, J.; Wang, X.C.Transgenic Arabidopsis overexpressing Mn-SOD enhanced salt-tolerance. *Plant Science.* **2004,** 167, 671-677.

124. Werner, K.; Heinz, R.; Andrea, P.Purification of two SOD isozymes and their sub-cellular localization in needles and roots of Norway spruce (*Piceaabies* L.). *Plant Physiology.* **1992,** 100, 334-40.

125. Willekens, H.;Inzé, D.; van Montagu, M.; Van Camp, W. Catalases in plants. *Molecular Breeding.* **1995,** 1, 207-228.

126. Willekens, H.;Villarroel, R.; van Montagu, M.;Inzé, D.; van Camp, W. Molecular identification of catalases from *Nicotianaplumbaginifolia* (L.). *FEBS Letters.* **1994,** 352, 79-83.

127. Wu, G.; Wei, Z.K.; Shao, H.B.The mutual responses of higher plants to environment: physiological and microbiological aspects. *Biointerfaces.* **2007,** 59, 113-119.

128. Xiang, C.; Werner, B.L.; Christensen, E.M.; Oliver, D.J. The biological functions of glutathione revisited in *Arabidopsis* transgenic plants with altered glutathione levels. *Plant Physiology.* **2001,** 126, 564-574.

129. Xu, W.F.; Shi, W.; Ueda, M.; Takabe, A.T.Mechanisms of salt tolerance in transgenic *Arabidopsis thaliana* carrying a peroxisomalascorbate peroxidase gene from barley. *Pedosphere.* **2008,** 18,486-495.

130. Yang, Y.; Han, C.; Liu, Q.; Lin, B.; Wang, J.Effect of drought and low light on growth and enzymatic antioxidant system of *Piceaasperata* seedlings. *ActaPhysiologiaePlantarum.* **2008,** 30, 433-40.

131. Yang, Z.; Wu, Y.; Li, Y.; Ling, H.Q.; Chu, C.OsMT1a, a type 1 metallothionein, plays the pivotal role in zinc homeostasis and drought tolerance in rice. *Plant Molecular Biology.* **2009,** 70, 219-29.

132. Yoon, H.S.; Lee, I.A.; Lee, H.; Lee, B.H.; Jo, J. Overexpression of a eukaryotic glutathione reductase gene from *Brassica campestris* improved resistance to oxidative stress in *Escherichia coli. Biochemical and Biophysical Research Communications.* **2005,** 326, 618-623.

133. Yousuf, P.Y.; Hakeem, K.U.R.;Chandna, R.; Ahmad, P.Role of glutathione reductase in plant abiotic stress. In: Ahmad, P., Prasad, M.N.V. (Eds.), Abiotic stress responses in plants: Metabolism, productivity and sustainability. Springer Science + Business Media, LLC. **2012,** 149-158.

Abiotic Stress Tolerance Mechanisms in Plants, Pages 203–265
Edited by: Gyanendra K. Rai, Ranjeet Ranjan Kumar and Sreshti Bagati
Copyright © 2018, Narendra Publishing House, Delhi, India

6

HEAT STRESS AND ITS EFFECTS ON PLANT GROWTH AND METABOLISM

Ashutosh Rai[1], Gyanendra K. Rai[2], and R. S. Dubey[1]

[1]Department of Biochemistry, Faculty of Science, BHU, Varanasi
[2]School of Biotechnology,
Sher-e- Kashmir University of Agricultural Sciences and Technology of Jammu-180009 (J&K)
E-mail: ashutoshraibhu@gmail.com

Abstract

High temperature (HT) stress is a major environmental stress that limits plant growth, metabolism, and productivity worldwide. Plant growth and development involve numerous biochemical reactions that are sensitive to temperature. Plant responses to HT vary with the degree and duration of HT and the plant type. HT is a major concern for crop production and development of HT stress tolerant crops is an important agricultural goal. Plants possess a number of adaptive, avoidance, or acclimatization mechanisms to cope with HT situations. In addition, major tolerance mechanisms that employ ion transporters, proteins, osmoprotectants, antioxidants, and other factors involved in signaling cascades and transcriptional control are activated to offset stress-induced biochemical and physiological alterations. Plant survival under HT stress depends on the ability to perceive the HT stimulus, generate and transmit the signal, and initiate appropriate physiological and biochemical changes. HT-induced gene expression and metabolite synthesis also substantially improve tolerance. The physiological and biochemical responses to heat stress are active research areas, and the molecular approaches are being adopted for developing HT tolerance in plants. This chapter deals with the recent findings on responses, adaptation, and tolerance to HT at the cellular, organellar, and whole plant levels and describes various approaches being taken to enhance thermotolerance in plants.

Keywords: Abiotic stress, antioxidant defense, high temperature, heat shock proteins, oxidative stress

1. Heat Stress

The phenomena of global warming, rise in atmospheric temperature day by day, unpredictable fluctuations in environmental temperature, have often led to exposure of crop plants to high atmospheric temperatures. Tropical plants need 25-30°C as optimal temperature range for their normal growth and development. When temperature exceeds much above optimal temperature, heat injury occurs to the plants. Heat stress is an important abiotic factor responsible for reducing tomato yield in Mediterranean and tropical countries. Tomato production under high temperature conditions, such as the summer in India, is often associated with reduced productivity and decline in quality and yield. For instance, low fruit setting, reduction in the flower fertilization rate, decrease in the lycopene content and high evaporation are all related to high temperature stress (Al-Khatib and Paulsen, 1999; Hall and Ziska, 2000; Sato *et al.*, 2000 and Hall, 2001).

Plants experience high temperature in many different ways and adaptation or acclimation to high temperature occurs depending on duration of exposure, intensity of exposure and stages of plant growth. Exposure to high temperature for long durations may lead to severe metabolic alterations in plants ultimately leading to decreased growth and yield. Different tissues and organs of the plants at different growth stages get damaged differently depending on the heat sensitivity of the key cellular processes that are active at the time of heat exposure. Sensitivity or tolerance of a crop species towards heat stress is a complex phenomenon involving many genes. During heat exposure complex perturbations take place in the metabolic machinery leading to alteration in cellular homeostasis.

2. Responses of Plants to Heat Stress

Abiotic stressful conditions of the environment cause extensive loss to tomato production worldwide (Boyer *et al.*, 1982). The stress conditions such as heat, salinity and drought have been the subject of intense research on tomato cultivation and productivity (Bray *et al.*, 2000; Cushman *et al.*, 2000). In field conditions, tomato and other vegetable crops are routinely subjected to a combination of different abiotic stresses (Craufurd *et al.*, 1993; Jiang *et al.*, 2001; Moffat *et al.*, 2002).

High temperatures during the entire growing season of vegetable crops have been reported to be detrimental to growth, reproductive development and yield (Hall, 1992; Hussain *et al.*, 2006; Singh *et al.*, 2007). However, in tomato high temperature exposure during reproductive developmental stage causes significant increment in flower drop (Hanna & Hernandez, 1982), marked decline in fruit setting (Weis and Berry, 1988; Sato *et al.*, 2000) consequently leading to severe decline in fruit yield. At high temperature, the reproductive parts of the flower get adversely affected. Stigma tube elongation, poor pollen germination, poor pollen tube growth and carbohydrate perturbations are the main reasons for poor fruit setting at high temperatures in tomato.

Recent studies have revealed that the molecular and metabolic responses of plants to a combination of drought and heat are unique and cannot be directly extrapolated from the response of plants to each of these two stresses applied individually (Pnueli *et al.*, 2002; Rizhsky *et al.*, 2002; Suzuki *et al.*, 2005). In addition, tolerance to a combination of these two different abiotic stresses is a well-known breeding target in corn and other crops (Craufurd *et al.*, 1993; Jiang *et al.*, 2001). Nevertheless, little is known about the molecular mechanisms underlying the acclimation of plants to a combination of two these two different stresses (Rizhsky *et al.*, 2002). The majority of abiotic stress studies performed under controlled conditions in the laboratory do not reflect the actual conditions that occur in the field, therefore a considerable gap exists between the knowledge gained by these studies and the knowledge required to develop crops plants with enhanced tolerance to stresses under field conditions. This gap might explain why some of the transgenic plants developed in the laboratory with enhanced tolerance to a particular stress condition failed to show enhanced tolerance when tested in the field (McKersie *et al.*, 1996; Gao *et al.*, 2000; Mohamed *et al.*, 2001). A focus on molecular, physiological and metabolic aspects of stress combination is needed to bridge this gap in order to facilitate the development of crop plants with enhanced tolerance to field stress conditions.

The complex plant response to an abiotic stress, such as heat stress involves many genes and biochemical-molecular mechanisms. The ongoing elucidation of the molecular control mechanisms of abiotic stress tolerance (such as heat tolerance), which may result in the use of molecular tools for engineering more tolerant plants, is based on the expression of specific stress-related genes. These genes include three major categories: (i) those involved in signaling cascades and in transcriptional control, such as MYC, MAP kinases and SOS kinase (Shinozaki and Yamaguchi-Shinozaki 1997; Munnik *et al.*, 1999), phospholipases (Frank *et al.*, 2000), and transcriptional factors such as HSF, and the CBF/DREB and ABF/ABAE families (Stochinger *et al.*, 1997; Choi *et al.*, 2000; Shinozaki and Yamaguchi-Shinozaki 2000); (ii) those that function directly in the protection of membranes and proteins, such as heat shock proteins (HSPs) and chaperones, late embryogenesis abundant (LEA) proteins (Vierling 1991; Ingram and Bartels 1996; Bray *et al.*, 2000), osmoprotectants, and free-radical scavengers (Bohnert and Sheveleva, 1998); (iii) those that are involved in water and ion uptake and transport such as aquaporins and ion transporters (Maurel 1997; Serrano *et al.*, 1999; Blumwald 2000). Plants vary greatly in their capacity to tolerate heat stress conditions; some of them are unable to endure heat stress so wilt and die (sensitive plants), while others can tolerate heat considerably by undergoing certain physiological and biochemical changes/adjustments in their tissues in order to maintain their cellular functions at normal level.

The work done on heat stress in tomato and some other crops is being reviewed under physiological, biochemical and molecular changes in responses in the following paragraphs.

2.1. Morpho-phenological Responses to Heat Stress

Temperatures 10°C above normal growth temperatures, affect a range of plant processes such as pollen meiosis, pollen germination, ovule development, ovule viability, development of the embryo (Peet *et al.,* 1998), seedling growth (Hong & Vierling 2001), etc. Successful flower development is critical for production of many agronomic and horticultural crops. High temperature exposures result in floral abortion in many species including *Lycopersicon esculentum* Mill. (Tomato; Levy *et al.,* 1978; Abdul-Baki, 1991), *Capsicum annum* L. (pepper; Rylski, 1986; Erickson *et al.,*2002), *Phaseolus vulgaris* L. (bean; Konsens *et al.,* 1991), *Vigna unguiculata* (L.) Walp. (cowpea; Craufurd *et al.,* 1998), *Pisum sativum* L. (pea; Guilioni *et al.,* 1997), *Gossypium hirsutum* L. (cotton; Reddy *et al.,* 1992), etc. In addition, a physiological disorder termed 'blindness' in roses is due to abortion of the flower at an early stage of development under low irradiance or high temperature conditions.

2.2. Physiological Responses of Heat Stress

2.2.1. *Canopy Temperature*

The high surface area of leaves makes plants most vulnerable to heat damage. A considerable amount of energy is spent by all plants to maintain the cellular temperature. The leaf and its canopy temperature in a plant system depends on transpiration, respiration and carbon dioxide level in leaf tissues. Direct relations with several physiological processes of plants make canopy temperature one of the best parameters for heat tolerance screening.

2.2.2. *Stomatal Conductance*

Under field conditions, when the plants are gradually exposed to heat or drought stress, the early response is the closure of stomata, which is thought to be in response to the migration of abscisic acid (ABA) synthesized in the root. This stomatal response has been linked more closely to the soil moisture content (Tardieu *et al.,* 1991; Stoll *et al.,* 2000). Due to heat, soil moisture decreases and water loss through transpiration increases. The first response of virtually all the plants to acute water deficit under heat and drought stress is the closure of their stomata to prevent the transpirational water loss. Closure of stomata may result from direct evaporation of water from the guard cells with no metabolic involvement.

Further evidence indicated that stomatal closure is likely to be mediated by chemical signals traveling from the dehydrating root to shoots. Abscisic acid has been identified (ABA) as one of the chemical signals involved in the regulation of stomatal functioning (Davis and Zang, 1991). Stomata respond directly to the rate of water supply, through, for example, changes in xylem conductance (Salleo *et al.*, 2000; Sperry 2000; Nardini *et al.*, 2001). Stomatal control of water loss has been identified as an early event in plant response to water deficit under field conditions leading to limitation of carbon uptake by the leaves (Chaves, 1991; Cornic and Massacci, 1996).

When heat stress is imposed slowly as case under field conditions, a reduction in the biochemical capacity for carbon assimilation and utilization may occur along with restriction in gaseous diffusion. For example, in grapevines grown in the field, CO_2 assimilation gets limited to a great extent due to stomatal closure as summer drought progresses. This is accompanied with proportional reduction in the activity of various enzymes of the reductive Calvin cycle (Maroco *et al.*, 2002 and Chaves *et al.*, 2002).

2.2.3. *Relative Water Content*

Leaf water status is intimately related to several leaf physiological variables, such as leaf turgor, growth, stomatal conductance, transpiration, photosynthesis and respiration. Water content and water potential have been widely used to quantify the effect of heat stress in leaf tissues. Alterations in these parameters occur when plants are exposed to heat stress. Leaf water content is a useful indicator of plant homeostasis, since it expresses the relative amount of water present in the plant tissues. On the other hand, water potential measures the energetic status of water inside the leaf cells (Slatyer & Taylor, 1960). Measurements of water content expressed on a tissue fresh or dry weight basis have been mostly replaced by measurements based on the maximum amount of water a tissue can hold. These measurements are referred to as Relative Water Content (Barrs, 1968; Boyer, 1968). Relative water content (RWC) is water content of a plant expressed as a function of its water content at full turgidity. The relative water content (RWC) of a plant tissue is expressed as

$$RWC (\%) = [(FW - DW) / (TW - DW)] \text{ X } 100$$

Where, FW, DW, and TW are the fresh weight, dry weight and turgid weight of the tissues, respectively.

This index may be useful for determining the plant leaf water status. When water uptake by roots is equal to transpiration, then RWC is about 85 to 95%. Critical RWC (below which tissue death occurs) varies amongst species and tissue types, it may be ~50% or less. A decline in RWC represents the severity of dehydration of plant tissues when it experiences water losses due to heat or other abiotic stresses (Morgan *et al.*, 1997).

Using different plant species, various studies have revealed that increasing duration and severity of stress decreased the RWC of plants (Sanchez *et al.*, 2006). Abdalla *et al.* (2007) reported that the RWC of wheat plants decreased progressively in tolerant and susceptible varieties compared to the untreated controls, although such decline was much pronounced in sensitive varieties (33.6%) than in the tolerant ones (28.6%) under similar level of stress treatment.

2.2.4. *Photosynthesis*

Photosynthetic CO_2 assimilation is known to be sensitive to environmental stresses (Dubey *et al.*, 1997). Drought and salinity, often occurring simultaneously with high temperature (heat) stress, are very common in arid and semi-arid areas leading to marked crop losses. In the recent years there has been considerable interest in the mechanism by which dehydration affects leaf CO_2 assimilation. There is supporting evidence for decline of net photosynthesis under drought-stress (Cornic *et al.*, 1992) principally due to stomatal closure; however non-stomatal responses have also been reported.

Photosynthesis in higher plants is known to decrease with the decrease in relative water content (RWC) and leaf water potential (Lawlor and Cornic, 2002).

Measurements of leaf photosynthesis rate have been shown to be reliable in distinguishing between drought tolerant and sensitive genotypes of some species such as sunflower (Gimenez *et al.*, 1992). Severe drought conditions results in limited photosynthesis due to decline in Rubisco activity (Bota *et al.*, 2004). The activity of photosynthetic electron chain is finely tuned to the availability of CO_2 in the plant and photosystem II (PS II) often declines in parallel as the drought conditions (Loreto, 1995).

The decrease in net CO_2 uptake under drought stress observed in many plant species is often associated with lowered internal CO_2 concentration that results in a limitation of photosynthesis at the acceptor site (Stuhlfauth *et al.*, 1988 and Cornic *et al.*, 1992) or due to direct inhibition of photosynthetic enzymes like Rubisco (Lawlor, 1995) or ATP synthase (Tezara *et al.*, 1999). The deficiency of CO_2, ATP, or RuBP as well as the inhibition of Rubisco would decrease the oxidation of NADPH in the Calvin cycle and, therefore, the primary acceptor for photosynthetic electrons, $NADP^+$, would not be sufficiently available. The conditions of high light absorption result in an over-reduction of the photosynthetic electron transport chain and damage to the photosynthetic apparatus.

2.2.5. *Transpiration*

The decline in cumulative transpiration under heat stress or drought conditions is usually linearly related to a reduction of dry matter production (Jones, 1992). Leaf expansion and transpiration for various crops show different sensitivities to declining plant available soil water. However, in general, these processes decrease when about a third of the transpirable soil water remains in the soil (Sadras and Milroy, 1996; Turner, 2001).

In a drying soil, the soil hydraulic conductivity declines as the volumetric water content decreases and the rate of water uptake by plants from the soil may get lowered. Consequently, stomatal conductance declines, and this reduces transpiration (Sinclair and Ludlow, 1986). In vegetable amaranth, transpiration declines in plants exposed to drying soil mainly due to reduction in stomatal conductance (Liu and Stutzel, 2002).

2.2.6. *Membrane Thermostability*

It is well known that membrane dysfunction is one of the key physiological processes of plants exposed to high temperatures (Levitt, 1980). When cellular membrane is injured due to high temperatures, cellular membrane permeability is increased, and electrolytes get diffused out of the cell (Marcum, 1998). Since the amount of electrolyte leakage is a function of membrane permeability, it can be an effective means of measuring cell thermostability. Electrolyte leakage has been regarded as an indicator of cell membrane thermostability (Kuo *et al.,* 1989). Membrane thermostabilty plays a significant role in conferring tolerance to a wide range of abiotic and biotic stresses. Plant tolerance to high temperature is a highly desirable agronomic trait. The processes involved in temperature acclimation of plants are initiated by the perception of temperature signals and transduction of these signals into series of biochemical processes that finally lead to the development of heat tolerance.

2.2.7. *Compatible Osmolytes and Hormones*

2.2.7.1. Proline

Proline is a proteinogenic amino acid with an exceptional conformational rigidity and is essential for plant primary metabolism (Szabados *et al.,* 2009). Proline (Pro) accumulates in plants in response to a wide range of environmental stresses including high temperatures. The role of high Pro levels in combating adverse environmental effects is still under debate. Most interpretations that support Pro contribution to tolerance against environmental stresses rely on the ability of Pro to mediate osmotic adjustment, stabilization of subcellular structures, scavenging of free radicals and its involvement in shuttling chemical energy required for the stress recovery processes. Proline is probably the most widely distributed osmolyte in plants and many other organisms (McCue and Hanson, 1990; Delauney, 1993). Besides osmotic adjustment other roles have been proposed for proline in osmotically stressed plant tissues, such as protection of plasma membrane integrity, a sink of energy or reducing power, a source for carbon and nitrogen (Ahmad and Hellebust, 1988), or hydroxyl radical scavenger (Smirnoff and Cumbes, 1989; Hong et al., 2000).

Accumulation of proline has been advocated as a criterion for selection of stress tolerant crop species (Yancy et al., 1982). Proline accumulation is often observed when plants encounter abiotic stresses such as high temperature, drought and salinity (Sairam et al., 2002). High levels of proline enable the plants to maintain low water potentials. By lowering water potentials, the accumulation of compatible osmolytes involved in osmoregulation allows additional water to be taken up from the environment, thus buffering the immediate effect of water shortages within the organism (Kumar et al., 2003).

Concerning the effect of heat stress on proline content, it was documented by many workers that it accumulated intensively in all stressed organs of plants especially in leaves as a consequence of increasing breakdown of proteins with simultaneous decline in its synthesis in addition to conversion of some of the amino acids such as ornithine, arginine and glutamic and to proline (Yoshiba et al., 1997; Chaitante et al., 2000).

In plants proline is synthesised in the cytosol and mitochondria from glutamate via Δ^1-pyrroline-5carboxylate (P_5C) by two successive reductions catalyzed by P_5C synthetase (PC_5S) and P_5C reductase (PC_5R), respectively (Hare et al., 1999). Genes encoding these enzymes are shown to be up regulated under heat or osmotic stress involving both ABA dependent and ABA independent signaling cascades (Hare et al., 1999). In addition, there is evidence suggesting that accumulation of products of proline synthesis and catabolism (namely glutamine and P_5C) can selectively increase the expression of several stress regulated genes in rice (Iyer and Caplant, 1998).

Accumulation of high proline content in cells has been associated with prevention of protein denaturation, preservation of enzyme structure and activity (Samuel et al., 2000) and protection of membranes from damage by ROS produced under many abiotic stresses including heat (Saradhi et al., 1995, Hamilton and Heckathorn, 2001). These effects of proline accumulation may become even more important than its role in osmotic adjustment (Hare et al., 1999).

2.2.7.2. Glycine Betaine

Like proline another compatible solute is glycine betaine which accumulates in plants exposed to salt stress and cold. Glycine betaine is synthesized from choline. In many plants, it is synthesized through a two-step oxidation of choline, by choline monooxygenase and betaine aldehyde dehydrogenase. The two enzyme genes have been cloned from plants, and attempts have been made to introduce them separately into plants to synthesize glycinebetaine, but sufficient quantities of the choline substrate and the intermediate betaine aldehyde could not be obtained within the cell. For this and other reasons, generating a sufficient quantity of glycinebetaine in vivo to contribute to a significant stress tolerance has not been possible. However, when a transgenic rice plant containing the betaine aldehyde dehydrogenase gene was supplied with sufficient exogenous betaine aldehyde to artificially generate a considerable quantity of glycinebetaine, it showed improved tolerance against both salt and low temperatures.

2.2.7.3. Gibberellic Acid

Gibberellins (GAs) are endonenous plant growth regulators (PGRs) involved in the regulation of many aspects of plant growth and development such as fermentation of roots, seed germination, extension of growth, flowering, etc. (Vreugdenhil *et al.,* 1999; Blazquez *et al.,* 2000 and Heuvel *et al.,* 2001). The multiple role of GAs in stress signaling have been investigated using a number of approaches, including the application of specific GAs and GA biosynthesis inhibitors to whole plants, tissues and plant cells and the identification of mutants defective in GA biosynthesis.

2.2.7.4. Salicylic Acid

The role of salicylic acid (SA) in the signal transduction process of heat tolerance has been widely recognized. The level of endogenous bound and free SA increases during the heat acclimatization in many plant species (Dat *et al.,* 1998b). When tobacco plants were grown *in vitro* for 4 weeks on medium containing SA, low concentration of SA (0.01 mM) was found to increase heat tolerance, while a concentration of 0.1 mM no longer had any protective effect (Dat *et al.,* 2000). The endogenous glucosylated SA content was enhanced in the shoots of plants grown on 0.01 or 0.1 mM SA, while free SA level was also enhanced in plants grown on 0.1 mM SA. In a similar manner, 0.01 mM acetyl SA was able to improve the heat tolerance of potato in tissue culture, while there was an increase in the endogenous H_2O_2 level in the plants (Lopez-Delgado *et al.,* 1998, 2004). It was also observed that potato microplants grown from explants incubated for 1 h in 0.1- 50 mM H_2O_2 exhibited a concentration-dependent decrease in stem height, but were significantly more thermo tolerant than the controls, even more than a month after the H_2O_2 treatment, showing that not only SA but also H_2O_2 is able to increase the heat tolerance of potato plants. SA treatment also reduced the oxidative damage caused by heat stress in *Arabidopsis* plants. In addition to SA, ethylene, ABA and calcium have also been shown to play a role in the development of tolerance to high temperatures (Larkindale and Knight, 2002).

2.2.8. *Secondary Metabolites*

2.2.8.1. Polyamines, Polyamine Oxidases and Flavonoids

Nitric oxide (NO), polyamines (PAs), diamine oxidase (DAO) and polyamine oxidases (PAO) play important roles in many biological processes such as breakup of dormancy, root development, flowering, senescence and in defense responses against abiotic and biotic stresses (Wendehenne *et al.*, 2005; Gill *et al.*, 2010; Wimalasekera *et al.*, 2011). The antioxidative effect of polyamines is due to a combination of their anionic and cationic-binding properties in radical scavenging, inhibiting lipid peroxidation and metal-catalyzed oxidative reaction and production of H_2O_2 by DAO and PAO (Groppa *et al.*, 2008). Increase in polyamines concentrations in known to occur as adaptive response to stress caused by UV radiation. It is well demonstrated that plants can protect themselves against ozone damage and ozone-derived oxidative damage by modulation of polyamine levels (Groppa *et al.*, 2008). H_2O_2 produced by polyamine catabolism may promote activation of anti-oxidative defense responses in oxidative stress conditions. Phenylpropanoid- polyamine conjugates can act as antioxidants against ROS and reactive nitrogen species in response to stress conditions. Flavonoids comprise a large group of polyphenolic plant secondary metabolites (Koes *et al.,* 1994). In tomato, flavonoids play important roles in many biological processes such as pigmentation of flowers, fruits and vegetable, plant-pathogen interactions, fertility and protection against abiotic stresses (Bovy *et al.,* 2007).

3. Heat, Oxidative Stress and Antioxidants

It has been observed that high temperature exposure or heat stress causes overproduction of reactive oxygen species (ROS) in plants. The ROS in plant cells is a normal and continuously occurring phenomenon mainly in chloroplasts, mitochondria and peroxisomes. ROS are generated by the stepwise monovalent reduction of molecular oxygen (O_2) due to single electron transfer reaction or due to incomplete oxidation of water by mitochondrial or chloroplastic electron transfer chains (Foyer *et al.*, 1994). The ROS include the superoxide anion $(O_2{}^{\cdot-})$, hydrogen peroxide (H_2O_2), hydroxyl radical (OH·), singlet oxygen ($^1O_2$), etc. Some of these ROS, mainly H_2O_2 and O_2^- are considered as the first line of defense (directly or indirectly) against pathogen attacks (Apostol et al., 1989). ROS have dual role in plants. At low concentrations ROS act as signaling molecules and are involved in many growth and developmental processes in plants, whereas at higher concentrations ROS cause oxidative damage to lipids, proteins, nucleic acids, etc. Due to overproduced ROS in the cells, when biomolecules are being oxidatively damaged, the cell are said to be in the state of 'Oxidative Stress'.

To counter the oxidative damage caused due to ROS under stress, plants have evolved a complex antioxidative defense system comprising of both non-enzymatic and enzymatic components (Sairam et. al., 1998, 2002). Responses of plants to extreme temperature stress are often correlated with responses typically observed from increased oxidative stresses. Oxidative stress appears to be a major component in expression of heat induced injury in plants. ROS are highly reactive, and can react with proteins, nucleic acids etc., lipids, potentially causing oxidative modifications, denaturation, mutagenesis, etc. (McKersie, 1996).

3.1. Sites of Production of Reactive Oxygen Species

The reduction of molecular oxygen to form $O_2^{·-}$, H_2O_2 and ·OH is the principal mechanism of oxygen activation in biological systems to produce ROS. Chloroplasts, mitochondria, endoplasmic reticulum, microbodies, plasma membranes, and cell walls are the major sites of production of ROS in plant cells (McKersie, 1996).

As described by Elstner (1991), there are at least four sites within the chloroplast that can activate oxygen. Firstly, the reducing site of PSI (Photosystem I) was thought to contribute significantly to the monovalent reduction of oxygen under conditions where $NADP^+$ is limiting. Secondly, under conditions that prevent the captured light energy from being utilized in the electron transport systems, the excitation energy of photoactivated chlorophyll can excite oxygen from the triplet to singlet form. Thirdly, leakage of electrons from the oxidizing side of PSII to molecular oxygen, or release of partially reduced oxygen products contributes to ROS production. Lastly, due to photorespiration glycolate is formed, and its subsequent metabolism in the peroxisomes leads to the generation of activated oxygen.

The main superoxide generators in the mitochondria are the ubiquinone radical and NADH dehydrogenases. Superoxide is formed by the auto-oxidation of the reduced components of the respiratory chain (Dat et al., 2000). The various Fe-S proteins have also been implicated as possible sites of superoxide and hydrogen peroxide formation in mitochondria (Turrens et al., 1982).

Various oxidative processes occurring on the smooth endoplasmic reticulum involve the transfer of oxygen into an organic substrate using NAD(P)H as the electron donor. Superoxide is produced by microsomal NAD(P)H dependent electron transport involving cytochrome P_{450} (Winston and Cederbaum, 1983). Cytochrome P_{450} reacts with its organic substrate (RH) and forms a radical intermediate (cytP450-R) that can readily react with triplet oxygen (forming cytP$_{450}$-ROO·). This oxygenated complex may be reduced by cytochrome b or occasionally the complex may get decomposed releasing superoxide (McKersie, 1996).

Peroxisomes and glyoxysomes contain enzymes involved in β-oxidation of fatty acids and the glyoxylic acid cycle including glycolate oxidase, catalase, and various peroxidases. Glycolate oxidase produces H_2O_2 in a reaction involving two electron transfers from glycolate to oxygen (Lindqvist *et al.,* 1991). Xanthine oxidase, urate oxidase, and NAD(P)H oxidase generate superoxide as a consequence of oxidation of their substrates (McKersie, 1996).

A superoxide-generating NAD(P)H oxidase activity has been clearly identified in plasmalemma enriched fractions. Wounding, heat shock and xenobiotics transiently activate this superoxide generating enzyme, and consequently, it has been proposed that these superoxide generating reactions may serve as a signal in plant cells to elicit responses to biological, physical, or chemical stress (Doke *et al.,*1991). In the cell wall, such mechanism is also thought to be present. Some biosynthetic reactions (such as lignin biosynthesis) and oxidative enzymes (diamine oxidase, NADH oxidase) also lead to production of ROS on the cell wall which may induce oxidative stress (McKersie, 1996).

3.2. Damage caused due to ROS

The reactions of ROS with organic substrates are complex in biological systems due to the surface properties of membranes, electrical charges, binding properties of macromolecules, and compartmentalization of enzymes, substrates, and catalysts. Thus, various sites even within a single cell differ in the nature and extent of the reactions of ROS (McKersie, 1996). The mechanisms by which ROS cause damage to membrane lipids are well accepted, and consequently oxidative damage of membrane lipids is exclusively associated with peroxidation reactions occurring in fatty acids. ROS also cause oxidative damage to proteins and nucleic acids, depending on severity these reactions may also become lethal to cells (McKersie, 1996).

There are two common sites of attack of ROS on the phospholipids molecule; the unsaturated double bonds of the fatty acid and the ester linkage between the glycerol and the fatty acid. Hydroxyl radicals attack on the double bonds and initiate the peroxidation reaction by abstracting a single hydrogen atom. This creates a carbon radical product that is capable of reacting with the ground state oxygen in a chain reaction. The resulting molecule is ready to react with another phospholipid, and the reaction propagates. The basis for the hydroxyl radical's extreme reactivity in lipid systems is that at very low concentrations it initiates a chain reaction involving triplet oxygen, the most abundant form of oxygen in the cell (McKersie, 1996).

Oxidative attack on proteins results in site-specific amino acid modifications, fragmentation of the peptide chain, aggregation of cross-linked reaction products, altered electrical charge and increased susceptibility to proteolysis. Sulphur containing amino acids, and thiol groups specifically, are very susceptible sites. Although the oxidation of thiol groups can be reversed by various enzymes, some forms of free radical attack on proteins are not reversible. For example, the oxidation of iron-sulphur centers by superoxide destroys enzymatic function (Gardner and Fridovich, 1991). The oxidative degradation of proteins is enhanced in the presence of metal cofactors that are capable of redox cycling, such as iron. In these cases, the metal binds to divalent cation binding site on the protein. The metal then reacts with hydrogen peroxide in a Fenton reaction to form a hydroxyl radical that rapidly oxidizes an amino acid residue at or near the cation-binding site of the protein (Stadtman, 1986).

DNA is an obvious weak link in a cell's ability to tolerate ROS. First, it seems that DNA is effective in binding metals that are involved in Fenton reactions, and secondly less damage can be tolerated in DNA than other macromolecules. ROS induce numerous lesions in DNA that cause deletions, mutations, and other lethal effects. Both sugar and the base moieties are susceptible to oxidation, causing base degradation, single strand breakage, and crosslinking to protein (Imlay and Linn, 1986).

3.3. Antioxidative Defense System in Plants

The presence of oxygen in the environment and various cellular locations where ROS are produced render oxidant scavengers necessary for plant growth and survival. Plants have antioxidative defense system comprising of non enzymatic and enzymetic antioxidants located in different plant cell compartments to scavenge ROS and to prevent the deleterious effects of ROS (Dat *et al.*, 2000). Elevation in activities and concentrations of these antioxidant molecules under abiotic stresses confers tolerance to plants.

3.3.1. *Non-Enzymatic Antioxidants*

The non-enzymatic defense mechanism includes the molecules such as glutathione, ascorbate (vitamin C), carotenoids, Q-tocopherol (vitamin E), and various phenylpropanoid derivatives (phenolic compounds) such as flavonoids, lignans, tannins and lignins.

The tripeptide glutathione (GSH, γ-Glu- Cys-Gly) is the major low molecular weight thiol in plants. GSH is found in cells and subcellular compartments of all higher plants. At subcellular level, GSH concentration is highest in the chloroplasts, but significant quantities also accumulate in the cytosol. The antioxidant function of GSH is mediated by sulfhydryl group of cysteine, which forms a disulfide bond with a second molecule of GSH to form oxidized glutathione (GSSG) upon oxidation (Hausladen and Alscher, 1993). GSH can react chemically with singlet oxygen, superoxide, and hydroxyl radicals; therefore, it functions directly as a free radical scavenger. Also, GSH may stabilize membrane structure by removing acyl peroxides formed by lipid peroxidation reactions (Price *et al.,* 1990).

L-ascorbic acid (vitamin C) is an important antioxidant in both animal and plant tissues. Ascorbate is synthesized in the cytosol of higher plants primarily from the conversion of D-glucose to ascorbate. It is abundant in plant protoplasts, chloroplasts and certain fruits. Ascorbate has been shown to have an essential role in several physiological processes in plants, including growth, differentiation, and metabolism. It functions as a reductant for many free radicals, thereby minimizing the damage caused by oxidative stress. Ascorbate is the terminal electron donor in the processes, which scavenges the free radicals in the hydrophilic environments of plant cells. It scavenges hydroxyl radicals at diffusion-controlled rates (McKersie, 1996). The reaction with superoxide may serve a physiologically similar role to superoxide scavenging enzyme, superoxide dismutase (SOD):

$$2O_2^- + 2H^+ + \text{ascorbate } 2H_2O_2 + \text{dehydroascorbate}$$

Ascorbate also reacts non-enzymatically with H_2O_2 at a significant rate, producing water and monodehydroascorbate. The reaction is catalyzed by ascorbate peroxidase (APX) in the chloroplast and cytosol of higher plants.

In plants, H_2O_2 detoxification in chloroplasts is extremely important because photosynthesis is highly sensitive to very low concentrations of H_2O_2. Because chloroplasts lack catalase, which scavenges hydrogen peroxide in the peroxisomes, ascorbate has a central importance in eliminating the H_2O_2 from chloroplasts. In addition to its role as a primary antioxidant, ascorbate has a significant secondary antioxidant function. The ascorbate pool represents a reservoir of antioxidant potential that is used to regenerate other membrane-bound antioxidants such as α-tocopherol and zeaxanthin, which scavenge lipid peroxides and singlet oxygen, respectively (Foyer, 1993).

Carotenoids are C_{40} isoprenoids or tetraterpenes which are located in the plastids of both photosynthetic and non-photosynthetic plant tissues. In addition to their function as accessory pigments in light harvesting, they detoxify various forms of activated oxygen and triplet chlorophyll that are produced as a result of excitation of photosynthetic complexes by light. In terms of their antioxidant properties, carotenoids can protect the photosystems one of four ways: by reacting with lipid peroxidation products to terminate chain reactions (Burton and Ingold, 1984), by scavenging singlet oxygen and dissipating the energy as heat (Mathis and Kleo, 1973), by reacting with triplet or excited chlorophyll molecules to prevent formation of singlet oxygen, or by the dissipation of excess excitation energy through the xanthophyll cycle (McKersie, 1996). Tocopherol, another family of antioxidants, has been found in all higher plants in both photosynthetic and non-photosynthetic tissues.

3.3.2. *Enzymatic Antioxidants*

The antioxidative enzymes include superoxide dismutase (SOD; EC 1.15.1.1), catalase (CAT; EC 1.11.1.6), peroxidase (POD; EC 1.11.1.7) and the enzymes of Ascorbate Glutathione Cycle or Halliwet- Asada pathway namely ascorbate peroxidase (APX; EC 1.1.1.11), glutathione reductase (GR; EC 1.6.4.2), monodehydroascorbate reductase (MDHAR; EC 1.6.5.4) and dehydroascorbate reductase (DHAR; EC 1.8.5.1). These enzymes constitute enzymic antioxidants in cells which deirectly scavange ROS and remove ROS from the cells and also catalyze synthesis, degradation, and recycling of antioxidant molecules in the cells.

3.3.2.1. Superoxide Dismutase

The enzyme superoxide dismutase (SOD), originally discovered by McCord and Fridovich (1969), catalyzes the dismutation of superoxide to hydrogen peroxide and oxygen:

$$2O_2^{\cdot -} + 2H^+ \longrightarrow H_2O_2 + O_2$$

This enzyme is unique in that its activity determines the concentrations of O_2 and H_2O_2, the two Haber-Weiss reaction substrates, which generate the most reactive hydroxyl radicals. The importance of SOD has been established by the demonstration that SOD-deficient mutants of *Escherichia coli* (Carlioz and Touati, 1986) and yeast (Van Loon *et al.,* 1986) that are hypersensitive to oxygen. The enzyme serves as the first line of defense against ROS, because this enzyme is primarily involved in scavenging $O_2^{\cdot -}$ which produced immediately after one electron reduction of molecular oxygen. SOD is present in all aerobic organisms and in most subcellular compartments that generate ROS. The three known isoform types of SOD are classified depending on their metal cofactor: copper/zinc (Cu/ZnSOD), manganese (MnSOD), and iron (FeSOD) forms. Experimentally these three different types can be identified by their differential sensitivities to KCN and H_2O_2. Cu/Zn SOD is characterized as being sensitive to both H_2O_2 and KCN; FeSOD is sensitive only to H_2O_2, while MnSOD is resistant to both inhibitors. Subcellular fractionation studies have been performed in many plant species and in general plants contain a mithochondrial matrix localized MnSOD and a cytosolic Cu/ZnSOD, with FeSOD and/or Cu/ZnSOD present in the chloroplast stroma (Bowler *et al.,* 1992). Moreover, there are findings that MnSOD has also glyoxysomal and peroxisomal isozymes in certain plant species (Bowler *et al.,* 1994).

3.3.2.2. Peroxidase

Peroxidases are a family of heme containing enzymes which are primarily responsible for removal and reduction of H_2O_2 to water in the cells. Among peroxidases, the most common are guaiacol peroxidases (GPX, EC 1.11.1.7), heme containing proteins, which preferably oxidize aromatic electron donor such as guaiacol and pyragallol at the expences of H_2O_2. GPX is widely found in animals, plants and microbes. These enzymes have four conserved disulfide bridges and contain two structural Ca^{2+} ions. Many isoenzymes of GPX exist in plant tissues localized in vacuoles, the cell wall and the cytosol (Shame *et al.* 2010). GPX is associated with many important biosynthetic processes, including lignifications of cell wall, degradation of IAA, biosynthesis of ethylene, wound healing and defense against abiotic and biotic stresses. GPXs are widely regarded as "stress enzyme". GPX can fuction as effective quencher of reactive intermediary form of O_2 and peroxy radicals under stressed conditions. Various abiotic stressful conditions of the environment have been shown to induce the activity of GPX (Dubey, 2010).

3.3.2.3. Catalase

Catalase is a heme-containing enzyme that catalyzes the dismutation of hydrogen peroxide into water and oxygen:

$$2H_2O_2 \longrightarrow 2H_2O + O_2$$

The enzyme is found in all aerobic eukaryotes in different subcellular compartments and is important in the removal of H_2O_2 generated in peroxisomes (microbodies), the glyoxylate cycle (photorespiration) and purine catabolism (McKersie, 1996). Various isoforms of catalase have been described in many plant species. In maize, there are three isozymes localized separately in peroxisomes, cytosol and mitochondria (Scandalios, 1993). Catalase is very sensitive to light and has a rapid turnover rate. Stress conditions, which reduce the rate of protein turnover, such as salinity, heat shock or cold, cause depletion of catalase activity (Hertwig *et al.*, 1992). Detoxification of H_2O_2 is mediated by catalase. However, catalase possesses a low affinity for H_2O_2 and its activity is either extremely low or not detectable in the cytosol, mitochondria and chloroplasts (Halliwell, 1981). In plant cells, an alternative and more effective detoxification mechanism against H_2O_2 also exists, operating both in chloroplast and cytosol, called ascorbate-glutathione or Halliwell- Asada cycle (Foyer and Halliwell, 1976; Asada and Takahashi, 1987). The pathway is likely to be a major H_2O_2 detoxification system in cytosol, chloroplast as well as in mitochondria. This cycle is also important for the maintenance of ascorbate and glutathione pools in their reduced states.

Enzymes of Halliwell- Asada Cycle

3.3.2.4. Ascorbate Peroxidase

Ascorbate peroxidase (APX) is the first enzyme of Halliwell-Asada cycle and catalyses the reduction of H_2O_2 to water and has high specificity and affinity for ascorbate as reductant (Asada, 1999). Its sequence is distinct from other peroxidases. Different forms of APX occur in the chloroplast, cytosol, mitochondria, peroxisomes and glyoxysomes (Jimenez *et al.*, 1997; Leonardis *et al.*, 2000). Membrane bound APX occur in the peroxisomes and thylakoid membranes. By the ascorbate-glutathione cycle, H_2O_2 is effectively scavenged while the ascorbate level is maintained at a constant level.

The oxidation of ascorbic acid occurs in two sequential steps, in the first instance, monodehydroascorbate, and subsequently dehydroascorbate is formed. Monodehydroascorbate is either directly reduced to ascorbate by the action of NAD (P) H-dependent monodehydroascorbate reductase (MDHAR) or spontaneously disproportionates to dehydroascorbate. Dehydroascorbate is also very unstable at pH values greater than 6.0. The carbon chain is cleaved to products such as tartrate and oxalate, and may decompose to yield toxic derivatives. To prevent loss of ascorbate pool following oxidation, the chloroplast contains efficient mechanisms of recycling both monodehydroascorbate and dehydroascorbate, and to ensure that the ascorbate pool is maintained largely in reduced form. Reduced glutathione (GSH), which is present in chloroplast in millimolar concentrations, will non-enzymatically reduce dehydroascorbate back to ascorbate at pH values greater than 7.0. This reaction is catalyzed by dehydroascorbate reductase (DHAR), which is present in high activities in leaves, seeds and other tissues. DHAR uses reduced glutathione (GSH) as an electron donor for the reduction of dehydroascorbate to ascorbate. Thiol groups are involved in the catalysis, as SH-reagents deactivate the enzyme (Foyer, 1993).

3.3.2.5. Glutathione Reductase

The last enzyme of Halliwell-Asada cycle, glutathione reductase (GR), catalyses the NADPH-dependent reduction of oxidized glutathione. Several isozymes of GR are present in plant tissues. Subcellular fractionation studies have shown GR to occur in chloroplast, cytosol and also in mitochondria (Hausladen and Alscher, 1993). GR is the rate-limiting enzyme in H_2O_2 scavenging pathway and it is involved in the maintenance of high ratio of GSH/GSSG, which is required for the regeneration of ascorbate (Sudhakar *et al.*, 2001).

4. Heat Stress Specific Proteins

4.1. Heat Shock Proteins (HSPs) and Chaperons

Because of the temperature sensitivity of the forces responsible for protein folding (Pace *et al.*, 1996), proteins are easily denatured at high temperatures. Biological organisms have a suite of proteins that are made in response to high temperatures that appear to be designed to prevent or reverse the effects of heat. These are called heat shock proteins (HSPs) because these are made in abundance when organisms are subjected to potentially damaging high temperatures. HSPs comprise an evolutionary highly conserved group of proteins, which are present in all living organisms, including plants. HSPs, mainly molecular chaperones also have house-keeping functions and are essential for cellular homeostasis even at physiological temperature (Walter and Buchner, 2002). Many of these proteins are also synthesized in response to other stresses, notably osmotic and oxidative stress. Thus, an adequate basal expression level of these chaperones is important under normal growth conditions. When exposed to stress, cells require an increased chaperone capacity due to increase in the amount of partially unfolded proteins. This need is satisfied either by enhanced transcription of the respective chaperone genes (as in bacteria) or by the additional expression of close homologues specifically dedicated to stress response (as in most eukaryotes). Typical examples for this kind of regulation are the HSP90 and HSP70 families. Under normal conditions, HSP70s facilitate folding of newly synthesized proteins, preventing undesirable interactions. Members of HSP70 family play distinct roles in plant growth and development. Developmentally regulated HSP70 genes have been described in tomato (Duck *et al.*, 1989) and Arabidopsis (Wu *et al.*, 1988).

It is believed that these proteins help other proteins fold correctly or refold after damage by heat stress (Boston *et al.*, 1996), although recent data has established unique roles for each of the families of HSPs. HSPs were discovered as proteins that were not expressed or not very prominent in assays of plants grown at moderate temperatures but which became prominent when organisms were subjected to a heat shock (Lindquist, 1986). The HSPs are divided into different families based on their molecular weight. For example, most organisms have heat-inducible genes with a molecular weight between 100 and 104; these are called HSP100. The different HSP families appear to have different functions. Some are constitutive, expressed at a somewhat higher level in response to heat stress, while others are nearly completely inducible, with no protein detectable in unstressed plants and lots of protein in stressed plants. The following is a quick tour of some of the important aspects of each class of heat shock proteins.

HSP-100: This class of HSP has been shown to be directly related to thermotolerance (Queitsch *et al.*, 2000). Plants lacking Hsp101 can grow at normal temperatures and reproduce, but unlike wild type plants, a pretreatment at moderately high temperature does not induce thermotolerance. HSP101 is synthesized in Arabidopsis seedlings and confers tolerance to heat stress (Hong and Vierling, 2000; Queitsch *et al.*, 2000).

Expression of heat shock proteins and heat shock factors (proteinaceous factors that stimulate expression of heat shock proteins as well as other heat tolerance genes) results in increased thermo tolerance in many plants (Sun *et al.*, 2002). HSP101 is highly inducible and plants that have not experienced high temperature have only very low level, if any, HSP101. The HSP100 family has sequence similarities to caseinolytic protease B (ClpB) and is often referred to as HSP100/ clpB (Agarwal *et al.*, 2002). These proteins have nucleotide binding mo- tifs and appear to recruit other HSPs to help repair damaged proteins. In yeast, HSP104 can reverse the effect of high temperature on protein structure. Yeast HSP104 acts with other HSPS (HSP40 and HSP70) to undo damage to proteins caused by heat (Glover and Lindquist, 1998).

HSP-90: In animals HSP90 is best known as a component of a complex that helps fold the glucocoticoid receptor. In plants, HSP90 participates in a similar multiprotein complex that also includes HSP70 and a small acidic protein known as p23 (Pratt *et al.*, 2001). HSP90 complexes appear to be involved not only in protein folding, but also protein transport along microtubules inside the cell, especially transport of proteins into peroxisomes (Pratt *et al.*, 2001). HSP90 is essential for growth and it gets prominently expressed at high temperature. HSP90 has been called a "capacitor" for genetic variation.

HSP-70: Like HSP90, HSP70 is required for plant growth and is essential for processes such as protein import into chloroplasts and mitochondria (Marshall *et al.*, 1990). HSP70 proteins appear to be involved in virus infectivity and perhaps in allowing viruses to move between cells of plants through plasmadesmata (Alzhanova *et al.*, 2001). HSP60 – Rubisco folding requires a "rubisco binding protein" that appears to be a member of the HSP60 family. As many as 14 HSP60 protein chains will aggregate to make a functional complex to fold Rubisco (Roy and Andrews, 2000). HSP60s are essential for folding of many proteins in plants and it is presumed that these proteins are needed for refolding when proteins are denatured by heat.

Low Molecular Weight HSPs

Low molecular weight HSPs or small molecular weight HSPs (smHSPs) have molecular weight 17 to 28 kD and represent a much bigger family of HSPs in plants and other organisms (Vierling, 1991). It has been hypothesized that the rapid divergence of small heat shock proteins in plants into five or six families may have been an important step in the evolution of plants onto land required to allow plants to cope with the large temperature fluctuations that occur in air with its low heat capacity relative to water (Waters, 1995, 2003). The families of small heat shock proteins in plants are located in compartments, such as chloroplast, endoplasmic reticulum, cytosol, etc. It has been shown that there are two small HSP families in the mitochondrion (Scharf *et al.*, 2001). The plant small HSPs appear to play a role in thermotolerance of a number of processes (Basha *et al.*, 2004). Modulation of thermotolerance in carrot cell cultures has been reported to occur by modulating the abundance of a small HSP (Malik *et al.*, 1999). The chloroplast small heat shock proteins have been associated with increased thermotolerance (Wang and Luthe, 2003) and the mechanism has been suggested to be due to protection of photosystem II (Heckathorn *et al.*, 1998, 2002). The small HSPs have a motif that is closely related to the α-crystallin protein in the lens of animal eyes. Another family of smHSPs in *Arabidopsis* also contain α-crystallin domains and are called Acd proteins but their role in heat tolerance is unknown. EvidenceS indicates that there are heat shock factors with DNA binding domains that may be important components in the transduction pathway between high temperature stress and gene expression leading to accumulation of HSPs (Nover *et al.*, 2001). There appear to be two classes of heat shock transcription factors in Arabidopsis (Czarnecka-Verner *et al.*, 2000). In summary, heat stress may affect membrane function and protein folding.

4.2. Heat Sensing and Signaling Molecules

Although the responses of plants to stress signals have been extensively studied at the physiological and the biochemical levels, the perception and the intracellular transmission mechanisms are largely unknown. Progress in understanding signal transduction in animals and yeast showed that reversible protein phosphorylation plays a pivotal role in many signaling cascades. Increasing evidence indicates that protein kinase pathways are involved in signal transduction in plants. Changes in protein phosphorylation patterns were observed after exposure of plant cells to fungal elictors and to ethylene, and during establishment of freezing tolerance. Furthermore, a continuously growing number of genes coding for protein kinases in plants have been reported. A highly conserved group of serine/threonine protein kinases, which show sequence homology to MAP kinases, have been isolated from different plant species.

N-terminal kinase/p38 is activated by various types of stress. JNK and p38 can functionally replace the HOG1 yeast MAP kinase that is necessary for adaptation to high extracellular osmolarity. Genetic and biochemical studies in yeast revealed several distinct MAP kinase cascades in signaling different extracellular stimuli. These kinase cascades are important regulators in pheromone response, pseudohyphal differentiation, and osmolarity responses. Activation of MAP kinases requires tyrosine and threonine phosphorylation. The highly conserved threonine and tyrosine residues are located close to kinase domain VIII and seem to be also important for activation of plant MAP kinases. Phosphorylation of these crucial residues is performed by a dual specific MAP kinase kinase that in turn has to be activated by a serine/threonine MAP kinase kinase kinase. These kinase cascades seem to be conserved in modular form throughout evolution, mediating distinct signal transduction pathways. Isolation of homologous upstream kinases from plants indicates the presence of similar biochemical modules for extracellular signal transmission. Despite the fact that all of the components of MAP kinase modules have been identified in plants, little is known about their functions. An Arabidopsis MAP kinase was proposed to be involved in auxin signal transduction. Genetic studies and the isolation of the CTR1 gene suggest that a MAP kinase cascade may also be involved in mediating ethylene responses. Increased transcript levels of genes encoding a MAP kinase module has been taken as evidence for the involvement of a MAP kinase pathway in signaling touch, cold, salt, and water stress.

5. Conventional Breeding Approach for Heat Tolerance

Breeding for heat tolerance is still in its infancy stage and warrants more attention than it has been given in the past (Ortiz *et al.*, 2008; Ashraf, 2010). It is unfortunate that the literature contains relatively little information on breeding for heat tolerance in different crop species. During the past two decades, the use of marker-assisted selection (MAS) approaches has contributed greatly to a better understanding of the genetic bases of plant stress-tolerance in tomato and maize (Liu *et al.*, 2006; Sun *et al.*, 2006; Momcilovic and Ristic, 2007) and, in some cases, led to the development of plants with enhanced tolerance to abiotic stress. Because of the general complexity of abiotic stress tolerance and the difficulty in phenotypic selection for tolerance, MAS has been considered as an effective approach to improve plant stress tolerance (Foolad, 2005). Comparatively, however, limited research has been conducted to identify genetic markers associated with heat tolerance in different plant species. Use of SSR markers revealed mapping of 2 QTLs for grain filling duration under heat stress in wheat (Yang *et al.,* 2002).

6. Expressed Sequence Tags: A Molecular Approach to Study Growth Processes and Stress Mechanisms

Because of the relatively large size of most plant genomes and the associated high cost of sequencing, it is unlikely to have the full genomic sequence for many plant species in the near future. A less expensive alternative is to sequence or partially sequence cDNA clones, which can reveal a substantial portion of the expressed genes of a genome at a fraction of the cost of genomic sequencing. As a result, extensive efforts on expressed sequence tags (EST) are under way in a wide variety of plant species (National Science Foundation Plant Genome Research Program [http://www.nsf.gov/bio/dbi/dbi_pgr.htm]; Pennisi, 1998; Adam, 2000; Paterson *et al.,* 2000). ESTs can provide a basis for gene discovery and the determination of gene function.

ESTs are unedited, automatically processed, single read sequences produced from cDNAs (small DNA molecules reverse transcribed from cellular mRNA population). The concept of using cDNAs as a rout to expedited in early 1980s (Putney *et al.,* 1983).

Expressed sequence tags are created by sequencing the 5' and/or 3' ends of randomly isolated gene transcripts that have been converted into cDNA (Adams *et al.,* 1991). Despite the fact that a typical EST represents only a portion (approximately 200–900 nucleotides) of a coding sequence, the partial sequence data is of substantial utility. For example, EST collections are a relatively quick and inexpensive route for discovering new genes (Bourdon *et al.,* 2002; Rogaev *et al.,* 1995), confirm coding regions in genomic sequence (Adams *et al.,* 1991), create opportunities to elucidate phylogenetic relationships (Nishiyama *et al.,* 2003), facilitate the construction of genome maps (Paterson *et al.,* 2000), can sometimes be interpreted directly for transcriptome activity (Ewing *et al.,* 1999; Ogihara *et al.,* 2003; Ronning *et al.,* 2003), and provide the basis for development of expression arrays also known as DNA chips (Chen *et al.,* 1998; Shena *et al.,* 1995).

In addition, high-throughput technology and EST sequencing projects can result in identification of significant portions of an organism's gene content and thus can serve as a foundation for initiating genome sequencing projects (Van der Hoeven *et al.,* 2002). Currently there are 60,923,778 ESTs in the NCBI public collection, 290,973 of which derived from tomato (http://www.ncbi.nlm. nih.gov/dbEST/). With many large-scale EST sequencing projects in progress and new projects being initiated, the number of ESTs in the public domain will continue to increase in the coming years. The sheer volume of this sequence data has and will continue to require new computer- based tools for systematic collection, organization, storage, access, analysis, and visualization of this data. Not surprisingly, despite the relative youth of this field, an impressive diversity of bioinformatics resources exists for these purposes.

As sequence and annotation data continue to accumulate, public databases for genomic analysis will become increasingly valuable to the plant science community. The Arabidopsis Information Resource (TAIR; http://www. arabidopsis.org/home. html), the Salk Institute Genomic Analysis Laboratory (SIGnAL; http://signal.salk. edu/), the Solanaceae Genomics Network (SGN; http://sgn.cornell.edu/), and GRAMENE (http://www.gramene.org/) serve well as examples of these on-line resources.

By clustering genes according to their relative abundance in various EST libraries, expression patterns of genes across various tissues were generated and genes with similar patterns were grouped. In addition, tissues themselves were clustered for relatedness based on relative gene expression as a means of validating the integrity of the EST data as representative of relative gene expression.

EST collections from other species (e.g. Arabidopsis) were also characterized to facilitate cross-species comparisons where possible (http://ted.bti.cornell.edu/). With the rapid expansion of available EST data (e.g. http://www.arabidopsis.org; http://www.gramene.org; http://www.medicago.org; http:// www.sgn.cornell.edu; http://www.tigr.org), opportunities for digital analysis of gene expression will continue to expand. Expressed sequence tag collections also have limitations when being used for genomic analysis from the perspectives of accurate representation of genome content, gene sequence, and as windows into transcriptome activity. The fact that ESTs reflect actively transcribed genes makes it difficult to use EST sequencing alone as a means of capturing the majority of an organism's gene content. Despite many limitations, it has been shown that EST databases can be a valid and reliable source of gene expression data (Ewing *et al.,* 1999; Ogihara *et al.,* 2003; Ronning *et al.,* 2003).

Defining the transcriptome of a complex, multicellular eukaryote is, however, a daunting challenge. The two most widely used and comprehensive approaches are whole genome sequencing coupled with application of gene prediction algorithms (Mathe *et al.,* 2002) and single pass sequencing of cDNAs to obtain expressed sequence tags (ESTs; Adams *et al.,* 1991). Among newer approaches that have not yet been used widely are targeted sequencing of gene-rich regions, identified either as being hypomethylated (Rabinowicz *et al.,* 1999; Bedell *et al.,* 2005) or enriched in single-copy sequences (Peterson *et al.,* 2002), and serial analysis of gene expression (Velculescu *et al.,* 1995). However, gene prediction algorithms are as yet imperfect (Mathe *et al.,* 2002), while other methods are in a practical sense incapable of identifying every potentially expressed gene. Ultimately, a combination of strategies employed in parallel will be required to provide a near-complete description of any complex transcriptome.

7. Transcriptome Studies in Plants at Various Developmental Stages

The diversity of behavior and responses that a cell undergoes through differentiation and growth is the product of the highly coordinated and regulated expression of different genes. The identification of genes that are differentially expressed in a particular tissue or developmental stage or under stress condition is critical to understand a unique function played for such a tissue and/or the molecular ground of the differentiation process.

Silva *et al.*(2005) analysed and annotated 146,075 expressed sequence tags from *Vitis* species. The majority of these sequences were derived from different cultivars of *Vitis vinifera*, comprising an estimated 25,746 unique contig and singleton sequences that survey transcription in various tissues and developmental stages and during biotic and abiotic stresses. Putatively homologous proteins were identified for over 17,752 of the transcripts, with 1,962 transcripts further subdivided into one or more Gene Ontology categories. A simple structured vocabulary was developed to describe the relationship between individual expressed sequence tags and cDNA libraries which provided query terms to facilitate data mining within the context of a relational database.

Fernandeza *et al.* (2003) generated differential organ-specific Sunflower ESTs from leaf, stem, root and flower bud at two developmental stages (R1 and R4). The use of different sources of RNA as tester and driver cDNA for the construction of differential libraries was evaluated as a tool for detection of rare or low abundant transcripts. Organ-specificity ranged from 75 to 100% of non-redundant sequences in the different cDNA libraries. Sequence redundancy varied according to the target and driver cDNA used in each case. Comparison with sunflower ESTs on public databases showed that 197 of non-redundant sequences (60%) did not exhibit significant similarity to previously reported sunflower ESTs. This approach helped to successfully isolate a significant number of new reported sequences putatively related to responses to important agronomic traits and key regulatory and physiological genes.

Ray *et al.* (1988) isolated a cDNA clone for tomato pectin esterase, an enzyme also implicated in cell wall softening. They reported the structure of this cDNA and compare it with the structure of pectin esterase derived from amino acid sequence experiments.

7.1. Transcriptome Studies under Abiotic Stresses

Differences in gene expression during abiotic stresses such as drought, salinity, cold and high temperature varies to the type and extent of stress. In the model plant Arabidopsis deeper insights were gained into functional genomics aspects of multiple stress interactions. Using 1300 full-length clones (Seki *et al.,* 2001) and 7000 full-length clone inserts (Seki *et al.,* 2002a,b) multi-stress interactions of abiotic stress treatments were studied to overlapping responses as well to identify genes of potential interest to salt, drought and cold responses. By using 1300 full-length clones, Seki *et al.* (2001) identified a set of only 44 and 19 genes, which were induced either by drought or cold stress response, respectively. By using 7000 full-length inserts, 299 drought-inducible genes, 213 high salinity-stress-inducible genes, 54 cold-inducible genes and 245 ABA-inducible genes were identified (Seki *et al.,* 2002a,b).

Genes responding to a particular stress vary between species and even genotypes due to the fact that certain genotypes have efficient signal perception and transcriptional changes that lead to successful adaptations and eventually tolerance. Kawasaki *et al.* (2001) reported large-scale gene expression profiling in the salt-tolerant rice variety Pokkali as well as in the salt sensitive variety IR29 at 15-min to 7-day time intervals under control and high salinity conditions. These authors concluded that the tolerant cultivar responded at the level of transcription after 15 min of salt stress and displayed up-regulation of many genes encoding glycine-rich proteins, ABA and stress-induced proteins, metallothionein-like proteins, glutathione S-transferase, ascorbate peroxidase, water channel protein isoforms, subtilisin inhibitor, tyrosine inhibitor and others. Many transcripts that were up-regulated in the tolerant cultivar responded more slowly in the sensitive cultivar during salt treatment.

Brautigam *et al.* (2005) generated 9792 single-pass ESTs sequences from coldacclimated oat, *Avena sativa.* The library was prepared from pooled RNA samples isolated from leaves of four-week old *Avena sativa* plants incubated at +4°C for 4, 8, 16 and 32 hours. Clustering and assembly identified a set of 2800 different transcripts in which 398 displayed significant homology to genes previously reported to be involved in cold stress related processes, 107 novel oat transcription factors were also identified, out of which 51 were similar to genes previously shown to be cold induced. Four oat CBF sequences were found which play a major role in regulating cold acclimation, belonging to the monocot cluster of DREB family ERF/AP2 domain proteins. Finally in the total EST sequence data (5.3 Mbp) approximately 400 potential SSRs were found, a frequency similar to what has previously been identified in *Arabidopsis* ESTs.

Lee *et al.* (2005) characterized an abiotic stress-inducible dehydrin Gene, OsDhn1, in Rice (*Oryza sativa* L.). The tomato ripening-associated membrane protein (TRAMP) was reported to be expressed in numerous tissues, including roots, and was upregulated in tomato stems during water stress (Fray *et al.*, 1994).

Ji *et al.* (2005) constructed a full-length cDNA library of *Glycine soja* (50109) leaf treated with 150 mM NaCl, using the SMART technology. Random expressed sequence tag (EST) sequencing of 2,219 clones produced 2,003 ESTs for gene expression analysis. The average read length of cleaned ESTs was 454 bp, with an average GC content of 40%. These ESTs were assembled using the PHRAP program to generate 375 contigs and 696 singlets. The EST sequences of *Glycine soja* were compared to that of Glycine max by using the blastn algorithm.

By differential library screening, many dehydration responsive genes including RD29A were identified in Arabidopsis (Yamaguchi- Shinozaki and Shinozaki., 1993). The transcription factors known as dehydration-responsive element binding factors (DREB) that bind to dehydration-responsive elements (DRE) in the RD29A promoter were later identified using a yeast one-hybrid system (Liu *et al.*, 1998). Using similar methods, a set of cold responsive genes was isolated from Arabidopsis (Thomashow, 1999). Among them is the C-repeat binding factor (CBF) family, which turned out to be the DREB1 family. Constitutive expression of some of these transcription factors improves both freezing and drought tolerance by inducing a cold acclimated-like state (Jaglo-Ottosen *et al.*, 1998; Liu *et al.*, 1998; Kasuga *et al.*, 1999; Gilmour *et al.*, 2000).

To understand the molecular basis of plant responses to salt stress better, suppression subtractive hybridization (SSH) and microarray approaches were combined to identify the potential important or novel genes involved in the early stage of tomato responses to severe salt stress. First, SSH libraries were constructed for the root tissue of two cultivated tomato (*Solanum lycopersicum*) genotypes: LA2711, a salt-tolerant cultivar, and ZS-5, a salt-sensitive cultivar, to compare salt treatment and non-treatment plants. Then a subset of clones from these SSH libraries were used to construct a tomato cDNA array and microarray analysis was carried out to verify the expression changes of this set of clones upon a high concentration of salt treatment at various time points compared to the corresponding non-treatment controls. A total of 201 non-redundant genes that were differentially expressed upon 30 min of severe salt stress either in LA2711 or ZS-5 were identified from microarray analysis; most of these genes have not previously been reported to be associated with salt stress (Ouyang *et al.*, 2007).

Several tomato genes (*Solanum lycopericum*) have been isolated (e.g. cDNA *le4, le16, le20, le25*) whose expression is increased by treatment with abscisic acid (ABA) or under by drought (Cohen and Bray, 1990; Khan *et al.*, 1993). These genes were identified as a result of their elevated expression in detached, water stressed leaves of wild type plants compared to ABA-deficient *Flacca* plants.

Further, Seki *et al.* (2002a) employed a full-length cDNA microarray containing 7000 independent Arabidopsis cDNAs to identify cold, drought and salinity-induced target genes and stress-related transcription factor family members such as DREB, ERF, WRKY, MYB, bZIP, helix-loop-helix and NAC. These results indicated that there was a greater cross-talk between salt and drought stress signaling processes in comparison to salt and cold stresses.

Similarly, transcriptome response to dehydration, salinity and ABA has been monitored in sorghum seedlings and identified approximately 22 transcription factors (Buchanan *et al.*, 2005). These regulators include ABF from bZIP factors, DREB from AP2/EREBP family, HD-ZIP and MYB factors, which are also known to be stress-responsive in other model species such as Arabidopsis and rice. Also, there is a greater need to verify the roles that these transcription factors play in the networks for better designing plants that can tolerate a variety of environmental stresses.

Because plant responses to stress are complex, the functions of many of the genes are still unknown (Cushman and Bohnert, 2000; Bray, 2002). In contrast, not all stress responsive genes are involved in cellular adaptive processes; some of them are simply involved in short term deleterious responses. Many of the traits that explain plant adaptation to drought are those determining plant development and shape, such as phenology, the size and depth of root system, xylem properties or the storage of reserves. These traits are mostly constitutive rather than stress inducible (Blum, 1984; Passioura, 2002).

Genome-wide transcriptome analysis has identified hundreds of genes encoding transcription factors that are induced or repressed by many environmental stresses (Chen and Zhu, 2004). The expression patterns of these transcription factors are highly complex and tolerance and resistance are controlled at the transcriptional level by an extremely intricate gene regulatory network.

Also, there is a greater need to verify the roles that these transcription factors play in the networks for better designing plants that can tolerate a variety of environmental stresses.

7.2. Expression Profiling Studies Based on Microarray Experiments

Vast amounts of DNA sequence information's are currently being produced that require accurate annotation based on bioinformatics algorithms and solid experimental evidence. The new area of functional genomics benefits from recent advances in high throughput technologies that allow the description of biological processes from a global genetic perspective (Somerville and Somerville, 1999). One such technology, DNA microarray, first described by Schena *et al.*(1995) in *Arabidopsis thaliana*, allows the simultaneous analysis of transcription profiles of thousands of genes in parallel. It has since been applied to most other model organisms including *Escherichia coli*, yeast, *Caenorhabditis elegans*, *Drosophila melanogaster*, mouse and human (Alon *et al.*, 1999; Jelinsky and Samson, 1999; Richmond *et al.*, 1999; White *et al.*, 1999; Ross *et al.*, 2000; Kim *et al.*, 2001). On a whole genome level, microarrays provide a high-throughput platform to measure gene expression and thereby generate functional data for many genes simultaneously.

A prerequisite of microarray technology is a large collection of cDNA clones or oligonucleotides representing the individual genes of the organism under investigation. These can be either in the form of expressed sequence tags (ESTs) that are usually derived from cDNA libraries, or of synthetic oligonucleotides matching to the genes (sometimes called DNA chips; Lipshutz *et al.*, 1999). These are spotted and immobilised in a high-density array on a solid phase such as a nylon membrane or glass slide. These arrays can then be used for the hybridization of fluorescently-labelled cDNA representing the mRNA levels from the biological samples under investigation. To compare the relative mRNA abundance in different cells, tissues or physiological states, mRNA from these samples is extracted, converted to cDNA, labelled with different fluorescent dyes and hybridised simultaneously onto the array. The amount of each labelled target bound to each spot on the array is then quantified. Skewed ratios of signal intensities between control and test cDNA represent induction or repression of transcription activities related to the two different biological samples.

Microarray platforms for crop species have been developed as well. Because the research communities for some of these species are smaller, several projects have been organized as consortia to provide a microarray expression platform for these species, either from Affymetrix or as synthesized long oligonucleotide sets. Because most of these microarrays have become available only recently, little data is publicly available. However, studies using arrays with crop species have been reported for barley(close et al 2004, Oztur *et al.*, 2002), grape (Waters *et al.*, 2005), maize (Fernandes *et al.*, 2002, Cho *et al.*, 2002, Wang *et al.*, 2003; Yu *et al.*, 2003), pine (Watkinson, 2003), poplar (Moreau *et al.*, 2005), potato (Ducreux *et al.*, 2005; Restrepo *et al.*, 2005; Rensink *et al.*, 2005), tomato (Baxter *et al.*, 2005; Errikson *et al.*, 2004), soybean (Moy *et al.*, 2004, Vodkin *et al.*, 2004; Thibaud-Nissen *et al.*, 2003) and wheat (Clarke *et al.*, 2005).

7.3. Use of cDNA Microarray

Microarray technology has become a useful tool for the analysis of genome-scale gene expression (Schena *et al.*, 1995; Eisen and Brown, 1999). This DNA chip-based technology arrays cDNA sequences on a glass slide at density >1000 genes/ cm^2. These arrayed sequences are hybridized simultaneously to a two-color fluorescently labeled cDNA probe pair prepared from RNA samples of different cell or tissue types, allowing direct and large-scale comparative analysis of gene expression. This technology was first demonstrated by analyzing 48 Arabidopsis genes for differential expression in roots and shoots (Schena *et al.*, 1995). Microarrays were used to study 1000 randomly chosen clones from a human cDNA library for identification of novel genes responding to heat shock and protein kinase C activation (Schena *et al.*, 1996). In another study, expression profiles of inflammatory disease-related genes were analyzed under various induction conditions by this chip-based method (Heller *et al.*, 1997). Furthermore, the yeast genome of >6000 coding sequences has been analyzed for dynamic expression by the use of microarrays (DeRisi *et al.*, 1997; Wodicka *et al.*, 1997). In plant science, several reports of microarray analyses have been published (Aharoni *et al.*, 2000; Reymond *et al.*, 2000).

Several biological processes have been studied in *Arabidopsis* using DNA microarrays, including seed development (Girke *et al.*, 2000), comparisons of gene expression between roots and leaves (Schena *et al.*, 1995; Ruan *et al.*, 1998), between light-grown and dark grown *Arabidopsis* seedlings (Desprez *et al.*, 1998), between two different *Arabidopsis* accessions (Kehoe *et al.*, 1999) and under different nitrate treatments (Wang *et al.*, 2000). Novel diurnal and circadian-regulated genes in *Arabidopsis* were also recently identified using microarrays (Schaffer *et al.*, 2001). Other studies monitored gene expression changes of *Arabidopsis* genes following inoculation with pathogens such as *Alternaria brassicicola* (Schenk *et al.*, 2000) and *Peronospora parasitica* (Maleck *et al.*, 2000).

DNA microarray analyses were also carried out following treatments with the plant defense signalling compounds salicylic acid, methyl jasmonate and ethylene (Schenk *et al.*, 2000) or chemical analogues benzothiadiazol and 2,6 dichloroisonicotinic acid (Maleck *et al.*, 2000). Other examples of plants used for microarrays analysis include strawberry, petunia and maize (Aharoni *et al.*, 2000;

Baldwin *et al.,* 1999). The expression profiles of genes with no known function obtained by microarray analyses have been useful for assigning putative roles (e.g. studies by Grant and Mansfield, 1999). However, our current understanding of gene function still does not match recent advances in genome sequencing and gene discovery. For example, in *Arabidopsis* putative functions of less than half of its 25,498 predicted genes are known based on sequence homology to characterised genes (The Arabidopsis Genome Initiative, 2000). Depending on the biological question asked, DNA microarrays can be used mainly for two purposes: the 'discovery' of individual genes with important features and the identification of gene clusters with similar expression profiles.

Expression profiling has become an important tool to investigate how an organism responds to environmental changes. Plants, being sessile, have the ability to dramatically alter their gene expression patterns in response to environmental changes such as temperature, water availability or the presence of deleterious levels of ions. Sometimes these transcriptional changes are successful adaptations leading to tolerance while in other instances the plant ultimately fails to adapt to the new environment and is labeled as sensitive to that condition.

Expression profiling can define both tolerant and sensitive responses. These profiles of plant response to environmental extremes (abiotic stresses) are expected to lead to regulators that will be useful in biotechnological approaches to improve stress tolerance as well as to new tools for studying regulatory genetic circuitry. Finally, data mining of the alterations in the plant transcriptome will lead to further insights in to how abiotic stress affects plant physiology.

Gene expression profiling using cDNA macroarrays (Sreenivasulu *et al.,* 2006) or microarrays (Chen *et al.,* 2002) are novel approaches to identify higher number of transcripts and pathways related to stress tolerance mechanisms than before. There are several studies reported related to abiotic stress transcriptome profiling in model species such as Arabidopsis and rice that have revealed several new stress-related pathways in addition to the previously well described stress-related genes (Desikan *et al.,* 2001; Chen *et al.,* 2002; Kreps *et al.,* 2002; Seki *et al.,* 2002a; Oh *et al.,* 2005).

To obtain an overall view on transcript modification during the cherry tomato fruit responding to biocontrol yeast *Cryptococcus laurentii,* Feng *et al.* (2009) performed a microarrary analysis, using Affymetrix Tomato Genechip arrays, representing approximately 10,000 genes. The results showed that 194 and 312 genes were up- or downregulated, respectively, more than ten time fold in biocontrol yeast treated tomato fruit as compared with control fruits. Those up-regulated included genes involved in metabolism, signal transduction, and stress response. Conversely, genes related to energy metabolism and photosynthesis was generally down-regulated.

7.4. Limitations in the Use of Microarrays

There are some limitations in using cDNA microarrays. Firstly, these are usually to be developed for individual crop species, although in some cases a cDNA microarray of one species can be used for studying the expression patterns of another species. Further, among the same species, it is desirable to develop a cDNA microarray for a specific purpose. For example, specific cDNA microarrays have to be developed for monitoring the gene expression patterns for developing seeds (Girke *et al.*, 2000; Ohiorogge *et al.*, 2000), salt tolerance (Kawashaki *et al.*, 2001), and plant defense mechanisms (Schenk *et al.*, 2000, Maleck *et al.*, 2000). Otherwise, one would have to possess a cDNA microarray that contains all the ESTs that are available in a particular species. This could be achieved from the complete genome sequence, using bioinformatics, to identify all ESTs. The higher the number of relevant probes represented in the cDNA microarray, the more information is developed on expression patterns. Although global EST analysis is desirable, the current bottle neck is the availability of bioinformatics tools and expertise to conclusively analyze the large amount of data. A subtractive cDNA library, representational difference analysis (RDA), and suppressive subtractive hybridization (SSH) techniques could also be adopted to specifically enrich the transcripts in one set of conditions and use them to make a cDNA microarray.

Another major constraint is the requirement for a high amount of target (1-2 ng of mRNA or 50-200 ng of total RNA) for each hybridization experiment. This could be difficult to obtain from specific tissues or cell types during development. This approach has been demonstrated using 1 µg of total RNA (target) before amplification and was successfully used in identifying gene expression patterns in developing xylem and phloem of Populus (Hertzberg *et al.*, 2001). Since the microarray technique can monitor the expression pattern at mRNA levels, gene regulation at the post transcriptional level cannot be studied. In such cases, one has to gather information from other resources, such as proteomics, in assigning the functional role of different genes. Improvement in microarray technology is on fast pace to solve various limitations and is now emerging as an ideal tool for studying gene expression.

8. Conclusion and Future Perspectives

Heat is one of the most widespread environmental stresses and affects almost all plant functions (Yamaguchi-Shinozaki *et al.*, 2002). Response to heat stress involves many structural and metabolic alterations in plants (Bohnert and Sheveleva, 1998). Some of these alterations include in root to shoot ratios, leaf anatomical changes, accumulation of many compatible metabolites, alterations in carbon and nitrogen metabolism, synthesis of many novel proteins, etc. (Pinheiro *et al.,* 2001; Chartzoulakis *et al.,* 2002). A variety of physiological key processes such as stomatal conductance, transpiration, electron transport, photosynthesis, respiration are adversely affected due to heat, which ultimately determine the yield of crops (Qing *et al.,* 2001). Atmospheric temperature is closely associated with photosynthetic activity, dry matter production and yield in many species (Tollenaar and Aguilera, 1992; Qing *et al.,* 2001). To maintain growth and productivity under high temperature environments, plants must adapt to stress conditions and exercise specific tolerance mechanisms.

Different abiotic stresses such as heat, drought, salinity and oxidative stress are often interconnected, and may induce similar cellular damage. For example, heat and/or salinization are manifested primarily as osmotic stress, resulting in the disruption of homeostasis and ion distribution in the cell (Serrano *et al.,* 1999; Zhu, 2001a). Oxidative stress, which frequently accompanies high temperature, salinity, or drought stress, may cause denaturation of functional and structural proteins (Smirnoff, 1998). As a consequence, these diverse environmental stresses often activate similar cell signaling pathways (Shinozaki and Yamaguchi-Shinozaki, 2000; Knight and Knight, 2001 and 2002) and cellular responses, such as the production of stress proteins, up-regulation of anti-oxidants and accumulation of compatible solutes (Cushman and Bohnert 2000).

The complex plant response to abiotic stress, involves many genes and biochemical-molecular mechanisms. The ongoing elucidation of the molecular control mechanisms of abiotic stress tolerance, which may result in the use of molecular tools for engineering more tolerant plants, is based on the expression of specific stress-related genes. These genes include three major categories: (i) those that are involved in signaling cascades and in transcriptional control, such as MYC, MAP kinases and SOS kinase (Shinozaki and Yamaguchi-Shinozaki 1997; Munnik *et al.,* 1999), phospholipases (Frank *et al.,* 2000), and transcriptional factors such as HSF, and the CBF/DREB and ABF/ ABAE families (Stochinger *et al.,* 1997; Choi *et al.,* 2000; Shinozaki and Yamaguchi-Shinozaki 2000); (ii) those that function directly in the protection of membranes and proteins, such as heat shock proteins (HSPs) and chaperones, late embryogenesis abundant (LEA) proteins (Vierlintg 1991; Ingram and Bartels 1996; Bray *et al.,* 2000), osmoprotectants, and free-radical scavengers (Bohnert and Sheveleva, 1998); (iii) those that are involved in water and ion uptake and transport such as aquaporins and ion transporters (Maurel 1997; Serrano *et al.,* 1999; Blumwald, 2000). In order to produce transgenic plants with enhanced capacity for heat tolerance, it is essential to examine the expression of stress related genes and biochemical changes that occur in plants exposed to heat and to identify the candidate genes with their specific functions which are associated with heat stress tolerance.

9. References

1. Abdalla, M.M.; Khoshiban, E.The Influence of Water Stress on Growth, Relative Water Content, Photosynthetic Pigments, Some Metabolic and Hormonal Contents of two *Triticium aestivum*cultivars. *J Appl Sci Res.* **2007**, 3, 2062–2074.

2. Abdul-Baki, A.A. Tolerance of tomato cultivars and selected germplasm to heat stress.*J Am Soc Horticult Sci.* **1991**,116,1113–1116.

3. Adams, M.D.; Dubnick, M.; Kerlavage, A.R.; Moreno, R.; Kelley, J.M.; Utterback, T.R.; Nagle, J.W.; Fields, C.; Venter, J.C. Sequence identification of 2,375 human brain geBuchanan ar nes. *Nature.* **1992**, 355, 632-634.

4. Adams, M.D.; Kelley, J.M.; Gocayne, J.D.; Dubnick, M.; Polymeropoulos, M.H.; Xiao, H.; Merril, C.R.; Wu, A.; Olde, B.; Morenco, R.F.; Kerlavage, A.R.; McCombie, W.R.; Venter, J.C. Complementary DNA sequencing: Expressed sequence tags and human genome project. *Science.* **1991**, 252, 1651–1656.

5. Adamska, I.; Kloppstech, K. Evidence for the localization of the nuclear coded 22-kDa heat shock protein in a subfraction of thylakoid membranes. *Eur J Biochem.* **1991**, 198,375–381.

6. Adams, P.L.; Barry, C.; Giovannoni, J. Signal transduction systems regulating fruit ripening. *Trends in Plant Science.* **2004**, 9, 331–338.

7. Agrawal, S.; Kandimalla, E.R.; Yu, D.; Ball, R.; Lombardi, G.; Lucas, T.; Dexter, D.L.; Hollister, B.A.; Chen, S.F. GEM 231, a second-generation antisense agent complementary to protein kinase A RIalpha subunit, potentiates antitumor activity of irinotecan in human colon, pancreas, prostate and lung cancer xenografts. *Int J Oncol.* **2002**, 21,65–72.

8. Aharoni, A. DNA microarrays for functional plant genomics, Oscar Vorst. *Plant Molecular Biology.* **2001**, 48, 99-118.

9. Ahmad,I.; Hellebust, J.A. The relationship between inorganic nitrogen metabolism and proline accumulation in osmoregulatory responses of two euryhaline microalgae. *Plant Physiol.* **1988**, 88,348–354.

10. Ahmed, S.; Nawata, E.; Hosokawa, M.; Domae, Y.; Sakuratani, T. Alterations in photosynthesis and some antioxidant enzymatic activities of mungbean subjected to waterlogging.*Plant Sci.* **2002**, 163,117–123.

11. Ahn, Y.J.; Zimmerman, J.L. Introduction of the carrot HSP17.7 into potato (Solanum tuberosum L.) enhances cellular membrane stability and tuberization in vitro. *Plant Cell Environ.* **2006**, 29, 95–104.

12. Alexander, H.; Daniela, B.; Enrico, S.; Klaus, D.S. Crosstalk between Hsp90 and Hsp70 Chaperones and Heat Stress Transcription Factors in Tomato. *Plant Cell.* **2011**, 23(2),741–55.

13. Alfonso, M.; Yruela, I.; Almarcegui, S.; Torrado, E.; Perez, M.A.; Picorel, R. Unusual tolerance to high temperatures in a new herbicide-resistant D1 mutant from *Glycine max* (L.) Merr. Cell cultures deficient in fatty acid desaturation. *Planta.* **2001**, 212,573–582.

14. Al-Khatib, K.; Paulsen, G.M. High-temperature effects on photosynthetic processes in temperate & tropical cereals. *Crop Sci.* **1999**, 39,119–125.

15. Allakhverdiev, S.I.; Kreslavski, V.D.; Klimov, V.V.; Los, D.A.; Carpentier, R.; Mohanty, P. Heat stress: an overview of molecular responses in photosynthesis. *Photosynth Res.* **2008**, 98,541–550.

16. Almeselmani, M.; Deshmukh, P.S.; Sairam, R.K.; Kushwaha, S.R.; Singh, T.P. Protective role of antioxidant enzymes under high temperature stress. *Plant Science.* **2006**, 171, 382–388.

17. Alon, U.; Barkai, N.; Notterman, D.A.; Gish,K.; Ybarra, S.; Mack, D.; Levine, A.J. Broad patterns of gene expression revealed by clustering analysis of tumor and normal colon tissues probed by oligonucleotide arrays *Proc Natl Acad Sci USA.* **1999**, 96(12),6745-6750.

18. Alsadon, A.A.; Wahb-allah, M.A.; Khalil, S.O. Heat Tolerance Studies in Tomato (*Lycopersicon esculentum* Mill.). *Int J Agri Biol.* **2007**, 9(4), 649–652.

19. Alzhanova,D.V.; Napuli, A.J.; Creamer, R.; Dolja, V.V. Cell to cell movement and assembly of a plant closterovirus: Roles for the capsid proteins and Hsp70 homolog. *EMBO Eur Mol Biol Organ J.* **2001**, 20, 6997-7007.

20. Andersen, P.P. Food security: definition and measurement. *Food Sci.* **2009**, 1,5–7.

21. Apostol, I.; Heinstein, P.F.; Low, P.S. Rapid stimulation of an oxidative burst during elicidation of cultured plant cells. In: Role in defense and signal transduction. *Plant Physiol.* **1989**, 90, 106—116.

22. Arumuganathan, K.; Earle, E.D. Nuclear DNA content of some important plant species. *J Plant Mol Biol Rep.* **1991**, 9,208–218.

23. Asada, K. The water-water cycle in chloroplasts: scavenging of active oxygens and dissipation of excess photons. *Annu Rev Plant Physiol. Plant Mol Biol.* **1999**, 50, 601–639.

24. Asada, K.; Takahashi, M. Production and scavenging of active oxygen in photosynthesis. In: Kyle, D.J., Osmond, C.B., Arntzen, C.J.(Eds.),Photo inhibition. Elseiver Science Publishers, Amsterdam. **1987**, pp. 227- 287.

25. Ashburner, M.; Ball, C. A.; Blake, J. A.; Botstein, D.; Butler, H.; Cherry, J. M.; Davis, A. P.; Dolinski, K.; Dwight, S.S.; Eppig, J. T. Gene ontology: tool for the unification of biology. The Gene Ontology Consortium. *Nat Genet,* **2000**, 25, 25–29.

26. Ashraf, M.; Harris, P.J.C. Potential biochemical indicators of salinity tolerance in plants. *Plant Science.* **2004**, 166,3–16.

27. Aspinall, D.; Paleg, L.G. Proline accumulation-physiological aspects. In: Paleg, L.G., Aspinall, D. (Eds.), Physiology and Biochemistry of Drought Resistance in Plants. Academic Press, Sydney. **1981**, pp. 205.

28. Baker, B.; Zambryski, P.; Staskawicz, B.; Dinesh-Kumar, S.P. Signaling in plant-microbe interactions. *Science.* **1997**, 276,726–733.

29. Baniwal, S.K.; Bharti, K.; Chan, K.Y.; Fauth, M.; Ganguli, A.; Kotak, S.; Mishra, S.K.; Nover, L.; Port, M.; Scharf, K.D.; Tripp, J.; Weber, C.; Zielinski, D.; von Koskull-Döring, P. Heat stress response in plants: a complex game with chaperones and more than twenty heat stress transcription factors. *J Biosci.* **2004**, 29,471–487.

30. Barrs, H.D. Determination of water deficits in plant tissue. In: Kozlowski, T.T. (Ed.), Water deficits and plant growth. Academic Press 1, New York. **1968**, pp.235-368.

31. Basha, M.R.;Wei, W.; Reddy, G.R.; Zawia, N.H. Zinc finger transcription factors mediates perturbations of brain gene expression elicited by heavy metals. In: Zawia, N.H.,(Ed.), Molecular neurotoxicology, environmental agents, and transcription-transduction coupling. Boca Raton, FL: CRC. **2004**, pp. 43–64.

32. Bedell, V.M.; Yeo, S.Y.; Park, K.W.; Chung, J.; Seth, P.; Shivalingappa,V.; Zhao, J.; Obara, T.; Sukhatme, V.P.; Drummond, I.A.; Li, D.Y.; Ramchandran, R. Roundabout is essential for angiogenesis in vivo. *Proc Natl Acad Sci, USA.* **2005**, 102(18), 6373-6378.

33. Bharti, K.; Döving, K.V.P.; Bharti, S.; Kumar, P.; Körbitzer, A.T.; Treuter, E.; Nover, L. Tomato heat stress transcription factor HsfB1 represents a novel type of general transcription coactivator with a histone-like motif interacting with the plant CREB binding protein ortholog HAC1. *The Plant Cell.* **2004**, 16, 1521–1535.

34. Blaazquez, M.A.; Weigel, D. Integration of Floral inductive signals in *Arabidopsis. Nature.* **2000**, 404, 889-892.

35. Blumwald, E. Sodium transport and salt tolerance in plants.*Curr Opin Cell Biol.* **2000**, 12,431–434.

36. Bohnert, H.J.; Sheveleva, E. Plant stress adaptations making metabolism move. *Curr Opin Plant Biol.* **1998**, *1*, 267–274.

37. Boscheinen, O.; Lyck, R.; Queitsch, C.; Treuter, E.; Zimarino, V.; Scharf, K.D. Heat stress transcription factors from tomato can functionally replace Hsf1 in the yeast Saccharomyces cerevisiae. *Mol Gen Genet.* **1997**, 255,322–331.

38. Boston, R.S.; Viitanen, P.V.; Vierling, E. Molecular chaperones and protein folding in plants. *Plant Mol Biol.* **1996**, 32,191-222.

39. Bota, J.; Flexas, J.; Medrano, H. Is photosynthesis limited by decreased Rubisco activity and RuBP content under progressive water stress? *New Phytol.* **2004**, 162,671–681.

40. Bourdon, V.; Naef, F.; Rao, P.; Reuter, V.; Mok, S.; Bosl, G.; Koul, S.; Murty, V.; Kucherlapati, R.; Chaganti, R. Genomic and expression analysis of the 12p11-p12 amplicon using EST arrays identifies two novel amplified and over expressed genes. *Cancer Res.* **2002**, 62, 6218–6223.

41. Bovy, A.; Schijlen, E.; Hall, R.D. Metabolic engineering of flavonoids in tomato (*Solanum lycopersicum*): the potential for metabolomics. *Metabolomics.* **2007**, 3, 399–412.

42. Bowler, C.; Fluhr, R. The role of calcium and activated oxygens as signals for controlling cross-tolerance. *Trends Plant Sci.* **2000**, 5, 241–245.

43. Bowler, C.; Van Camp, W.; Van Montagu, M.; Inzé, D. Superoxide dismutase in plants. *Critical Review in Plant Science.* **1994**, 13,199–218.

44. Bowler, C.M.; Montagu, V.; Inze, D. Superoxide dismutase and stress tolerance, Annu. Rev. *Plant Physiol Plant Mol Biol.* **1992**, 43, 83-116.

45. Boyer, J.S. Measurement of the water status of plants. *Ann Review of Plant Physio.* **1968**, 9, 351-363.

46. Boyer, J.S. Plant productivity and environment. *Science.* **1982**, 218,443–448.

47. Boyes, D.C.; Zayed, A.M.; Ascenzi, R.; McCaskill, A.J.;Hoffman, N.E.; Davis, K.R.; Gorlach, J. Growth stage-based phenotypic analysis of Arabidopsis a model for high throughput functional genomics in plants. *Plant Cell.* **2001**, 13, 1499–1510.

48. Bräutigam, M.; Lindlöf, A.; Zakhrabekova, S.; Gharti-Chhetri, G.; Olsson, B.; Olsson, Olof. Generation and analysis of 9792 EST sequences from cold acclimated oat, *Avena sativa. BMC Plant Biol.* **2005,** 5, 18.

49. Bray, E.A. Abscisic acid regulation of gene expression during water deficit stress in the era of the Arabidopsis genome. *Plant Cell and Environ.* **2002**, 25, 153-161.

50. Bray, E.A.; Bailey-Serres, J.; Weretilnyk, E. Responses to abiotic stresses. In: Gruissem W, *et al.,*(Eds.), Biochemistry and Molecular Biology of Plants. American Society of Plant Physiologists. **2000**, pp, 158–1249.

51. Brazma, A.; Hingamp, P.; Quackenbush, J.; Sherlock, G.;Spellman, P.; Stoeckert, C.; Aach, J.; Ansorge, W.; Ball, C.A.; Causton, H.C.; Gaasterland, T.; Glenisson, P.; Holstege, F.C.; Kim, I.F.; Markowitz, V.; Matese, J.C.; Parkinson, H.; Robinson, A.; Sarkans, U.; Schulze-Kremer, S.; Stewart,J.; Taylor, R.; Vilo, J.; Vingron, M. Minimum information about a microarray experiment (MIAME)-toward standards for microarray data. *Nat Genet.* **2001**, 29, 365–371.

52. Brummell, D.; Harpster, M. Cell wall metabolism in fruit softening and quality and its manipulation in transgenic plants. *Plant Molecular Biology.* **2001**, 47, 311–340.

53. Buchanan, C.; Lim, S.; Salzman, R.; Kagiampakis, I.; Morishige, D.; Weers, B.; Klein, R.; Pratt, L.; Cordonnier-Pratt, M.; Klein, P. Sorghum bicolor's Transcriptome Response to Dehydration. High Salinity and ABA. *Plant Mol Biol.* **2005**, 58(5), 699-720.

54. Burton, G.W.; Ingold, K.U. β -carotene: an unusual type of lipid antioxidant. *Science.* **1984**, 224, pp. 569-573.

55. Camejo, D.; Rodríguez, P.; Morales, M.A.; Dell'amico, J.M.; Torrecillas,A.; Alarcón, J.J. High temperature effects on photosynthetic activity of two tomato cultivars with different heat susceptibility. *J Plant Physiol.* **2005**, 162,281–289.

56. Casano L.; Gomez, L.; Lascano, C.; Trippi, V. Inactivation and degradation of Cu/ZnSOD by active oxygen species in wheat chloroplasts exposed to photooxidative stress. *Plant Cell Physiol.* **1997**, 38,433-440.

57. Chaitante, D.; Di, A.I.; Maiuro, L.; Scippa, S.G. Effect of water stress on root meristems in woody and herbaceous plants during the first stage of development. *Form Fun Physio.* **2000**, 245-258.

58. Charles, J.B.; Mohammed, S.W.; Paul, Q.; Lee. Sweet love Comparison of changes in fruit gene expression in tomato introgression lines provides evidence of genome-wide transcriptional changes and reveals links to mapped QTLs and described traits. J of Exp Bot. **2005**, 56(416),1591–1604.

59. Chartzoulakis, K.; Patakas, A.; Kofidis, G.; Bosabalidis, A.; Nastou, A. Water stress affects leaf anatomy, gas exchange, water relations and growth of two avocado cultivars. *Scien Hort.* **2002**, 95, 39-50.

60. Chaves, M.M. Effects of water deficits on carbon assimilation. *J exp Bot.* **1991**, 42,1–16.

61. Chaves, M.M.; Pereira, J.S.; Maroco, J.; Rodrigues, M.L.; Ricardo, C.P.; Osorio, M.L.;Carvalho, I.; Faria, T.; Pinheiro, C.How plants cope with water stress in the field, Photosynthesis and growth. *Annals Bot.* **2002**, 89, 907–916.

62. Chen, J.; Wu, R.; Yang, P.C.;Huang, J.Y.;Sher, Y.P.;Han, M.H.;Kao, W.C.;Lee, P.J.;Chiu, T.F.;Chang, F.;Chu, Y.W.;Wu, C.W.; Peck, K. Profiling expression patterns and isolating differentially expressed genes by cDNA microarray system with colorimetry detection. *Genomics.* **1998**, 51,313–324.

63. Chen, S.; Gollop, N.; Heuer, B. Proteomic analysis of salt-stressed tomato (Solanum lycopersicum) seedlings: effect of genotype and exogenous application of glycinebetaine. *J Exper Bot.* **2009**, 60(7),2005–2019.

64. Chen, W.; Provart, N.J.; Glazebrook, J.; Katagiri, F.; Chang, H.S.; Eulgem, T.; Mauch, F.; Luan, S.; Zou, G.; Whitham, S.A. Expression profile matrix of Arabidopsis transcription factor genes suggests their putative functions in response to environmental stresses. *Plant Cell.* **2002**, 14, 559–574.

65. Chen, W.Q.J.; Zhu, T. Networks of transcription factors with roles in environmental stress response, *Trends Plant Sci.* **2004**, 9, 591–596.

66. Cho, Y.; Fernandes, J.; Kim, S.H.; Walbot, V. Gene-expression profile comparisons distinguish seven organs of maize. *Genome Biol.* **2002**, 3(9),45.1–45.16.

67. Choi, W.Y.; Kang, S.Y.; Park, H.K.; Kim, S.S.; Lee, K.S.; Shin, H.T.; Chai, S.Y. Effects of water stress by PEG on growth and physiological traits in rice seedlings. *Korean J Crop Sci.* **2000**, 45(2),112–117.

68. Chollet, R.; Vidal, J.; O'Leary, M.H. Phosphoenolpyruvate carboxylase: a ubiquitous, highly regulated enzyme in plants. *Ann Rev Plant Phys Plant Mol Biol.* **1996**, 47, 273–298.

69. Chrispeels, M.J.; Holuigue, L.; Latorre, R.; Luan, S.; Orellana, A.; Pena-Cortes, H.; Raikhel, N.V.; Ronald, P.C.; Trewavas, A. Signal transduction networks and the biology of plant cells. *Biol Res.* **1999**, 32, 35–60.

70. Close, T.J.; Wanamaker, S.I.; Caldo, R.A.; Turner, S.M.; Ashlock, D.A.; Dickerson, J.A.; Wing, R.A.; Muehlbauer, G.J.; Kleinhofs, A.; Wise, R.P. A new resource for cereal genomics: 22K barley Gene Chip comes of age. *Plant Physiol.* **2004**, 134, 960–968.

71. Cohen, A.; Bray, E.A. Characterization of three mRNAs that accumulate in wilted tomato leaves in response to elevated levels of endogenous abscisic acid. *Planta.* **1990**, 182,27-33.

72. Cornic, G.; Ghasghaie, J.; Genty, B.; Briantais, J.M. Leaf photosynthesis is resistant to a mild drought stress. *Photosynthetica.* **1992**, 27, 295–309.

73. Cornic, G.; Massacci, A.Leaf photosynthesis under drought stress. In: Baker, N.R., (Ed.), Photosynthesis and the Environment. The Netherlands, Kluwer Academic Pub. **1996**, pp.230.

74. Cortina, C.; Francisco, A.; Culianez-Macia. Tomato abiotic stress enhanced tolerance by trehalose biosynthesis. *Plant Science*. **2005**, 169, 75–82.

75. Costa, M.A.; Collins, R.E.; Anterola, A.M.; Cochrane, F.C.; Davin, L.B.; Lewis, N.G. An in silico assessment of gene function and organization of the phenylpropanoid pathway metabolic networks in Arabidopsis thaliana and limitations thereof. *Phytochemistry*. **2003**, 64, 1097–1112.

76. Craufurd, P.Q.; Bojang, M.; Wheeler, T.R.; Summerfield, R.J. Heat tolerance in cowpea: effect of timing and duration of heat stress. *Annals of Applied Biology*. **1998**, 133, 257–267.

77. Craufurd, P.Q.; Peacock, J.M. Effect of heat and drought stress on sorghum. *Exp Agric*. **1993**, 29,77–86.

78. Cushman, J.C.; Bohnert, H.J. Genomic approaches to plant stress tolerance. *Curr Opin Plant Biol*. **2000**, 3, 117–124.

79. Czarnecka-Verner, E.; Yuan, C.X.; Scharf, K.D.; Englich, G.; Gurley, W.B. Plants contain a novel multi-member class of heat shock factors without transcriptional activation potential. *Plant Mol Biol*. **2000**, 43,459-471.

80. Darwin, C. The Origin of Species by means of Natural Selection. In: Murray.,(6th Edn.), London, UK. **1875**.

81. Dat, J.; Vandenabeele, S.; Vranova, E.; Van, M.M.; Inze, D.; Van Breusengem, F. Dual action of the active oxygen species during plant stress responses. *Cell and Mol Life Sci*. **2000**, 57,779-795.

82. Dat, J.F.; Foyer, C.H.; Scott, I.M. Changes in salicylic acid and antioxidants during induced thermotolerance in mustard seedlings. *Plant Physiol*. **1998**, 118, 1455–1461.

83. Daveis, W.J.; Zhang, J. Root signals and the regulation of growth and development of plants in drying soil. *Ann Review of Plant Physiol*. **1991**, 42, 55–76.

84. Davletova, S.; Schlauch, K.; Coutu, J.; Mittler, R. The zinc-finger protein Zat12 plays a central role in reactive oxygen and abiotic stress signaling in Arabidopsis. *Plant Physiol*.**2005**, 139, 847–56.

85. Delauney, A.; Verma, D.P.S.) Proline biosynthesis and osmoregulation in plants. *Plant J*. **1993,** 4, 215-223.

86. DellaPenna, D.; Alexander, D.C.; Bennett, A.B. Molecular cloning of tomato fruit polygalacturonase: analysis of polygalacturonase mRNA levels during ripening. *Proceedings of the National Academy of Sciences of the USA*. **1986**, 83, 6420–6424.

87. DeRisi, J.; Penland, L.; Brown, P.; Bittner, M.; Meltzer, P.; Ray, M.; Chen, Y.; Su, Y.; Trent, J. Use of a cDNA microarray to analyze gene expression patterns in human cancer. *Nat Genet.* **1996**, 14,457–460.

88. Desikan, R.; Mackerness, S.A.H.; Hancock, J.T.; Neill, S.J. Regulation of the Arabidopsis transcriptome by oxidative stress. *Plant Physiol.* **2001**, 127, 159–172.

89. Desprez, T.; Amselem, J.; Caboche, M.; Höfte, H. Differential gene expression in Arabidopsis seedlings monitored using cDNA arrays. *Plant J.* **1998**, 14,643-652.

90. Dill, A.; Jung, H.S.; Sun, T.P. The DELLA motif is essential for gibberellin-induced degradation of RGA. *Proc Natl Acad Sci USA.* **2001**, 98, 14162–14167.

91. Doke, N.; Miura, Y.; Chai, H.B.; Kawakita, K. Involvement of Active Oxygen in induction of plant defense against infection and injury. In: Pelt, E.J., Steffen, K.L., (Eds.), Active Oxygen/Oxidative Stress and Plant Metabolism. American Soc. Plant Physiol. Rockville, M.D. **1991**, pp. 84-96.

92. Don, B.; Virginia, C.; Douglas, R. A comparison of gel-based, nylon filter and microarray techniques to detect differential RNA expression in plants. *Curr Opin in Plant Biol.* **1999**, 2,96-l03.

93. Dong, X. SA, JA, ethylene, and disease resistance in plants. *Curr Opin Plant Biol.* **1998**, 1, 316–323.

94. Dubey, R.S. Photosynthesis in plants under stressful conditions, In: Pessarakli, M., (Ed.), Handbook of Photosynthesis, Marcel Dekker, New York, NY. **1997**, 859–875.

95. Duck, N.; McCormick, S.;Winter, J. Heat shock protein hsp70 cognate gene expression in vegetative and reproductive organs of Lycopersicon esculentum. *Proc Natl Acad Sci. USA.* **1989**, 86, 3674–3678.

96. Ducreux, L.J.M.; Morris, W.L.; Hedley, P.E.; Shepherd, T.; Davies, H.V.; Millam, A.; Taylor, M.A. Metabolic engineering of high carotenoid potato tubers containing enhanced levels of b-carotene and lutein. *J Exp Bot.* **2005**, 56, 81–89.

97. Elstner, E.F. Mechanisms of oxygen activation in different compartments of plant cells, *Curr Top In Plant Physio*, **1991,** 6:13-25.

98. Emma, M.; Eriksson, A.B.; Ken, M.; John, A.; Jacquie, D.S.; Gregory, A.T.; Graham, B.S. Effect of the Colorless non-ripening Mutation on Cell Wall Biochemistry and Gene Expression during Tomato Fruit Development and Ripening. *Plant Physiol.* **2004**, 136, 4184–4197.

99. Erickson, A.N.;Markhart, A.H. () Flower developmental stage and organ sensitivity of bell pepper (*Capsicum annuum* L.) to elevated temperature. *Plant, Cell and Environment.* **2002**, 25, 123–130.

100. Etterson, J.R.; Shaw, R.G. Constraint to Adaptive Evolution in Response to Global Warming. *Science.* **2001**, *294*, 151.

101. Ewing, R.; Kahla, A.; Poirot, O.; Lopez, F.; Audic, S.; Claverie, J. Large-scale statistical analyses of rice ESTs reveal correlated patterns of gene expression. *Genome Res.* **1999**, 9, 950–959.

102. Falcone, D.L.; Ogas, J.P.; Somerville, C.R. Regulation of membrane fatty acid composition by temperature in mutants of Arabidopsis with alterations in membrane lipid composition. *BMCPlant Biol.* **2004**, 4, 17.

103. Feder, M.E.; Hoffman, G.E. Heat-shock proteins, molecular chaperones, and the stress response: evolutionary and ecological physiology. *Annu Rev Physiol.* **1999**, 61, 243–282.

104. Fei, Z.; Tang, X.; Alba, R.; White, J.; Ronning, C.; Martin, G.; Tanksley, S.; Giovannoni, J. Comprehensive EST analysis of tomato and comparative genomics of fruit ripening. *Plant J.* **2004**, 40(1), 47–59.

105. Feng, J.; Xiaodong, Z.; Jishuang, C. Microarray analysis of gene expression profile induced by the biocontrol yeast Cryptococcus laurentii in cherry tomato fruit. *Gene.* **2009**, 430, 12–16.

106. Fernandez, A.; Kardos, J.; Scott, L.R.; Goto, Y.; Berry, R.S. Structural defects and the diagnosis of amyloidogenic propensity. *Pro Natl Acad Sci USA Phys Rev Lett.* **2003**, 91, 8-102.

107. Foolad, M.R. Breeding for abiotic stress tolerances in tomato. In: Ashraf, M., Harris, P.J.C., (Eds.), Abiotic Stresses: Plant Resistance Through Breeding and Molecular Approaches. The Haworth Press Inc., New York, USA. **2005**, pp. 613–684.

108. Fowler, S.; Thomashow, M.F. Arabidopsis transcriptome profiling indicates that multiple regulatory pathways are activated during cold acclimation in addition to the CBF cold response pathway. *Plant Cell.* **2002**, 14, 1675–1690.

109. Foyer, C.H. Ascorbic acid. In: Alscher, R.G., Hess, J.L., (Eds.), Antioxidants in higher plants. CRC Press, Boca Raton. **1993**, pp. 31-58.

110. Foyer, C.H.;Descourvieres, P.; Kunert, K.J. Protection against oxygen radicals: An important defense mechanism studied in transgenic plants. *Plant Cell Environ.***1994**, 17, 507–523.

111. Foyer, C.H.; Halliwell, B. The presence of glutathione and glutathione reductase in chloroplasts: a proposed role in ascorbic acid metabolism. *Planta.* **1976**, 133, 21–25.

112. Frank, W.; Munnik, T.; Kerkmann, K.; Salamini, F.; Bartels, D. Water deficit triggers phospholipase D activity in the resurrection plant Craterostigma plantagineum. *Plant Cell.* **2000**, 12,111–124.

113. Fray, R.G.; Grierson, D. Identification and genetic analysis of normal and mutant phytoene synthase genes of tomato by sequencing, complementation and co-suppression. *Plant Mol Biol.* **1993**, 22, 589–602.

114. Fray, R.G.; Grierson, D.) Molecular genetics of tomato fruit ripening. *Trends in genetics.* **1993**, 9(12), 438–443.

115. Fray, R.G.; Wallace, A.; Grierson, D.; Lycett, G.W. Nucleotide sequence and expression of a ripening and water stress related cDNA from tomato with homology to the MIP class of membrane channel proteins. *Plant Mol Biol.* **1994**, 24, 539–543.

116. Gallego, S.M.; Benavides, M.P.; Tomaro, M.L. Effect of heavy metal ion excess on sunflower leaves: evidence for involvement of oxidative stress. *Plant Sci.* **1996**, 121, 151–159.

117. Gao, A.G.; Hakimi, S.M.; Mittanck, C.A.; Wu, Y.; Woerner, B.M.; Stark, D.M.; Shah, D.M.; Liang, J.; Rommens, C.M. Fungal pathogen protection in potato by expression of a plant defensin peptide. *Nat Biotechnol.* **2000**, 18, 1307–1310.

118. Gardner, P.R.; Fridovich, I. Superoxide sensitivity of Escerichia coli 6-phosphogluconate dehydratose. *J of Biol Chem.* **1991**, 266, 1478-1483.

119. Gebhardt, C.; Valkonen, J.P. Organization of genes controlling disease resistance in the potato genome. *Annu Rev Phytopathol.* **2001**, 39, 79–102.

120. Gibson, G.; Weir, B. The quantitative genetics of transcription. *Trends in Genetics.* **2005**, 21(11), 616–623.

121. Gill, S.S.; Tuteja, N. Ployamines and abiotic stress tolerance in plants. *Plant Sign Behavior.* **2010**, 5(1), 26–33.

122. Gilmour, S.J.; Sebolt, A.M.; Salazar, M.P.; Everard, J.D.; Thomashow, M.F. () Overexpression of the Arabidopsis CBF3 transcriptional activator mimics multiple biochemical changes associated with cold acclimation. *Plant Physio.* **2000**, 124(4), 1854-1865.

123. Gimenez, C.; Mitchell, V.J.; Lawlor, D.W. Regulation of photosynthetic rate of two sunflower hybrids under water stress. *Plant Physiol.* **1992**, 96, 635-643.

124. Girke, T.; Todd, J.; Ruuska, S.; White, J.; Benning, C.; Ohirogge. *J Plant Physiol.* **2000**, 124, 1570-1581.

125. Glazebrook, J.Genes controlling expression of defense responses in Arabidopsis. *Curr Opin Plant Biol.* **1999**, 2, 280–286.

126. Glover, J.R.; Lindquist, S. Hsp104, Hsp70, and Hsp40: a novel chaperone system that rescues previously aggregated proteins. *Cell.* **1998**, 94, 73–82.

127. Gong, M.; Li, X.J.; Dai, X.; Tian, M.; Li, Z.G. Involvement of calcium and calmodulin in the acquisition of HS induced thermotolerance in maize seedlings. *J Plant Physiol.* **1997**, 150, 615–621.

128. Grant, J.J.; Loake, G.J. Role of active oxygen intermediates and cognate redox signaling in disease resistance. *Plant Physiol.* **2000**, 124, 21–29.

129. Gray, J.; Picton, S.; Shabbeer, J.; Schuch, W.; Grierson, D. Molecular biology of fruit ripening and its manipulation with antisense genes. *Plant Mol Biol.* **1992**, 19, 69–87.

130. Groppa, M.D.; Benavides, M.P. Polyamines and abiotic stress: recent advances. *Amino Acids.* **2008**, 34, 35–45.

131. Grotewold, E. Plant metabolic diversity: a regulatory perspective. *Trends in Plant Science.* **2005**, 10(2), 57–62.

132. Gu, C.; Chen, S.; Liu, Z.; Shan, H.; Luo, H.; Guan, Z.; Chen, F. Reference gene selection for quantitative real-time PCR in *chrysanthemum* subjected to biotic and abiotic stress. *Mol Biotechnol.* **2011**, 49, 192–197.

133. Guilioni, L.; Wery, J.; Tardieu, F. Heat stress–induced abortion of buds and flowers in pea: is sensitivity linked to organ age or to relations between reproductive organs? *Annals of Botany.* **1997**, 80, 159–168.

134. Hall, A.E. Breeding for heat tolerance. *Plant Breed Rev.* **1992**, 10, 129–68.

135. Hall, A.E. Crop Responses to Environment. CRC Press LLC, Boca Raton, Florida. **2001**.

136. Hall, A.E.; Ziska, L.H. Crop breeding strategies for the 21st century. In: Reddy, K.R., Hodges, H.F. (Eds.), Climate Change and Global Crop Productivity, CABI Publishing. New York, USA. **2000**, pp:407–423.

137. Halliwell, B. *Bull. Eur. Physiopathol Respir.* **1981**, 17, 21-28.

138. Hameed, K.; Smit, A.; Boissinot, S. Molecular evolution and tempo of amplification of human LINE-1 retrotransposons since the origin of primates. *Genome Res January.* **2006**, 16(1), 78–87.

139. Hamilton, A.J.; Fray, R.G.; Grierson, D. Sense and antisense inactivation of fruit ripening genes in tomato. *Curr Topics in Microbiol and Immunol.* **1995**, 197, 77–89.

140. Hanks, S.K.; Quinn, A.M.; Hunter, T. The protein kinase family: conserved features and deduced phylogeny of the catalytic domains. *Science.* **1988**, 241, 42–52.

141. Hanna, H.Y.; Hernandez, T.P. Response of six tomato genotypes under summer and spring weather conditions in Louisiana. *Hort Science.* **1982**, 17, 758–759.

142. Hare, P.D.; Cress, W.A.; Van Staden, J. Proline synthesis and degradation: a model system for elucidating stress-related signal transduction. *J of Exp Bot.* **1999**, 50, 413-434.

143. Hare, P.D.;Cress, W. Metabolic implications of stress-induced proline accumulation in plants. *J Plant Growth Regul.* **1997**, 21, 79–102.

144. Hasegawa, P.; Bresson, R.; Zhu, J.K.; Bohnert, H.J. Plant cellular and molecular responses to salinity. *Annu Rev Plant Physiol Plant Mol Biol.* **2000**, 51, 463–499.

145. Hauser, M.T.; Aufsatz, W.; Jonak, C.; Luschnig, C. Transgenerational epigenetic inheritance in plants. *Biochimica et Biophysica Acta.* **2011**, 1809, 459–468.

146. Hausladen, A.; Alscher, R.G. Glutathione. In: Alscher, R.G., Hess, J.L. (Eds.), Antioxidants in Higher Plants. CRC Press, Boca Raton. **1993**, pp.1-23.

147. Heckathorn, S.A.; Downs, C.A.; Sharkey, T.D.; Coleman, J.S. The small methionine-rich chloroplast heat-shock protein protects photosystem II electron transport during heat stress. *Plant Physiol.* **1998**, 116, 439–444.

148. Heckathorn, S.A.; Ryan, S.L.; Baylis, J.A.; Wang, D.; Hamilton, E.W.; Cundiff, L.; Dawn, S.L. In vitro evidence from an Agrostis stolonifera selection genotype that chloroplast small heat-shock protein can protect photosystem II during heat stress. *Funct. Plant Biol.* **2002**, 29, 933–944.

149. Heerklotz, D.; Döring, P.; Bonzelius, F.; Winkelhaus, S.; Nover, L. The balance of nuclear import and export determines the intracellular distribution and function of tomato heat stress transcription factor HsfA2. *Mol Cell Biol.* **2001**, 21, 1759–1768.

150. Heller, R.A.; Schena, M.; Chai, A.; Shalon, D.; Bedilion, T.; Gilmore, J.; Woolley, D.E.; Davis, R.W. Discovery and analysis of inflammatory disease-related genes using cDNA microarrays. *Proc Natl Acad Sci USA.* **1997**, 94, 2150–2155.

151. Hertwig, B.; Streb, P.; Feierabend, J. Light dependence of catalase synthesis and degradation in leaves and the influence of interfering stress conditions. *Plant Physiol.* **1992**, 100, 1547–1553.

152. Hertzberg, M.; Sievertzon, M.; Aspeborg, H.; Nilsson, P.; Sandberg, G.; Lundeberg, J. *Plant J.* **2001**, 25, 585-591.

153. Heuvel, K.; Hulzink, J.; Barendse, G.; Wullems, G.) The expression of tgas118, encoding a defensin in Lycopersicon esculentum, is regulated by gibberellin. *Journal of Experimental Botany.* **2001**, 360, 1427-1436.

154. Heyne, E.G.; Brunson, A.M.) Genetic studies of heat and drought tolerance in maize. *J Am Soc Agro.* **1940**, 32,803–814.

155. Hong, S.W.; Vierling, E. Mutants of Arabidopsis thaliana defective in the acquisition of tolerance to high temperature stress. *Proc Natl Acad Sci USA.* **2000**, 97, 4392–4397.

156. Hong, S.W.; Vierling, E. Hsp101 is necessary for heat tolerance but dispensable for development and germination in the absence of stress. *Plant J.* **2001**, 27, 25–35.

157. Horváth, I.; Glatz, A.; Varvasovszki, V.; Török, Z.; Páli, T.; Balogh, G.; Kovács, E., Nádasdi, L.; Benkö, S.; Joó, F.; Vígh, L. Membrane physical state controls the signaling mechanism of the heat shock response in Synechocystis PCC 6803: identification of hsp17 as a 'fluidity gene'. *Proc Natl Acad Sci USA.* **1998**, 95, 3513–3518.

158. Howarth, C.J. Molecular responses of plants to an increased incidence of thermotolerance. *Plant Cell Environ.* **1991**, 14, 831–841.

159. Howarth, C.J. Genetic improvements of tolerance to high temperature. In: Ashraf, M., Harris, P.J.C. (Eds.), Abiotic Stresses: Plant Resistance Through Breeding and Molecular Approaches. Howarth Press Inc, New York. **2005**.

160. Hruz, T.; Laule, O.; Szabo, G.; Wessendorp, F.; Bleuler, S.; Oertle, L.; Widmayer, P.; Gruissem, W.; Zimmermann, P. Genevestigator V3: a reference expression database for the meta- analysis of transcriptomes. *Adv in Bioiformatics.* **2008**, 42074–77.

161. Hubbell, D.S. A morphological study of blind and flowering rose shoots, with special reference to flower-bud differentiation. *Journal of Agricultural Research.* **1934**, 48, 91–95.

162. Hugly, S.; Kunst, L.; Browse, J.; Somerville, C. Enhanced thermal tolerance of photosynthesis and altered chloroplast ultrastructure in a mutant of Arabidopsis deficient in lipid desaturation. *Plant Physiol.* **1989**, 90, 1134–1142.

163. Hussain, T.; Khan, I.A.; Malik, M.A.; Ali, Z. Breeding Potential for high temperature tolerance in corn (*Zea mays* L.). *Pakistan J Bot.* **2006**, 38, 1185–95.

164. Iba, K. Acclimative response to temperature stress in higher plants: approaches of gene engineering for temperature tolerance. *Annu Rev Plant Biol.* **2002**, 53, 225–245.

165. Ingram, J.; Bartels, D. The molecular basis of dehydration tolerance in plants. *Annu Rev Plant Biol.* **1996**, 47, 377–403.

166. IPCC. In: Penner, J.E., *et al.* (Eds.), Aviation and the Global Atmosphere: A Special Report of IPCC Working Groups I and III. Cambridge University Press, Cambridge, United Kingdom and New York, USA. **1999**, pp. 373.

167. Itoh, H.; Ueguchi-Tanaka, M.; Sato, Y.; Ashikari, M.; Matsuoka, M. The gibberellin signaling pathway is regulated by the appearance and disappearance of SLENDER RICE1 in nuclei. *Plant Cell.* **2002**, 14, 57–70.

168. Izui, K.; Matsumura, H.; Furumoto, T.; Kai, Y. Phosphoenolpyruvate carboxylase: a new era of structural biology. *Annu Rev Plant Biol.* **2004**, 55, 69–84.

169. Jaglo-Ottosen, K.R.; Gilmour, S.J.; Zarka, D.G.; Schabenberger, O.; Thomashow, M.F. Arabidpsis CBF1 overexpression induces COR genes and enhances freezing tolerance. *Science.* **1998**, 280, 104-106.

170. Jelinski, S.A.; Samson, L.D. Global response of Saccharomyces cerevisiae to an alkylating agent. *Proc Natl Acad Sci USA.* **1999**, 96, 1486–1491.

171. Ji, E.K.; Pretorius, D.H.; Newton, R.; Uyan, A.D.; Hollenbach, K.; Nelson, T.R. Effects of ultrasound on maternal-fetal bonding: A comparison of two- and three-dimensional imaging. *Ultrasound in Obstetrics and Gynecology.* **2005**, 25, 473-477.

172. Jiang, Y.;Huang, B. Drought and heat stress injury to two cool season turfgrasses in relation to antioxidant metabolism and lipid peroxidation. *Crop Sci.***2001**, 41, 436–442.

173. Jung, S. Variation in antioxidant metabolism of young and mature leaves of Arabidopsis thaliana subjected to drought. *Plant Science.* **2004**, 166, 459-466.

174. Kawasaki, S.; Borchert, C.; Deyholos, M.; Wang, H.; Susan Brazille,A.; Kiyoshi, Kawai.; David Galbraith,A.; Bohnert, H.J. Gene Expression Profiles during the Initial Phase of Salt Stress in Rice. *The Plant Cell.* **2001**, 13, 889–905.

175. Kehoe, D.M.; Villand, P.; Somerville, S. DNA microarrays for studies of higher plants and other photosynthetic organisms. *Trends Plant Sci.* **1999**, 4, 38–41.

176. Kim, J.M.; Lim, W.J.; Suh, H.J. Feather-degrading Bacillus species from poultry waste. *Process Biochemistry.* **2001**, 37(3), 287-291.

177. Kishor, P.; Sangam, S.; Amrutha, R.; Laxmi, P.; Naidu, K.; Rao, K.; Rao, S.; Reddy, K.; Theriappan, P.; Sreenivasulu, N. Regulation of proline biosynthesis, degradation, uptake and transport in higher plants: its implications in plant growth and abiotic stress tolerance. *Curr Sci.* **2005**, 88, 424.

178. Kishor, P.B.K.; Hong, Z.; Miao, G.H.; Hu, C.A.;Verma, D.P.S. Overexpression of 1-Pyrroline-5-Carboxylate Synthetase Increases Proline Production and Confers Osmotolerance in Transgenic Plants. *Plant Physiol.* **1995**, 108, 1387–1394.

179. Knight, H.; Knight, M.R. Abiotic stress signalling pathways: specificity and cross-talk. *Trends Plant Sci.* **2001**, 6, 262–267.

180. Koes, R.E.; Quattrocchio, F.; Mol, J.N.M. The flavonoin biosynthetic pathway in plants: Function and evolution. *BioEssays.* **1994**, 16, 123–132.

181. Konsens, I.; Ofir, M.; Kigel, J. The effect of temperature on the production and abscission of flowers and pods in snap bean (*Phaseolus vulgaris* L.). *Annals of Botany.* **1991**, 67, 391–399.

182. Kreps, J.A.; Wu, Y.; Chang, H.S.; Zhu, T.; Wang, X.; Harper, J.F. Transcriptome changes for Arabidopsis in response to salt, osmotic, and cold stress. *Plant Physio.* **2002**, 130, 2129-2141.

183. Kumar, S.G.; Mattareddy, A.; Sudhakar, C. NaCl effects on praline metabolism in two high yielding genotypes of mulberry (*Morus alba* L.) with contrasting salt tolerance. *Plant Sci.* **2003**, 165, 1245–1251.

184. Kuo, C.G.; Chen, H.M.; Sun, H.C. Membrane thermostability and heat tolerance of vegetable leaves. In: Kuo, C.G. (Ed.), Adaptation of Food Crops to Temperature and Water Stress Asian Vegetable Research and Development Center. Shanhua, Taiwan. **1993**, 160-168

185. Kurek, I.; Chang, T.K.; Bertain, S.M.; Madrigal, A.; Liu, L.; Lassner, M.W.; Zhu, G. Enhanced Thermostability of Arabidopsis Rubisco Activase Improves Photosynthesis and Growth Rates under Moderate Heat Stress. *The Plant Cell.* **2007**, 19, 3230–3241.

186. Kwon, Y.; Oh, J.E.; Noh, H.; Hong, S.W.; Bhoo, S.H.; Lee, H. The ethylene signaling pathway has a negative impact on sucrose-induced anthocyanin accumulation in Arabidopsis. *J Plant Res.* **2011**, 124, 193–200.

187. Lamb, C.; Dixon, R.A. The oxidative burst in plant disease resistance. *Annu Rev Plant Physiol Plant Mol Biol.* **1997**, 48, 251–275.

188. Lamb, C.J. Plant disease resistance genes in signal perception and transduction. *Cell.* **1994**, 76, 419–422.

189. Larkindale, J.; Knight, M.R. Protection against Heat Stress-Induced Oxidative Damage in Arabidopsis Involves Calcium, Abscisic Acid, Ethylene, and Salicylic Acid. *Plant Physiology.* **2002**, 128(2), 682-695.

190. Larkindale, J.; Mishkind, M.; Vierling, E. Plant responses to high temperature. In: Jenks, M.A., Hasegawa, P.M., (Ed.), Plant Abiotic Stress. Blackwell Publishing Ltd. **2005**, pp.100–132.

191. Larkindale, J.; Vierling, E. Core genome responses involved in acclimation to high temperature. *Plant Physiol.* **2008**, 146, 748–61.

192. Lawlor, D.W. Effects of water deficit on photosynthesis. In: Smirnoff, N., (Ed.), Environment and plant metabolism. Bios Scientific Publishers Ltd, Oxford. **1995**, 129–160.

193. Lawlor, D.W.; Cornic, G. Photosynthetic carbon assimilation and associated metabolism in relation to water deficits in higher plants. *Plant Cell Environ.* **2002**, 25, 275-294.

194. Lawrence, S.D.; Cline, K.; Moore, G.A. Chromoplast development in ripening tomato fruit: Identification of cDNAs for chromoplast-targeted proteins and characterization of a cDNA en-coding a plastid localized low molecular weight heat shock protein. *Plant Mol Biol.* **1997**, 33, 483–492.

195. Lee, S.C.; Lee, M.Y.; Kim, S.J.; Jun, S.H.; Gynheung, A.; Seong, R. Characterization of an Abiotic Stress-inducible Dehydrin Gene, OsDhn1, in Rice (Oryza sativa L.). *Mol.Cells.* **2005**, 19(2), 212-218.

196. Leonardis, S.D.; Dipierro, N.; Dipierro, S. Purification and characterization of an ascorbate peroxidase from potato tuber mitochondria. *Plant Physiology and Biochemistry.* **2000**, 38, 773–779.

197. Levitt, J. Responses of Plants to Environmental Stresses. New York, Academic Press. **1980**.

198. Levy, A.; Rabinowitch, H.D.; Kedar, N. Morphological and physiological characters affecting flower drop and fruit set of tomatoes at high temperatures. *Euphytica.* **1978**, 27, 211–218.

199. Li, L.; Steffens, J.C. Overexpression of polyphenol oxidase in transgenic tomato plants results in enhanced bacterial disease resistance. *Planta.* **2002**, 215, 239-247.

200. Lindquist, S. Varying patterns of protein synthesis in Drosophila during heat shock: implications for regulation. *Dev Biol.* **1980**, 77, 463–479.

201. Ling, J.; Wells, D.R.; Tanguay, R.L.; Dickey, L.F.; Thompson, W.F.; Gallie, D.R. Heat shock protein HSP101 binds to the Fed-1 internal light regulatory element and mediates its high translational activity. *Plant Cell.* **2000**, 12, 1213–1227.

202. Ling, Y.L.; Stanghellini, C.; Challa, H. Effect of electrical conductivity and transpiration on production of greenhouse tomato (*Lycopersicon esculentum* L.). *Sci Hort.* **2001**, 88(1), 11–29.

203. Wodicka, L.; Dong, Helin.; Mittmann, Michael.; Ho, M.H.; J Lockhart, D. Genome-wide expression monitoring in Saccharomyces cerevisiae. *Nature Biotechnology.* **1997**, 15, 1359-1367.

204. Liu, F.; Stützel, H. Leaf expansion, stomatal conductance, and transpiration of vegetable amaranth (*Amaranthus* spp.) in response to soil drying. *J Amer Soc Hort Sci.* **2002**, 127, 878-883.

205. Liu, J.; Shono, M. Molecular cloning the gene of small heat shock protein in the mitochondria and endoplasmic reticulum of tomato. *Acta Bot Sin.* **2001**, 43(2), 138–145.

206. Liu, Q.; Kasuga, M.; Sakuma, Yoh.; ABE Hiroshi.; MIURA, Setsuko.; YAMAGUCHI-SHINOZAKI, Kazuko.; SHINOZAKI, Kazuo. Two transcription factors, DREB1 and DREB2, with an EREBP/AP2 DNA binding domain separate two cellular signal transduction pathways in drought and low-temperature-responsive gene expression, respectively, in Arabidopsis. *The Plant Cell.* **1998**, 10(8), 1391-1406.

207. Liu, X.A.; Bush, D.R. Expression and transcriptional regulation of amino acid transporters in plants. *Amino Acids.* **2006**, 30, 113-120.

208. Livak, K.J.; Schmittgen, T.D. Analysis of relative gene expression data using real-time quantitative PCR and the 2[-Delta Delta C (T)] method. *Methods.* **2001**, 25, 402–408.

209. Loggini, B.; Scartazza, A.; Brugnoli, E.; Navari-Izzo, F. Antioxidant defense system, pigment composition and photosynthetic efficiency in two wheat cultivars subjected to drought. *Plant Physiol.* **1999**, 119, 1091–1099.

210. Long, S.P.; Ort, D.R. More than taking the heat: crops and global change. *Curr Opin Plant Biol.* **2010**, 13, 241–248.

211. Lopez-Delgado, H.; Dat, J.F.; Foyer, C.H.; Scott, I.M. Induction of thermotolerance in potato microplants by acetylsalicylic acid and H_2O_2. *Journal of Experimental Botany.* **1998**, 49(321), 713-720.

212. López-Delgado, H.; Mora-Herrera, M.E.; Zavaleta-Mancera, H.A.; Cadena-Hinojosa, M.; Scott, I.M. Salicylic acid enhances heat tolerance and potato virus X (PVX) elimination during thermotherapy of potato microplants. *American Journal of Potato Research.* **2004**, 81, 171-176.

213. Loreto, F.; Tricoli, G.D. Marco On the relationship between electron transport rate and photosynthesis in leaves of the C4 plant Sorghum bicolor exposed to water stress, temperature changes and carbon metabolism inhibition. *Aust J Plant Physiol.* **1995**, 22, 885–892.

214. Lyck, R.; Harmening, U.; Höhfeld, I.; Treuter, E.; Scharf, K.D.; Nover, L. Intracellular distribution and identification of the nuclear localization signals of two plant heat-stress transcription factors. *Planta.* **1997**, 202, 117–125.

215. Maestri, E.; Klueva, N.; Perrotta, C.; Gulli, M.; Nguyen, H.T.; Marmiroli, N. Molecular genetics of heat tolerance and heat shock proteins in cereals. *Plant Mol Biol.* **2002**, 48, 667–681.

216. Maleck, K.; Levine, A.; Eulgem, T.; Morgan, A.; Schmid, J.; Lawton, K.A.; Dangl, J.L.; Dietrich, R.A. The transcriptome of Arabidopsis thaliana during systemic acquired resistance. *Nat Genet.* **2000**, 26, 403–410.

217. Marcum, K.B. Cell membrane thermostability and whole-plant heat tolerance of kentucky bluegrass. *Crop Sci.* **1998**, 38, 1214–1218.

218. Mark, S.; Dari, S.; Renu, H.; Chai, A.; Brown, P.O.; Davis, R.W. Parallel human genome analysis: Microarray-based expression monitoring of 1000 genes. *ProcNatlAcadSci USA.* **1996**, 93, 10614-10619.

219. Maroco, J.P.; Rodriges, M.L.; Lopes, C.; Chaves, M.M. Limitation to leaf photosynthesis in grapevine under drought – metabolic and modelling approaches. *Functional Plant Physiol.* **2002**, 29, 1–9.

220. Marshall, B.J.; Barrett, L.J.; Prakash, C.; McCallum, R.W.; Guerrant, R.L. Urea protects Helicobacter (Campylobacter) pylori from the bactericidal effect of acid. *Gastroenterology.* **1990**, 99, 697-702.

221. Maurel, C. Aquaporins and water permeability of plant membranes. *Annu Rev Plant Biol.* **1997**, 48, 399–429.

222. McCabe, P.F.; Leaver, C.J. Programmed cell death in cell cultures. *Plant Mol Biol.* **2000**, 44, 359–368.

223. McCabe, P.F.; Levine, A.; Meijer, P.J.; Tapon, N.A.; Pennell, R.I. A programmed cell death pathway activated in carrot cells cultured at low cell density. *Plant J.* **1997**, 12, 267–280.

224. McCord, J.M.; Fridovich, I. Superoxide dismutase. An enzymic function for erythrocuprein (hemocuprein). *Journal of Biological Chemistry.* **1969**, 244, 6049-6055.

225. McCue, K.R.; Hanson, A.D. Drought and salt tolerance: towards understanding and application. *Ttetids Biotech.* **1990**, 8, 358-362.

226. McKersie, B.D.; Bowley, S.R.; Harjanto, E.; Leprince, O. Water deficit tolerance and field performance of transgenic alfalfa overexpressing superoxide dismutase. *Plant Physiol.* **1996**, 111, 1177–1181.

227. McKersie, B.D.; Bowley, S.R.; Jones, K.S. Winter survival of transgenic alfalfa overexpressing superoxide dismutase. *Plant Physiol.* **1999**, 119, 839–848.

228. McKersie, B.D.; Chen, Y.; de Beus, M.; Bowley, S.R.; Bowler, C.; Inze, D.; D'Halluin, K.; Botterman, J. Superoxide dismutase enhances tolerance of freezing stress in transgenic alfalfa (*Medicago sativa* L.). *Plant Physiol.* **1993**, 103, 1155–1163.

229. Moffat, A.S. Finding new ways to protect drought-stricken plants. *Science.* **2002**, 296,1226–1229.

230. Mohamed, R.; Meilan, R.; Ostry, M.E.; Michler, C.H.; Strauss, S.H. Bacterio-opsin gene overexpression fails to elevate fungal disease resistance in transgenic poplar (Populus). *Can J For Res.* **2001**, 31, 268–275.

231. Momcilovic, I.; Ristic, Z. Expression of chloroplast protein synthesis elongation factor, EF-Tu, in two lines of maize with contrasting tolerance to heat stress during early stages of plant development. *J Plant Physiol.* **2007**, 164, 90–99.

232. Moreau, C.; Aksenov, N.; Lorenzo, M.G.; Segerman, B.; Funk, C.; Nilsson, P.; Jansson, S.; Tuominen, H. A genomic approach to investigate developmental cell death in woody tissues of Populus trees. *Genome Biol.* **2005**, 6, 5-34.

233. Morgan, R.J.M.; Williams, F.; Wright, M.M. An early warning scoring system for detecting developing critical illness. *Clin Intensive Care.* **1997**, 8, 100.

234. Moy, P.; Qutob, D.; Chapman, B.P.; Atkinson, I.; Gijzen, M. Patterns of gene expression upon infection of soybean plants by Phytophthora sojae. *Mol Plant–Microbe Interact.* **2004**, 17, 1051–1062.

235. Munnik, T.; Ligterink, W.; Meskiene, I.; Calderini, O.; Beyerly, J.; Musgrave, A.; Hirt, H. Distinct osmo-sensing protein kinase pathways are involved in signaling moderate and severe hyper-osmotic stress. *Plant J.* **1999**, 20, 381–388.

236. Murmu, J.; Plaxton, W.C. Phosphoenolpyruvate carboxylase protein kinase from developing castor oil seeds: partial puriWcation, characterization, and reversible control by photosynthate supply. *Planta.* **2007**, 226, 1299–1310.

237. Nardini, A.; Tyree, Mt.; Salleo, S. Xylem cavitation in the leaf of Prunus laurocerasus and its impact on leaf hydraulics. *Plant Physiol.* **2001**, 125, 1700–1709.

238. Nimmo, H.G. Control of phosphoenolpyruvate carboxylase in plants. In: Plaxton, W.C., McManus, M.T. (Eds.), Control of primary metabolism in plants. Blackwell Publishing, Oxford. **2006**, pp. 219–233.

239. Nishiyama, T.; Fujita, T.; Shin, I.T.; Seki, M.;Nishide, H.;Uchiyama, I.;Kamiya, A.;Carninci, P.;Hayashizaki, Y.;Shinozaki, K.; Kohara, Y.;Hasebe, M. () Comparative genomics of Physcomitrella patens gametophytic transcriptome and Arabidopsis thaliana: implication for land plant evolution. *Proc Natl Acad Sci.* **2003**, 100, 8007–8012.

240. Nover, L.; Bharti, K.; Döring, P.; Mishra, S.; Ganguli, A.; Scharf, K.D. Arabidopsis and the HSF world: How many heat stress transcription factors do we need? *Cell Stress Chaperones.* **2001**, 6, 177–189.

241. Nover, L.; Scharf, K.D.; Neumann, D. Formation of cytoplasmic heat shock granules in tomato cell cultures and leaves. *Mol Cell Biol.* **1983**, 3, 1648–1655.

242. Ogihara, Y.; Mochida, K.; Nemoto, Y.; Murai, K.; Yamazaki, Y.; Shin, I.T.; Kohara, Y. () Correlated clustering and virtual display of geneexpression patterns in the wheat life cycle by large-scale statistical analyses of expressed sequence tags. *Plant J.* **2003**, 33, 1001–1011.

243. Oh, S.J.; Song, S.I.K.; Kim, Y.S.; Jang, H.J.; Kim, S.Y.; Kim, M.; Kim, Y.K.; Nahm, B.H.; Kim, J.K. Arabidopsis CBF3/DREB1A and ABF3 in Transgenic Rice Increased Tolerance to Abiotic Stress without Stunting. *Growth Plant Physiology.* **2005**, 138, 341-351.

244. Ohiorogge, J.; Pollard, M.; Bao, X.; Focke, M.; Girke, T.; Ruuska, S.; Mekhedov, S.; Benning, C. *Biochem Soc Trans.* **2000**, 28, 567-574.

245. Olszewski, N.; Sun, T.P.; Gubler, F. Gibberellin signaling: biosynthesis, catabolism, and response pathways. *Plant Cell.* **2002**, 14, S61–S80.

246. Ortiz, R.H.; Braun, J.; Crossa, J.H.; Crouch, G.; Davenport, J.; Dixon, S.; Dreisigacker, E.; Duveiller, Z.; He, J.; Huerta, A.K.; Joshi, M.; Kishii, P.; Kosina, Y.; Manes, M.; Mezzalama, A.; Morgounov, J.; Murakami, J.; Nicol, G.O.; Ferrara, I.; Ortiz-Monasterio, T.S.; Payne, R.J.; Pena, M.P.; Reynolds, K.D.; Sayre, R.C.; Sharma, R.P.; Singh, J.; Wang, M.; Warburton, H.; Wu, M.I. Wheat genetic resources enhancement by the International Maize and Wheat Improvement Center (CIMMYT*). Genetic Resources and Crop Evolution.* **2008**.

247. Örvar, B.L.; Sanger, L.; Omann, F.; Dindhsa, R.S. Early steps in cold sensing by plant cells: the role of actin cytoskeleton and membrane fluidity. *Plant J.* **2000**, 23, 785–794.

248. Ozturk, Z.N.; Talame, V.; Deyholos, M.; Michalowski, C.B.; Galbraith, D.W.; Gozukirmizi, N.; Tuberosa, R.; Bohnert, H.J. Monitoring large-scale changes in transcript abundance in drought- and salt-stressed barley. *Plant Mol Biol.* **2002**, 48, 551–573.

249. Pace, C.N.; Shirley, B.A.; McNutt, M.; Gajiwala, K. Forces contributing to the conformational stability of proteins. *FASEB J.* **1996**, 10, 75-83.

250. Panda, S.K.; Patra, H.K. () Does chromium (III) produce oxidative stress in excised wheat leaves? *J Plant Biol.* **2000**, 27, 105–110.

251. Pascual, L.; Blanca, J.M.; Canizares, J.; Nuez, F. Analysis of gene expression during the fruit set of tomato: A comparative approach. *Plant Science.* **2007**, 173, 609–620.

252. Passioura, J.B. Environmental biology and crop improvement. *Functional Plant Biol.* **2002**, 29, 537-546.

253. Pastori, G.M.; Foyer, C.H. () Common components, networks, and pathways of cross-tolerance to stress. The central role of "redox" and abscisic acid-mediated controls. *Plant Physiol.* **2002**, 129, 460–468.

254. Paterson, A.; Bowers, J.; Burow, M.; Draye, X.; Elsik, C.G.; Chun-Xiao, J.; Katsar, C.S.; Lan, T.H.; Lin, Y.R.; Ming, R.; Wright, R.J. () Comparative genomics of plant chromosomes. *Plant Cell.* **2000**, 12, 1523–1540.

255. Patrick, O.; Brown, D.B. Exploring the new world of the genome with DNA microarrays. *Nature Genetics Supplement.* **1999**, 21, 33-37.

256. Pedley, K.F.; Martin, G.B. Molecular basis of Pto-mediated resistance to bacterial speck disease in tomato. *Annu Rev Phytopathol.* **1998**, 41, 215–243.

257. Peet, M.M.; Sato, S.; Gardner, R.G. Comparing heat stress effects on male-fertile and male-sterile tomatoes. *Plant Cell and Environment.* **1998**, 21, 225–231.

258. Peet, M.M.; Willits, D.H.; Gardner, R.G. Response surface analysis for post-pollen production processes in male-sterile tomatoes under sub-acute high temperature stress. *J Exp Bot.* **1997**, 48, 101–112.

259. Penninckx, I.A.; Eggermont, K.; Terras, F.R.; Thomma, B.P.; De Samblanx, G.W.; Buchala, A.; Metraux, J.P.; Manners, J.M.; Broekaert, W.F. Pathogen-induced systemic activation of a plant defensin gene in Arabidopsis follows a salicylic acid-independent pathway. *Plant Cell.* **1996**, 8, 2309–2323.

260. Penninckx, I.A.; Thomma, B.P.; Buchala, A.; Metraux, J.P.; Broekaert, W.F. Concomitant activation of jasmonate and ethylene response pathways is required for induction of a plant defensin gene in Arabidopsis. *Plant Cell.* **1998**, 10, 2103–2113.

261. Pennisi, E. A closer look at SNPs suggests difficulties. *Science.* **1998**, 281, 1787-1789.

262. Petretto, E.; Mangion, J.; Dickens, N.J.; Cook, S.A.; Kumaran, M.K.; Lu, H.; Fischer, J.; Maatz, H.; Kren, V.; Pravenec, M.; Hubner, N.; Aitman, T.J. Heritability and tissue specificity of expression quantitative trait loci. *PLoS Genet.* **2006**, 2(10),e172.

263. Pieterse, C.M.; van Wees, S.C.; Hoffland, E.; van Pelt, J.A.; van Loon, L.C. Systemic resistance in Arabidopsis induced by biocontrol bacteria is independent of salicylic acid accumulation and pathogenesis-related gene expression. *Plant Cell.* **1996**, 8, 1225– 1237.

264. Pinheiro, C.; Chaves, M.M.; Ricardo, C.P. Alterations in carbon and nitrogen metabolism induced by water deficit in the stems and leaves of Lupinus albus L. *J Exp Bot.* **2001**, 52, 1063–1070.

265. Pnueli, L.; Herr, E.H.; Rozenberg, M.; Cohen, M.; Goloubinoff, P.; Kalan, A.; Mitler, R. () Molecular and biochemical mechanisms associated with dormancy and drought tolerance in the desert legume Retama raetam. *Plant J.* **2002**, 31, 319–330.

266. Porter, J.R. Rising temperature is likely to reduce crop yields. *Nature.* **2005,** 436, 174.

267. Pratt, S.C.; Brooks, S.E.; Franks, N.R. The epic project: developing national evidence-based guidelines for preventing healthcare associated infections. Phase 1: guidelines for preventing hospital-acquired infections. *J of Hospital Infec.* **2001,** 47, S1-S82.

268. Price, A.; Lucas, P.W.; Lea, P.J. Age dependent damage and glutathione metabolism in ozone fumigated barley: A leaf section approach. *J of Exp Bot.* **1990,** 41, 1309-1317.

269. Putney, J.W.; Jr, Burgess, G.M.; Halenda, S.P.; McKinney, J.S.; Rubin, R.P. *Biochem. J.* **1983,** 212, 473-482.

270. Qing, Z.M.; Jing, L.G.; Kai, C.R. Photosynthesis characteristics in eleven cultivars of sugarcane and their responses to water stress during the elongation stage Proc ISSCT. **2001,** 24, 642-643.

271. Queitsch, C.; Hong, S.W.; Vierling, E.; Lindquist. S. Heat shock protein 101 plays a crucial role in thermotolerance in Arabidopsis. *Plant Cell.* **2000,** 12, 479-492.

272. Rabbani, M.A.; Maruyama, K.; Abe, H.; Khan, M.A.; Katsura, K.; Ito, Y.; Yoshiwara, K.; Seki, M.; Shinozaki, K.; Yamaguchi-Shinozaki, K. Monitoring expression profiles of rice genes under cold, drought, and high-salinity stresses and abscisic acid application using cDNA microarray and RNA gel-blot analyses. *Plant Physiol.* **2003,** 133, 1755–1767.

273. Rabinowitz, A.R.; Myint, T.; Khaing, S.T.; Rabinowitz, S. Description of the leaf deer (Muntiacus putaoensis), a new species of muntjac from northern Myanmar. *J Zool.* **1999,** 249, 427-435.

274. Ray, T.C.; Callow, J.A.; Kennedy, J.F. Composition of root mucilage polysaccharide from Lepidium sativum. *J Exp Bot.* **1988,** 39, 1249–1261.

275. Reddy, K.R.; Hodges, H.F.;Reddy, V.R. Temperature effects on cotton fruit retention. *Agronomy Journal.* **1992,** 84, 26–30.

276. Ren, X.; Kong, Q.; Wang, P.; Jiang, F.; Wang, F.; Yu, T.; Zheng, X. Molecular cloning of a PR-5 like protein gene from cherry tomato and analysis of the response of this gene to abiotic stresses. *Mol Biol Rep.* **2009,** 38, 801–807.

277. Rensink, W.A.; Iobst, S.; Hart, A.; Stegalkina, S.; Liu, C.R.; Buell, J. Gene expression profiling of potato responses to cold, heat, and salt stress. *Funct Integr Genomics.* **2005,** 5, 201–207.

278. Restrepo, S.; Myers, K.L.; Del Pozo, O.; Martin, G.B.; Hart, A.L.; Buell, C.R.; Fry, W.E.; Smart, C.D. Gene profiling of a compatible interaction between Phytophthora infestans and Solanum tuberosum suggests a role for carbonic anhydrase. *Mol Plant–Microbe Interact.* **2005,** 18, 913–922.

279. Reymond, P.; Weber, H.; Damond, M.; Farmer, E.E. Differential gene expression in response to mechanical wounding and insect feeding in Arabidopsis. *Plant Cell.* **2000**, 12, 707–720.

280. Richmond, B.J.; Oram, M.W.; Wiener, M.C. Response features determining spike times. *Neural Plasticity.* **1999**, 6, 133-145.

281. Rick, C.M. Potential genetic resources in tomato species: clues from observation in native habitats, In: Srb, A.M. (Ed.), Genes, Enzymes and Populations. Plenum Press, New York. **1973**, pp 255–269.

282. Rick, C.M. Tomato-like nightshades: affinities, auto-ecology, and breeders opportunities. *Economic Botany.* **1988**, 42, 145–154.

283. Rizhsky, L.; Liang, H.; Mittler, R. The combined effect of drought stress and heat shock on gene expression in tobacco. *Plant Physiol.* **2002**, 130, 1143–1151.

284. Rizhsky, L.; Liang, H.; Shuman, J.; Shulaev, V.; Davletova, S.; Mittler, R. When defense pathways collide: the response of Arabidopsis to a combination of drought and heat stress. *Plant Physiol.* **2004**, 134, 1683–1696.

285. Robertson, A.J.; Ishikawa, M.; Gusta, L.V.; MacKenzie, S.L. () Abscisic acid-induced heat tolerance in Bromus inermis Leyss cell-suspension cultures. Heat-stable, abscisic acid-responsive polypeptides in combination with sucrose confer enhanced thermostability. *Plant Physiol.* **1994**, 105, 181–190.

286. Rogaev, E.I.; Sherrington, R.; Rogaeva, E.A.; Levesque, G.; Ikeda, M.; Liang, Y.; Chi, H.; Lin, C.; Holman, K.; Tsuda, T.; Mar, L.; Sorbi, S.; Nacmias, B.; Piacentini, S.; Amaducci, L.; Chumakov, I.; Cohen, D.; Lannfelt, L.; Fraser, P.E.; Rommens, J.M.; George-Hyslop, P.H. Familial Alzheimer's disease in kindreds with missense mutations in a gene on chromosome 1 related to Alzheimer's disease type 3 gene. *Nature.* **1995**, 376, 775–778.

287. Ronning, C.; Stegalkina, S.; Ascenzi, R.A.;Bougri, O.;Hart, A.L.;Utterbach, T.R.;Vanaken, S.E.;Riedmuller, S.B.;White, J.A.;Cho, J.;Pertea, G.M.;Lee, Y.;Karamycheva, S.;Sultana, R.;Tsai, J.;Quackenbush, J.;Griffiths, H.M.;Restrepo, S.;Smart, C.D.;Fry, W.E.;Van, D.H.R.;Tanksley, S.;Zhang, P.;Jin, H.;Yamamoto, M.L.;Baker, B.J.;Buell, C.R. Comparative analyses of potato expressed sequence tag libraries. *Plant Physiol.* **2003**, 131, 419–429.

288. Ross, D.T.; Scherf, U.; Eisen, M.B.; Perou, C.M.; Rees, C.; Spellman, P.; Iyer, V.; Jeffrey, S.S.; de Rijn, M.V.; Waltham, M.; Pergamenschikov, A.; Lee, J.C.; Lashkari, D.; Shalon, D.; Myers, T.G.; Weinstein, J.N.; Botstein, D.; Brown, P.O. Systematic variation in gene expression patterns in human cancer cell lines. *Nat Genet.* **2000**, 24(3), 227-235.

289. Rounsley, S.D.; Glodeck, A.; Sutton, G.; Adams, M.D.; Somerville, C.R.; Venter, J.C.; Kerlavage, A.R. The construction of Arabidopsis expressed sequence tag assemblies. *Plant Physiol.* **1996**, 112, 1177-1183.

290. Roy, H.; Andrews, T.J. Rubisco: assembly and mechanism. In: Leegood, R., Sharkey, T., von Caemmerer, S. (Eds.), Photosynthesis: physiology and metabolism. Vol.9. Springer, Dordrecht, The Netherlands. **2000**, pp. 53–83.

291. Rozen, S.; Skaletsky, H.J. Primer3.http://www-genome.wi.mit.edu/genome_software/other/primer3.html. **1996, 1997, 1998.**

292. Ryals, J.; Uknes, S.; Ward, E. Systemic acquired resistance. *Plant Physiol.* **1994**, 104, 1109–1112.

293. Rylski, I. Pepper (Capsicum): In: Monselise, S.P. (Ed.), Handbook of Fruit Set and Development. CRC Press, Boca Raton, FL, USA. **1986**, pp. 341–354.

294. Saccardy, K.B.; Pineau, O.; Roche, C.G. Photo chemical efficiency of photosystem and xanthophyll cycle components in Zea mays leaves exposed to water stress and high light. *Photosy Res.* **2002** ,56, 57-66.

295. Sadras, V.O.; Milroy, S.P. Soil water thresholds for the responses of leaf expansion and gas exchange: a review. *Field Crops Res.* **1996**, 47, 253–266.

296. Saeed, A.; Hayat, K.; Khan, A.A.; Iqbal, S. Heat Tolerance Studies in Tomato (Lycopersicon esculentum Mill.). *Int J Agri Biol.* **2007**, 9(4), 649–652.

297. Sairam, R.K.; Deshmukh, P.S.; Saxena, D.C. Role of antioxidant systems in wheat genotypes tolerance to water stress. *Biologia Plantarum.* **1998**, 41, 387–394.

298. Sairam, R.K.; Rao, K.V.; Srivastava, G.C. Differential response of wheat genotypes to long term salinity stress in relation to oxidative stress. *Plant Sci.* **2002**, 163, 1037–1046.

299. Sakuma, Y.; Liu, Q.; Dubouzet, J.G.; Abe, A.J.; Shinozaki, K.; Yamaguchi-Shinozaki, K. DNA-Binding Specificity of the ERF/AP2 Domain of Arabidopsis DREBs, Transcription Factors Involved in Dehydration- and Cold-Inducible Gene Expression. *Biochem. Biophys. Res. Comm.* **2002**, 290, 998–1009.

300. Salleo, S.; Nardini, A.; Pitt, F.; Lo Gullo, M.A. Xylem cavitation and hydraulic control of stomatal conductance in laurel (Laurus nobilis L.). *Plant Cell Environ.* **2000**, 23, 71–79.

301. Samuel, D.; Kumar, T.K.S.; Ganesh, G.; Jayaraman, G.; Yang, P.W.; Chang, M.M.; Trivedi, V.D.; Wang, S.L.; Hwang, K.C.; Chang, D.K.; Yu, C. Proline inhibits aggregation during protein refolding. *Protein Science.* **2000**, 9, 344–352.

302. Sanchez, B.; Fernandez, J.; Morales, T.A.; Morte, A.; Alarcon, J.J. () Variation in water stress, gas exchange, and growth in Rasmanrins officinalis plants infected with Glamus deserticola under drought conditions. *J Plant Physiol.* **2006**, 161, 675-682.

303. Saradhi, P.P.; Alia, A.S.; Prasad, K.V.S.K. Proline Accumulates in Plants Exposed to UV Radiation and Protects Them against UV-Induced Peroxidation. *Biochemical and Biophysical Research Communications.* **1995**, 209, 1-5.

304. Sasaki, A.; Itoh, H.; Gomi, K.; Ueguchi-Tanaka, M.; Ishiyama, K.; Kobayashi, M.; Jeong, D.H.; An, G.; Kitano, H.; Ashikari, M.; Matsuoka, M. Accumulation of phosphorylated repressor for gibberellin signaling in an F-box mutant. *Science.* **2003**, 299, 1896–1908.

305. Sato, S.; Peet, M.M.; Thomas, J.F. Physiological factors limit fruit set of tomato (Lycopersicon esculentum Mill.) under chronic, mild heat stress. *Plant Cell and Environment.* **2000**, 23, 719–726.

306. Scandalios, J.G. Oxygen stress and superoxide dismutase. *Plant Physiol.* **1993**, 101, 7–12.

307. Scharf, K.D.; Heider, H.; Höhfeld, I.; Lyck, R.; Schmidt, E.; Nover, L. () The tomato Hsf system: HsfA2 needs interaction with HsfA1 for efficient nuclear import and may be localized in cytoplasmic heat stress granules. *Mol Cell Biol.* **1998**, 18, 2240–2251.

308. Scharf, K.D.; Siddique, M.; Vierling, E. The expanding family of small Hsps and other proteins containing an α-crystallin domain. *Cell Stress Chaperones.* **2001**, 6, 225–237.

309. Schena, M.; Shalon, D.; Davis, R.; Brown, P. Quantitative monitoring of gene expression patterns with a complimentary DNA microarray. *Science.* **1995**, 270, 467–470.

310. Schöffl, F.; Prandl, R.; Reindl, A. Molecular responses to heat stress. In: Shinozaki, K., Yamaguchi-Shinozaki, K. (Eds.), Molecular Responses to Cold, Drought, Heat and Salt Stress in Higher Plants. RG Landes Co, Austin, Texas. **1999**, pp. 81–98.

311. Seki, M.; Narusaka, M.; Abe, H.; Kasuga, M.; Yamaguchi-Shinozaki, K.; Carninci, P.; Hayashizaki, Y.; Shinozaki, K. Monitoring the expression pattern of 1,300 Arabidopsis genes under drought and cold stresses by using a full-length cDNA microarray. *Plant Cell.* **2001**, 13, 61–72.

312. Seki, M.; Narusaka, M.; Ishida, J.; Nanjo, T.; Fujita, M.; Oono, Y.; Kamiya, A.; Nakajima, M.; Enju, A.; Sakurai, T.; Satou, M.; Akiyama, K.; Taji, T.; Yamaguchi-Shinozaki, K.; Carninci, P.; Kawai, J.; Hayashizaki, Y.; Shinozaki,

K. () Monitoring the expression profiles of 7000 Arabidopsis genes under drought, cold and high-salinity stresses using a full-length cDNA microarray. *Plant J.* **2002**, 31(3), 279–292.

313. Serrano, R.; Mulet, J.M.; Rios, G.; Marquez, J.A.; De Larrinoa, I.F.; Leube, M.P.; Mendizabal, I.; Pascual-Ahuir, A.; Proft, M.; Ros, R.; Montesinos, C.A. Glimpse of the mechanisms of ion homeostasis during salt stress. *J Exp Bot.* **1999**, 50, 1023–1036.

314. Shalon, D.; Smith, S.; Brown, P. A DNA microarray system for analyzing complex DNA samples using two-color fluorescent probe hybridization. *Genome Res.* **1996**, 6, 639–645.

315. Sharkey, T.D.; Chen, X.Y.; Yeh, S. Isoprene increases thermotolerance of fosmidomycin-fed leaves. *Plant Physiol.* **2001**, 125, 2001–2006.

316. Shinozaki, K.; Yamaguchi-Shinozaki, K. Gene expression and signal transduction in water-stress response. *Plant Physiol.* **1997**, 115, 327–334.

317. Shinozaki, K.; Yamaguchi-Shinozaki, K. Molecular responses to dehydration and low temperature: differences and cross-talk between two stress signalling pathways. *Curr Opin Plant Biol.* **2000**, 3, 217–23.

318. Shinozaki, K.; Yamaguchi-Shinozaki, K.; Seki, M.) Regulatory network of gene expression in the drought and cold stress responses. *Current Opinion in Plant Biology.* **2003**, 6, 410–17.

319. Siddique, M.; Port, M.; Tripp, J.; Weber, C.; Zielinski, D.; Calligaris, R.; Winkelhaus, S.; Scharf, K.D. Tomato heat stress protein Hsp16.1- CIII represents a member of a new class of nucleocytoplasmic small heat stress proteins in plants. *Cell Stress Chaperones.* **2003**, 8, 381–394.

320. Silva, F.G.D.; Iandolino, A.; Kayal, F.A.; Bohlmann, M.C.; Cushman, M.A.; Lim, H.; Figueroa, A.E.R.; Kabuloglu, E.K.; Osborne, C.; Rowe, J.; Tattersall, E.; Leslie, A.; Xu, J.; Baek, J.M.; Grant, R.C.; Cushman, J.C.; Douglas, R. Characterizing the Grape Transcriptome Analysis of Expressed Sequence Tags from Multiple Vitis Species and Development of a Compendium of Gene Expression during Berry Development. *Plant Physiology.* **2005**, 139, 574–597.

321. Sinclair, T.R.; Ludlow, M.M. Influence of water supply on the plant water balance of four tropical grain legumes. *Aust J Plant Physiol.* **1986**, 13, 329–341.

322. Singh, R.P.; Prasad, P.V.V.; Sunita, K.; Giri, S.N.; Reddy, K.R. Influence of high temperature and breeding for heat tolerance in cotton. *Adv Agron.* **2007**, 93, 313–85.

323. Slatyer, R.O.; Taylor, S.A. Terminology in plant-soil water relations. *Nature*. **1960**, 187, 922-924.

324. Slooten, L.; Capiau, K.; Van Camp, W.; Van Montagu, M.; Sybesma, C.; Inze, D. Factors affecting the enhancement of oxidative stress tolerance in transgenic tobacco over expressing manganese superoxide dismutase in the chloroplasts. *Plant Physiol.* **1995**, 107, 373–380.

325. Smertenko, A.; Draber, P.; Viklicky, V.; Opatrny, Z. () Heat stress affects the organization of microtubules and cell division in Nicotiana tabacum cells. *Plant Cell Environ.* **1997**, 20, 1534–1542.

326. Smirnoff, N. Plant resistance to environmental stress. *Curr Opin Biotech.* **1998**, 9, 214–219.

327. Smirnoff, N.; Cumbes, Q.J. Hydroxyl radical scavenging activity of compatible solutes. *Phytochemistry.* **1989**, 28, 1057–1060.

328. Smyth, G.K.; Yang, Y.H.; Speed, T. Statistical issues in cDNA microarray data analysis. *Methods Mol Biol.* **2003**, 224, 111–136.

329. Solomon, A.; Beer, S.; Waisel, Y.; Jones, G.P.; Paleg, L.G. Effects of NaCl on the Carboxylating Activity of Rubisco from Tamarix jordanis in the Presence and Absence of Proline-Related Compatible Solutes. *Physiol Plant.* **1994**, 90, 198–204.

330. Somerville, C.; Somerville, S. Plant functional genomics. *Science.* **1999**, 285(16), 380-383.

331. Song, S.Q.; Lei, Y.B.; Tian, X.R. Proline Metabolism and Cross-Tolerance to Salinity and Heat Stress in Germinating Wheat Seeds. *Russ J Plant Physiol.* **2005**, 5(6), 793–800.

332. Sperry, J.S. Hydraulic constraints on plant gas exchange. *Agric Forest Meteorol.* **2000**, 104,13–23.

333. Sreenivasulu, N.; Sopory, S.K.; Kishor, P.B.K. Deciphering the regulatory mechanisms of abiotic stress tolerance in plants by genomic approaches. *Gene.* **2007**, 388, 1-13.

334. Sreenivasulu, N.; Varshney, R.K.; Kishor, K.P.B.; Weschke, W.Tolerance to abiotic stress in cereals: a functional genomics approach. In: Gupta, P.K., Varshney, R.K. (Eds.), Cereal Genomic. **2004**, pp. 483–514.

335. Srivastava, S.; Dubey, R.S. Manganese-excess induces oxidative stress, lowers the pool of antioxidants and elevates activities of key antioxidative enzymes in rice seedlings. *Plant Growth Regul.* **2011**, 64, 1–16.

336. Stefanov, D.; Petkova, V.; Denev, I.D. Screening for heat tolerance in common bean (Phaseolus vulgaris L.) lines and cultivars using JIP-test. *Sci Hort.* **2011**, 128, 1–6.

337. Stockinger, E.J.; Gilmour, S.J.; Thomashow, M.F. Arabidopsisthaliana CBF1 encodes an AP2 domain-containing transcriptionalactivator that binds to the C-repeat/DRE, a cis-actingDNA regulatory element that stimulates transcription in responseto low temperature and water deficit. *Proc Natl Acad SciUSA.* **1997**, 94, 1035–1040.

338. Stoll, M.; Loveys, B.; Dry, P. Hormonal changes induced by partial rootzone drying of irrigated grapevine. *J Exp Bot.* **2000**, 51, 1627–1634.

339. Stuhlfauth, T.; Sultemeyer, D.F.; Weinz, S.; Fock, H.P. Fluorescence quenching and gas exchange in a water stressed C3 plant, Digitalis lanata. *Plant Physiol.* **1988**, 86, 246-250.

340. Sudhakar, C.; Lakshmi, A.; Giridara, K.S. Changes in the antioxidant enzyme efficacy in two high yielding genotypes of mulberry (*Morus alba* L.) under NaCl salinity. *Plant Sci.* **2001**, 161, 613-619.

341. Sun, A.; Yi, S.; Yang, J.; Zhao, C.; Liu, J. Identification and characterization of a heat-inducible ftsH gene from tomato (*Lycopersicon esculentum* Mill.). *Plant Sci.* **2006**, 170, 551–562.

342. Sun, W.; Montagu, M.V.; Verbruggen, N. Small heat shock proteins and stress tolerance in plants. *Biochim Biophys Acta.* **2002**, 1577, 1–9.

343. Sung, D.Y.; Kaplan, F.; Lee, K.J.; Guy, C.L. Acquired tolerance to temperature extremes. *Trends Plant Sci.* **2003**, 8, 179–187.

344. Suzuki, N.; Rizhasky, L.; Liang, H.; Shuman, J.; Shulaev, V.; Mittler, R. Enhanced tolerance to environmental stress in transgenic plants expressing the transcriptional co-activator Multiprotein Bridging Factor 1c. *Plant Physiol.* **2005**, 139, 1313–1322.

345. Swidzinski, J.A.; Sweetlove, L.J.; Leaver, C.J. A custom microarray analysis of gene expression during programmed cell death in Arabidopsis thaliana. *Plant J.* **2002**, 30, 431–446.

346. Szabados, L.; Savoure, A. Proline: a multifunctional amino acid. *Trends in Plant Science.* **2009**, 15(2), 89–97.

347. Tanksley, S.D. The genetic, developmental and molecular bases of fruit size and shape variation in tomato. *Plant Cell.* **2004**, 16, S181–S189.

348. Tardieu, F.; Katerji, N.; Bethenod, J.; Zhang, J.; Davies, W.J. Maize stomatal conductance in the field: its relationship with soil and plant water potentials, mechanical constraints and ABA concentration in the xylem sap. *Plant Cell and Environment.* **1991**, 14, 121–126.

349. Tezara, W.; Mitchell, V.J.; Driscoll, S.D.; Lawlor, D.W. Water stress inhibits plant photosynthesis by decreasing coupling factor and ATP. *Nature.* **1999**, 401, 914–917.

350. Thibaud-Nissen, F.; Shealy, R.T.; Khanna, A.; Vodkin, L.O. Clustering of microarray data reveals transcript patterns associated with somatic embryogenesis in soybean. *Plant Physiol.* **2003**, 132, 118–136.

351. Thomashow, M.F. So what's new in the field of plant cold acclimation? Lots! *Plant Physiol.* **2001**, 125, 89–93.

352. Tibbles, L.A.; Woodgett, J.R. The stress-activated protein kinase pathways. *Cell Mol Life Sci.* **1999**, 55, 1230–54.

353. Tollenaar, M.; Aguilera, A. Radiation use efficiency of an old and new maize hybrid. *Agron J.* **1992**, 84, 536-541.

354. Turner, N.C. Optimizing water use. In: Nösberger, J., Geiger, H.H., Struik, P.C. (Eds.), Proceedings of the Third Crop Science Congress on Crop Science Progress and Prospects. CABI International, Wallingford, UK. **2001**, pp 119–135.

355. Vallelian-Bindschedler, L.; Schweizer, P.; Mosinger, E.; Metraux, J.P. Heat induced resistance in barley to powdery mildew (Blumeria graminis f. sp. Hordei) is associated with a bust of AOS. *Physiol Mol Plant Pathol.* **1998**, 52, 185–199.

356. Van der Hoeven, R.; Ronning, C.; Giovannoni, J.; Martin, G.; Tanksley, S. Deductions about the number, organization, and evolution of genes in the tomato genome based on analysis of a large expressed sequence tag collection and selective genomic sequencing. *Plant Cell.* **2002**, 14, 1441–1456.

357. Van, H.R.B.; Cairns, W.L. Progress and prospects in the use of peroxidase to study cell-development. *Phytochemistry.* **1982**, 21, 1843–1847.

358. Varshney, R.K.; Bansal, K.C.; Aggarwal, P.K.; Datta, S.K.; Craufurd, P.Q. Agricultural biotechnology for crop improvement in a variable climate: hope or hype? *Trends in Plant Science.* **2011**, 16(7), 367-371.

359. Vierling, E. The roles of heat shock protein in plants. *Annu Rev Plant Physiol Plant Mol Biol.* **1991**, 42, 579–620.

360. Vierling, E.; Kimpel, J.A. Plant responses to environmental stress. *Curr Opin Biotech.* **1992**, 3, 164–170.

361. Wahid, A.; Gelani, S.; Ashraf, M.; Foolad, M.R. Heat tolerance in plants: an overview. *Environ Expt Bot.* **2007**, 61, 199–223.

362. Walley, J.W.; Coughlan, S.; Hudson, M.E.; Covington, M.F.; Kaspi, R.; Banu, G.; Harmer, S.L.; Dehesh, K. Mechanical stress induces biotic and abiotic stress responses via a novel cis-element. *PLoS Genet.* **2007**, 3(10), e172.

363. Walter, S.; Buchner, J. Molecular chaperones-cellular machines for protein folding. *Angew Chem Int Ed Engl.* **2002**, 41, 1098–1113.

364. Wang, D.; Luthe, D.S. Heat sensitivity in a bentgrass varient. Failure to accumulate a chloroplast heat shock protein isoform implicated in heat tolerance. *Plant Physiol.* **2003**, 133, 319–327.

365. Wang, H,Y.; Huang, Y.C.; Chen, S.F.; Yeh, K.W. Molecular cloning, characterization and gene expression of a water deficiency and chilling induced proteinase inhibitor I gene family from sweet potato (Ipomoea batatas Lam.) leaves. *Plant Science.* **2003**, 165(1), 191-203.

366. Wang, W.; Vinocur, B.; Shoseyov, O.; Altman, A. Role of plant heat-shock proteins and molecular chaperones in the abiotic stress response. *Trends Plant Sci.* **2004**, 9, 244–252.

367. Weckx, J.E.J.; Ciljsters, H.M. Zn phytotoxicity induces oxidative stress in primary leaves of *Phaseolus vulgaris* (L.). *Plant Physiol Biochem.* **1997**, 35, 405–410.

368. Weis, E.; Berry, J.A. Plants and high temperature stress. *Symp Soc Exp Biol.* **1988**, 42, 329–346.

369. Wells, D.R.; Tanguay, R.L.; Le, H.;Gallie, D.R. HSP101 functions as a specific translational regulatory protein whose activity is regulated by nutrient status. *Genes Dev.* **1998**, 12,3236–3235.

370. Wen, C.K.; Chang, C. Arabidopsis RGL1 encodes a negative regulator of gibberellin responses. *Plant Cell.* **2002**, 14, 87–100.

371. Wendehenne, D.; Gould, K.; Lamotte, O.; Durner, J.; Vandelle, E.; Lecourieux, D.; Courtois, C.; Barnavon, L.; Bentéjac, M.; Pugin, A. NO signaling functions in the biotic and abiotic stress responses. *BMC Plant Biol.* **2005**, 5(1),S35.

372. Wimalasekera, R.; Tebartz, F.;Scherer, G.F.E. Polyamines, polyamine oxidases and nitric oxide in development, abiotic and biotic stresses. *Plant Science.* **2011**, 181, 593–603.

373. Wu, C.H.; Caspar, T.; Browse, J.; Lindquist, S.; Somerville, C. Characterization of an HSP70 cognate gene family in Arabidopsis. *Plant Physiol.* **1988**, 88, 731–740.

374. Xiong, L.; Schumaker, K.S.; Zhu, J.K. Cell signaling during cold, drought, and salt stress. *Plant Cell.* **2002**, 14(1), S165–S183.

375. Xiong, L.; Zhu, J.K. Abiotic stress signal transduction in plants: molecular and genetic perspectives. *Physiol Plant.* **2001**, 112, 152–166.

376. Yancey, P.H. Compatible and counteracting solutes. Cellular and Molecular Physiology of Cell Volume Regulation, ed. Strange K, Austin, TX: CRC Press. **1994**, pp 81–109

377. Yang, J.; Sears, R.G.; Gill, B.S.; Paulsen, G.M. Quantitative and molecular characterization of heat tolerance in hexaploid wheat. *Euphytica.* **2002**, 126, 275–282.

378. Yi, S.Y.; Liu, L. Combinatorial interactions of two cis-acting elements, AT-rich regions and HSEs, in the expression of tomato Lehsp23.8 upon heat and non-heat stresses. *J. Plant Biol.* **2009**, 52, 560–568.

379. Yoshiba, Y.T.; Kiyosea, K.; Nakashima, K.; Yamaguchi-shinozaki, Shinozaki, K. Regulation of levels of proline as an osmolyte in plants under water stress. *Plant Cell Physiol.* **1997**, 38(10),1095-1102.

380. Yu, L.; Setter, T.L. Comparative transcriptional profiling of placenta and endosperm in developing maize kernels in response to water deficit. *Plant Physiol.* **2003**, 131, 568–582.

381. Zhang, J.; Kirkham, M.B. Drought-stress induced changes in activities of superoxide dismutase, catalase and peroxidases in wheat leaves. *Plant Cell Physiol.* **1994**, 35, 785–791.

382. Zhu, J.K. Cell signaling under salt, water and cold stresses. *Curr Opin Plant Biol.* **2001a**, 4, 401–406.

383. Zhu, J.K. Plant salt tolerance. *Trends Plant Sci.* **2001b**, 6, 66–71.

384. Zhu, J.K. Salt and drought stress signal transduction in plants. *Annu Rev Plant Biol.* **2002**, 53, 247–273.

Abiotic Stress Tolerance Mechanisms in Plants, Pages 267–290
Edited by: Gyanendra K. Rai, Ranjeet Ranjan Kumar and Sreshti Bagati
Copyright © 2018, Narendra Publishing House, Delhi, India

7

DEFENSE MECHANISM IN PLANTS AGAINST ABIOTIC STRESSES

Upama Mishra[1], Gyanendra K. Rai[2] and Sreshti Bagati[2]

[1]Indian Institute of Vegetable Research, Varanasi, India
[2]School of Biotechnology,
Sher-e- Kashmir University of Agricultural Sciences and Technology of Jammu-180009 (J&K)
E-mail: upama.mishra@gmail.com

Abstract

Abiotic stress is the major cause of crop loss worldwide, reducing average yield of plants by more than 50%. Plants are subjected to many abiotic stresses like drought, heat, low temperature and salinity stress. Plants have adopted to live in environments where they are often exposed to different stress factors in combination. They have developed specific mechanisms that allow them to sense particular environmental changes and act in response to complex stress conditions, reducing damage while conserving important resources for growth and development. Plant triggers a specific and unique stress response when subjected to different environmental stresses. The response to change in environment can be quick, depending on type of stress and can involve either adaptation mechanisms, or avoidance mechanism.

Keywords: Abiotic stress, Salinity, Environment, Response

1. Introduction

Plants are subjected to different abiotic stresses like cold, drought, heat, salinity and metal stress during their life cycle. Plant responses to different stresses are highly complex and involve changes at the, morphological physiological, biochemical, cellular and transcriptome levels. It has been reported that abiotic stress reduce average yields by more than 50% for most major crop plants (Wang *et al.,* 2003). Abiotic stress initiates a cascade of cellular and molecular response system implemented by the plant in order to avoid damage and ensure survival, but often results loss of growth and yield.

Plant triggers a specific and unique stress response when subjected to different environmental stresses (Rizhsky *et al.,* 2004b). To cope with the damaging effects of stress, plants have evolved many biochemical and molecular mechanisms. Understanding plant responses to abiotic stress is important for the improvement of crop productivity (Lawlor, 2013). During evolution, plants have developed different defence strategies against abiotic stresses like escape, avoidance and adaptation. For example during drought, plant tries to escape most sensitive stage of development (e.g. reproductive stage) from drought and allow occurring when the stress is over or less sensitive. Plants maintain high tissue water potential for avoidance of stress, while adaptation combines enhanced water acquisition using a deep root system with minimization of water loss by preventing transpiration. Mechanisms of drought tolerance include maintenance of turgor through osmotic adjustment, increased cell elasticity and decreased cell size as well as desiccation tolerance via protoplasmic tolerance. Several genes have been involved in drought tolerance; hence the plant drought stress responses, tolerance mechanisms and genetic control of tolerance are complex (Shinozaki and Yamaguchi- Shinozaki, 2007).

2. MicroRNAs

Plant MicroRNAs (miRNAs) are small, non-coding endogenous RNAs that are essential for plant growth and development. Reinhart *et al.* (2002) reported the existence of plant miRNAs. In plants they regulate gene expression at posttranscriptional level by degrading the target mRNA whereas in animal they act by blocking the protein translation through binding with 32 Untranslated Region of the target mRNA. Several reports indicate that plant miRNAs have been involved in various abiotic stress responses such as oxidative, mineral nutrient deficiency, drought, salinity, high temperature and low temperature. miRNA gene expression profiling reveals that miRNAs which are involved in the progression of plant growth and development are differentially expressed during abiotic stress responses. miRNA have an effect on many biological processes including development of organs such as leaves and flower parts, roots and stems (Bartel, 2004; Bian *et al.,* 2012; Chen, 2004; Chen *et al.,* 2011; Maizel and Jouannet, 2012) and many of the evidence suggest that miRNAs play important roles in plant responses to biotic and abiotic stresses.

miRNAs mediate the responses by adjusting the amount of themselves, mode of action of miRNA–protein complexes and the amount of mRNA targets which in turn changes the amount, timing and location of proteins expressed from other genes upon exposure to the stress. Upon exposure to stress these miRNAs mainly regulates gene expression at the post-transcriptional level (Ding *et al.,* 2013; Feng *et al.*, 2013; Floris *et al.*, 2009; Liu *et al.*, 2008, 2012; Ozhuner *et al.*, 2013; Sunkar *et al.*, 2006; Wang *et al.*, 2014; Xie *et al.*, 2012; Yang *et al.*, 2012; Zhang *et al.*, 2009, 2011, Ferdous *et al.*, 2015).

Many researchers have reported that miR393 is one of the important miRNAs during stress responses which showed its altered expression in *A. thaliana*, *Medicagotruncatula*, *Oryza sativa*, *Phaseolus vulgaris* and other plants under drought, chilling, salinity and aluminium stress (Sunkar and Zhu, 2004; Zhao *et al.*, 2007; Liu *et al.*, 2008; Arenas-Huertero *et al.*, 2009; Trindade *et al.*, 2010. Gao *et al.*, 2011 reported that mir393 play important role for salt tolerance stress in Arabidopsis.

The expression of the Rice nuclear transcription factor YA (*NF-YA*) genes under water stress is adjusted by members of the miR169 family (Zhao *et al.*, 2009). In *A. thaliana* plants over expressing miR169 are more sensitive to drought (Li *et al.*, 2008). On the contrary in tomato, plants over expressing miR169c, which targets a gene involved in the opening and closing of stomata, are more tolerant to drought (Zhang *et al.*, 2011a).

3. Mitogen Activated Protein Kinases

Plants have developed many signalling pathways like hormones, calcium ion and Mitogen-activated protein kinases (MAPK). MAPKs are protein kinases that are specific to the three amino acids serine, threonine, and tyrosine. They worked as signalling machinery for regulation of physiological and developmental responses such as cell growth, differentiation, hormone signalling, biotic and abiotic stresses. (Jonak *et al.*, 1999; Joshi *etal.*, 2011). MAPKs are involved in directing cellular responses to a diverse array of stimuli, such as abiotic and biotic stress. They regulate cell functions including proliferation, gene expression, differentiation, mitosis, cell survival, and apoptosis. MAPK3, MAPK4 and MAPK6 kinases of *Arabidopsis thaliana* are key mediators of responses to osmotic shock, oxidative stress, response to low temperature and involved in anti-pathogen responses (Sinha *et al.*, 2011, Rodriguez *et al.*, 2010). They are activated by phosphorylating the serine/threonine residue.

4. Antioxidants and Reactive Oxygen Species

Plants are frequently challenged by different biotic and abiotic stresses, which cause considerable losses in yield. It has been widely reported that different environmental stress like high temperature, drought and salinity conditions induce an oxidative stress.A common effect of such stresses is the accumulation of reactive oxygen species (ROS) and the establishment of a oxidative stress leading to metabolic damage and finally to cell death.

ROS are produced by all aerobic organisms and are usually balanced by the antioxidative mechanisms that exist in all living beings. Since ROS have an important signalling role in plants metabolism, their concentration must be carefully controlled through adequate pathways (Foyer andNoctor, 2003; Vranova *et al.*, 2002, Mittler, 2002). ROS can be produced during normal aerobic metabolic processes like respiration and photosynthesis and thus, the majority of ROS are produced in the chloroplast, mitochondria, peroxisomes, plasma membrane and apoplast (Ahmad *et al.*, 2008; Moller, 2001). Other sources of ROS production are NADPH oxidases, amine oxidases and cell-wall peroxidases (Mittler, 2002).

Under stressful condition these ROS are continuously produced as by product of normal aerobic metabolism. Antioxidative system (both non-enzymatic and enzymatic compounds) keeps tight control on ROS production, and adjusts intracellular ROS concentration, setting cellular redox homeostasis. The enzymatic antioxidant defence mechanism involve superoxide dismutase, catalase, ascorbateperoxidise, glutathione reductase, monodehydro-ascorbatereductase, dehydroascorbatereductase, glutathione peroxidase, guiacol peroxidase and glutathione-S-transferase (Gill and Tuteja, 2010). Among non-enzymatic scavengers, low molecular weight compounds, including ascorbic acid and glutathione are involved. In general ROS causes oxidations of polyunsaturated fatty acids in lipids, oxidations of amino acids in proteins, damage of DNA, and oxidatively deactivate specific enzymes by oxidation of co-factors.

ROS produced during plant metabolism are singlet oxygen, Superoxide (O_2^-), Hydrogen Peroxide (H_2O_2), Hydroxyl Radical (OH^-). Singlet oxygen is mainly produced in the chloroplasts at photosystem II but may also result from lipoxygenase activity and is a highly reactive species that have very short life span (Asada, 2006, Foyer *et al.*, 1994). The superoxide radical is mainly produced both in the chloroplasts (photo systems I and II) and mitochondria as sub products and in peroxisomes (del Rio *et al.*, 2006; Moller *et al.*, 2007; Rhoads *et al.*, 2006), has a half-life of 2-4 micro second and cannot cross phospholipid membranes (Gargand Manchanda, 2009) and so it is important that the cell has adequate *in situ* mechanism to scavenge this ROS. Superoxide dismutase can catalyse the conversion of superoxide radical into hydrogen peroxide. Superoxide radical can also be produced by NADPH oxidase in the plasma membrane (Moller *et al.*, 2007). Hydrogen peroxide is also results from the dismutation of superoxide (del Rio *et al.*, 2006, Rhoads *et al.*, 2006) but it is mainly formed in peroxisomes and mitochondria. It is not a radical and can easily cross membranes diffusing across the cell and has a half-life of around 1 milli second (Gargand Manchanda, 2009). Hydroxyl radical is the most reactive ROS and is formed from hydrogen peroxide via Fenton and Fenton-like reactions and, there are no known enzymatic systems able to degrade it (Freinbichler *et al.*, 2011). Under low concentration H_2O_2, act as a signal molecule but at higher concentration it leads to programme cell death.

Although the hydrogen peroxide and superoxide radical are not as reactive as other species they are produced in large amounts in the cell and can initiate other reactions that lead to more dangerous species (Noctorand Foyer, 1998). The superoxide radical can be converted by superoxide dismutase enzymes into hydrogen peroxide, and this can also be a problem as it causes the occurrence of Fenton reactions (Moller *et al.*, 2007). In order to improve tolerance against oxidative stress and finally to maintain the productivity of plants under different environmental stress conditions, fortification of the antioxidative mechanisms using genetic engineering has been widely reported.

Superoxide dismutases (SOD) are a class of enzymes that catalyze the dismutation of superoxide into oxygen and hydrogen peroxide. As such, they are an important antioxidant defense in nearly all cells exposed to oxygen. Catalase, which is concentrated in peroxisomes located next to mitochondria, reacts with the hydrogen peroxide to catalyze the formation of water and oxygen. Glutathione peroxidase reduces hydrogen peroxide by transferring the energy of the reactive peroxides to a very small sulfur-containing protein called glutathione. The sulphur contained in these enzymes acts as the reactive centre, carrying reactive electrons from the peroxide to the glutathione.

It is well studied that the stimulation of the cellular antioxidant machinery is important for defence against various stresses (Dalton *et al.,* 1999; Tuteja et al, 2007). Transformation of plants with a gene encoding for an antioxidant enzyme has been carried out in several plant species. Many researchers have reported that the enrichment of antioxidant metabolism enhances tolerance to oxidative stress and therefore provides stress tolerance. Faize *et al.* (2011) reported that transgenic tobacco plants over-expressing cytosolic Cu/Zn-SOD (cyt*sod*) and/or APX (cyt*apx*) show an increased tolerance to drought stress by enhancing antioxidant defences. Increases in the activity of a number of antioxidant enzymes was also observed in the chloroplast of these transgenic plants, suggesting a positive influence of cytosolic antioxidant machinery in protecting the chloroplast and underlining the complexity of the regulation network of plant antioxidant defenses during drought conditions.

5. Heat Shock Proteins (HSPs)

Heat shock proteins (HSP) are a family of proteins that are produced by cells in response to various stressful conditions. They were first described in relation to heat stress (Ritossa, 1962), but now well studied that these are also expressed during other stresses and also under unstressed conditions. Hsps are also called stress proteins or stress induced proteins (Lindquist and Crig, 1988; Morimoto *et al.,* 1994; Gupta *et al.,* 2010). All organisms respond to high temperature by modifying their gene expression pattern and turning on the heat shock gene. Heat stress activates the heat shock factors which in turn bind to heat shock elements/ promoters. Heat shock factors then up regulates the heat shock genes (Schoffl et al., 1998; von Koskull-Döring et al., 2007; Scharf et al., 2012). Heat shock proteins are the product of these heat shock genes.

Heat stress affects many basic physiological processes such as photosynthesis, respiration, and water relations (Wahid et al., 2007). Many researchers have suggested the five classes of HSPs characterized by their activities as molecular chaperones according to their approximate molecular weight viz., Hsp100, Hsp90, Hsp70, Hsp60, and small heat-shock proteins (Schlesinger, 1990; Schoffl et al., 1998; Kotak et al., 2007). Further Gupta et al. (2010) classified the heat-shock proteins into families according to their, amino acid sequence, molecular weight, homologies and functions: Hsp100 family, Hsp90 family, Hsp70 family, Hsp60 family, and the small Hsp family. It has been reported that HSPs function as molecular chaperone, regulating the folding and accumulation of proteins as well as localization and degradation in all plants and animal species (Feder and Hofmann, 1999 Schulze-Lefert, 2004; Panaretou and Zhai, 2008; Hu et al., 2009; Gupta et al., 2010). These HSPs, as chaperones, prevent the irreversible aggregation of other proteins and participate in refolding proteins during heat stress conditions (Tripp et al., 2009).

HSP60 is also known as chaperonin. It has been reported that HSP60 play important role in assisting rubisco (Wang et al., 2004). Some studies suggested that this class might participate in folding and aggregation of many proteins that were transported to organelles such as chloroplasts and mitochondria ((Lubben et al., 1989). HSP70 function as chaperones for newly synthesized proteins to prevent their accumulation as aggregates and folds in a correct way during their transfer to their final destination (Sung et al., 2001; Su and Li, 2008). In addition, Hsp70 and sHsps primarily act as molecular chaperone and play a crucial role in protecting plant cell from the damaging effects of heat stress (Rouch et al., 2004). Hsp90 binds with Hsp70 in many chaperone complexes and has important role in signaling protein function and trafficking (Pratt and Toft, 2003). One exclusive function of HSP100 class is the reactivation of aggregated proteins (Parsell and Lindquist, 1993) by resolubilization of non-functional protein aggregates and also helping to degrade irreversibly damaged polypeptides (Bosl et al., 2006; Kim et al., 2007). It has been suggested that HSP100 class also participates in facilitating the normal situation of the organism after severe stress (Gurley, 2000).

6. Osmolytes

Plants have evolved many biochemical and molecular mechanisms in response to adverse environmental effects. One of the well studied stress responses is accumulation of osmolytes in plant cell during stress. Osmolytes are compounds that accumulate themselves in higher concentration under stress condition and protect the plants from the adverse effect of stress. Osmolytes protect enzyme and membrane integrity, along with adaptive roles in mediating osmotic adjustment in plants grown under water stress conditions. Transgenic plants over expressing the genes involved in the biosynthesis of these osmolytes have improved the tolerance to osmotic stress in many crops.

Important compatible osmolytes are proline, glycine betaine, sugars and sugar alcohols, which play important role in abiotic stress tolerance (Bohnert and Jensen, 1996; Rajam *et al.*, 1998; Bohnert and Shen, 1999; Kumar *et al.*, 2006). These osmolytes do not have any adverse effect on normal cellular functions even when present at high concentration (Yancy *et al.*, 1982; Serraj and Sinclair, 2002). Accumulation of these molecules at higher concentration helps plants to preserve water within cells and protects cellular compartments from injury caused by dehydration or maintains turgor pressure during water stress. Furthermore, these molecules stabilize the structure and function of certain macromolecules, signaling functions or induction of adaptive pathways and scavenge reactive oxygen species (Hasegawa *et al.*, 2000; Chen and Murata, 2002). Though, the molecular and cellular interactions of these solutes are not completely understood.

6.1. Proline

Proline is an important osmolyte molecule that accumulates in many organisms, including plants, fungi and bacteria, in response to drought and salinity (Kumar *et al.*, 2003; Claussen, 2005). It is one of the most commonly found compatible osmolytes in water-stressed plants (Delauney and Verma, 1993; Yoshiba *et al.*, 1997). Pyrroline-5-carboxylate synthetase (P-5-CS) is an important enzyme of proline biosynthesis. Proline reduces the acidity of cell by scavenging the hydroxyl radical (Venekamp *et al.*, 1989, Smirnoff and Cumbes, 1989). Additionally, it may also function as an osmolyte (Kishor *et al.*, 1995) ROS scavenger and molecular chaperone, being able to protect protein integrity and enhance the activities of different enzymes (Iyer and Caplan, 1998,Szabados and Savoure, 2010).Glutamate is the precursor of proline, which is reduced to glutamate- semialdehyde (GSA) by D-1-pyrroline-5-carboxylate synthetase (P5CS). GSA can spontaneously convert to pyrroline-5-carboxylate (P5C), which is then further reduced by P5C reductase (P5CR) to proline. In mitochondria proline is degraded by proline dehydrogenase

(ProDH) and P5C dehydrogenase (P5CDH) to glutamate. Proline biosynthesis is enhanced during stress while proline catabolism is enhanced during recovery from stress. As represented in Table 1 overexpression of P5CS in many plants led to increased proline accumulation and enhanced salt and drought tolerance (Kishor et al., 1995; Hong et al., 2000; Yamada et al., 2005).

6.2. Gama Amino Butyric Acid (AGBA)

γ-amino butyric acid is a non protein amino acid, accumulates to high levels under different adverse environmental conditions (Shelp et al., 1999; Kinnersley and Turano, 2000; Kaplan and Guy, 2004; Kempa et al., 2008; Renault et al., 2010). Glutamate is the precursor for GABA synthesis. GABA is mainly synthesized in the cytosol by glutamate decarboxylase (GAD) and then transported to the mitochondria. GABA metabolism has been associated with carbon–nitrogen balance and ROS scavenging (Bouche and Fromm, 2004; Song et al., 2010; Liu et al., 2011). The activities of enzymes involved in GABA metabolism are enhanced under salt stress (Renault et al., 2010).

Table1: Proline biosynthesis in transgenic plants confers abiotic stress tolerance

S.No	Target plant	Stress tolerance	Transgene	Reference
1.	N. tabacum	Drought	P5CS	Kishor et al., 1995
2.	O. sativa	Salt	P5CS	Zhu et al., 1998
3.	Triticumaestivum	Salt	P5CS	Sawahel et al., 2002
4.	Daucuscarota	Salt	P5CS	Han et al., 2003
5.	Petunia	Drought	P5CS	Yamuda et al., 2005
6.	S. tuberosum	Salt	P5CS	Hmida-Sayari et al., 2005
7.	N. tabacum	Drought	P5CSF129A cDNA	Gubis et al ., 2007
8.	A. thaliana	Freezing, salt	Antisense ProDHcDNA	Namnjo et al., 2008
9.	C. cajan	Salt	P5CSF129A	Surekha et al., 2013
10.	N. tabacum	Abiotic stress tolerance	OsP5CS1 and OsP5CS2	Zhang et al.,2014

6.3. Polyamines

Polyamines are extensively present in living organism and play important role in the regulation of plant developmental and different physiological processes (Kusano *et al.*, 2007). Polyamines may also function as stress messengers in plant response to different stress signals (liu *et al.*, 2000, Liu *et al.*, 2007). Different poly amine like spermine, spermidin and putrescine play important role in plant development and different physiological processes. Polyamines are small polycations which play important role in modulation the defence response of plants to diverse environmental stresses which include drought (Yamaguchi *et al.*,2007), salinity (Duan *et al.*, 2008), chilling stress(Cuevas *et al.*,2008), metal toxicity (Gropa *et al.*, 2003) and oxidative stress (Rider *et al.*, 2007).

The polyamines 1, 4-diaminobutane or putrescine (Put), the triamine 1, 8 diamino-4- ozaoctane or spermidine (Spd) and the tetraamine 1, 12-diamino- 4, 9-diazadodecane or spermine (Spm; Cohen, 1998) are ubiquitous in plants. Less commonly found PAs are cadaverine (Cad), thermospermine (tSpm), norspermidine, norspermine, homocaldopentamine, homocaldohexamine, 1,3-diaminopropane and 4-aminobutylcadaverine, among others (Kuehn *et al.*, 1990; Fujihara *et al.*, 1995; Kuznetsov *et al.*, 2007).

Several genes are involved in biosynthesis of different amine like arginine decarboxylase (ADC), *S*-adenosylmethionine decarboxylase (SAMDC), Spermidined synthase (SPDS), ornithine decarboxylase (ODC), these amines lay important role in multiple abiotic stress tolerance (Wi *et al.*, 2006; Prabhavathi *et al.*, 2007; Kasukabe *et al.*, 2007; Kasukabe *et al.*, 2006; Wen X-P *et al.*, 2008. The first step in PAs biosynthesis is Putrescine formation, a process that may occur by two different pathways. First is through the decarboxylation of arginine by the enzyme arginine decarboxylase (ADC, EC 4.1.1.19), a reaction that generates agmatine and *N*-carbamoylputrescine as intermediates, by the action of the corresponding enzymes agmatineiminohydrolase (EC 3.5.3.12) and *N*-carbamoylputrescineamidohydrolase (EC 3.5.1.53). The second pathway starts with the decarboxylation of ornithine by ornithine decarboxylase (EC 4.1.1.17), which yields Put directly. Both ODC and ADC enzymes use pyridoxal 5'-phosphate as cofactor. Other higher polyamines, Spermidined and Sperminem are synthesized by successive addition of aminopropyl groups to putrescine, through the activity of Spd synthase (EC 2.5.1.16) and Spm synthase (EC 2.5.1.22). S-adenosylmethionine is considered the major regulatory enzyme involved in higher polyamines biosynthesis and plays an essential role in modulating ethylene production in plants (Bagni and Tassoni, 2001). Basu *et al.*(2010) reported that drought resistant cultivars of rice had higher free Spermidine and free Spermine in the leaves than drought-susceptible ones during the whole period of withholding of water.

Different abiotic stresses like salinity, drought and cold induces oxidative stress in plants (Santos *et al.*, 2001; Hajiboland and Joudmand, 2009; Janmohammadi *et al.*, 2012) and produce reactive oxygen species (ROS) these free radicals can cause lipid peroxidation, protein denaturation and DNA mutation (Gutteridge and Halliwell, 1990; Bowler *et al.*, 1992). To overcome the damaging effect of ROS, plants have evolved an array of antioxidant compounds, whose presence in several plant species has been shown to correlate with the ability to tolerate those stresses (Gossett *et al.*, 1994). Many authors suggested that polyamines act as antioxidants under salinity and other environmentally adverse conditions. For example, under NaCl-induced stress, a higher level of lipid peroxidation was observed in the salt-sensitive, relative to the salt-tolerant cultivar in wheat (El-bassiouny 118 P.I. Calzadilla and Bekheta, 2005) and rice (Roychoudhury *et al.*, 2008), along with augmented Spermidined and Spermine levels, not observable in the salt-sensitive cultivars. However, the precise role of polyamines as antioxidants is still a matter of discussion (Groppa and Benavides, 2008).

6.4. Glycine Betaine

Glycine betaine is a quaternary ammonium compound that occurs in a wide variety of plants, but its distribution among plants is random (Cromwell and Rennie, 1953). Many plant species do not accumulate glycinbetain. In some plants abiotic stress, enhances glycine betaine accumulation (Rhodes and Hanson, 1993; Chen and Murata, 2011). As represented in Table 2, introduction of the glycine betaine gene in to biosynthetic pathway genes of non-accumulators improved their ability to tolerate abiotic stress conditions (Hayashi *et al.*, 1997; Alia Holmstrom *et al.*, 2000; Park *et al.*, 2004; Waditee *et al.*, 2005; Bansal *et al.*, 2011), pointing to the beneficial role of GB in stress tolerance. Glycine betaine also accumulate up- to osmotically significant levels in salt tolerant plant species (Rhodes and Hanson, 1993).

It has been reported that transgenic plants accumulated 2.8–3.8 times higher GB in reproductive organs than in the leaves. A number of transgenic plants with GB biosynthetic genes have been tested for GB accumulation and the resultant salt, drought and temperature tolerance.Transgenic plants such as Arabidopsis, eucalyptus, tobacco, rice, tomato, potato and wheat with GB biosynthetic genes have showed increased GB accumulation and stress tolerance (Ahmad *et al.*, 2008; Goel *et al.*, 2011;Sulpice *et al.*, 2003).

6.5. Sugars

Soluble sugars are highly responsive to abiotic stresses, which act on the supply of carbohydrates from source organs to sink ones. Sucrose and hexoses both play dual functions in gene regulation as demonstrated by the upregulation of growth-related genes and downregulation of stress-related genes. Although coordinately regulated by sugars, these growth- and stress-related genes are upregulated or downregulated through HXK-dependent and/or HXK-independent pathways. Sucrose is readily broken down in to glucose and fructose, while these hexoses lead to sucrose synthesis (Roitsch *et al.*, 1999). Furthermore, these interconvertions are strongly affected by environmental stresses (Podazza *et al.*, 2006, Rosa *et al.*, 2009). Like hormones, soluble sugars also act as primary messengers and regulate signals that control the expression of different genes involved in plant growth and metabolism (Rolland *et al.*, 2006, Chen *et al.*, 2007). Differential source-sink effects on metabolism induced by unfavourable environmental conditions lead to a differential expression of several proteins related to carbohydrate metabolism (Stitt *et al.*, 1999, Coruzzi *et al.*, 2001) e.g., enzymes related to starch biosynthesis (AGPase, ADP-Glcpyrophosphorylase) and sucrose metabolism (SuSy, sucrose synthase; SPS, sucrose phosphate synthase).

Table 2: **Metabolic engineering of plants for GB biosynthesis confers tolerance to various abiotic stresses**

S.No	Target plant	Stress tolerance	Transgene	Reference
1.	S. lycopersicum	Chilling	codA	Park et al., 2004
2.	A. thaliana	Salt	GSMT/SDMT	Waditee et al., 2005
3.	O. sativa	Salt	COX	Su et al., 2006
4.	O. sativa	Salt	CMO (spinach)	Shirasawa et al., 2006
5.	Gossypiumhir-sutum	Drought	betA (CDH)	Lv et al., 2007
6.	N. tabacum	Salt	BADH (spinach)	Yang et al., 2008
7.	S. tuberosum	Salt, drought	codA	Ahmad et al., 2008
8.	N. tabacum	Salt	BADH/SeNHX1	Zhou et al., 2008
9.	Eucalyptus globulus	Salt	codA	Yu et al., 2009
10.	T. aestivum	Drought	beta	He et al., 2010
11.	S. lycopersicum	Salt, drought	codA	Goel et al., 2011
12.	S. tuberosum	Salt, drought	codA	Cheng et al., 2013
13.	N. tabacum	Salt	codA	Jing et al., 2013

6.6. ABA and Metabolic Adjustments

Plant hormone abscisic acid plays an important role in plant growth and development including seed germination and development and abiotic stress tolerance, particularly drought resistance. ABA is the central regulator of abiotic stress resistance in plant and coordinates a complex regulatory network enabling plant to cope with drought stress. ABA content significantly increases under drought and salinity stress condition and stimulates alterations in gene expression, accumulation of osmolytes and stomatal closure to survive under stress condition (Cutler *et al.*,2010; Kim *et al.*,2010).

ABA mediated signalling plays crucial role in plant responses to environmental stress. Water or any osmotic stress promotes the synthesis of the ABA which then turns on a major change in gene expression and adaptive physiological responses (Seki *et al.*, 2002; Shinozaki and Yamaguchi-Shinozaki, 2007; Yamaguchi-Shinozaki and Shinozaki, 2006). The major signalling complexes that recognize ABA and transmit signal to downstream events have been also been reported and well studied (Fujii *et al.*, 2009; Ma *et al.*, 2009; Park *et al.*, 2009). In Arabidopsis thaliana, ABA sensing and signalling are mediated by three classes of protein (I) pyrabactin resistance 1(PYR/PYR -1 like) regulatory component of ABA receptor (II) protein phosphates 2C(PP2C) and (III)sucrose nonfermenting-1(SNF-1) related protein kinase 2(SnRK2). Under unstressed conditions when ABA levels are low, PPC2 interact with subclass III SnRK2s and inhibit their kinase activities. Under water stress condition, when ABA accumulates it is sensed by PYR/PYR-1 receptors that inhibit the phosphatase activities of PP2Cs.Ternary complexes composed of ABA–PYR/PYL/RCAR–PP2C enable the activation of SnRK2s, resulting in the phosphorylation of downstream substrates, such as transcription factors and membrane channel proteins (Nishimura *et al.*,2010).

7. References

1. Ahmad, R.; Kim, M.D.; Back, K.H.; Kim, H.S.; Lee, H.S.; Kwon, S.Y.; Norio, M.; Chung Won II.; Kwak, S.K. Stress-induced expression of choline oxidase in potato plant chloroplasts confers enhanced tolerance to oxidative, salt, and drought stresses. *Plant Cell Rep.* **2008**, 27, 687–698.

2. Ahmad, P.; Sarwat, M.; Sharma, S. Reactive oxygen species, antioxidants and signalling in plants. *Journal of plant biology.* **2008**, 51, 3, 167-173.

3. Alia.; Hayashi, H.; Sakamoto, A.; Murata, N. Enhancement of the tolerance of Arabidopsis to high temperatures by genetic engineering of the synthesis of glycinebetaine. *The Plant Journal.* **1998**, 16, 155–161.

4. Arenas-Huertero, C.; Perez, B.; Rabanal, F.; Blanco-Melo, D.; De la Rosa, C.; Estrada-Navarrete, G.; Sanchez, F.; Covarrubias, A.A.; Reyes, J.L. Conserved and novel miRNAs in the legume Phaseolus vulgaris in response to stress. *Plant Mol Biol.* **2009**, 70, 385-401.

5. Bagni, N.; Tassoni, A. Biosynthesis, oxidation and conjugation of aliphatic polyamines in higher plants. *Amino Acids.* **2001**, 20, 301–317.

6. Bansal, K.C.; Goel, D.; Singh, A.K.; Yadav, V.; Babbar, S.B.; Murata, N. Transformation of tomato with a bacterial codA gene enhances tolerance to

salt and water stresses. *Journal of Plant Physiology.* **2011**, 168, 1286–1294.

7. Bartel, D.P. MicroRNAs: genomics, biogenesis, mechanism, and function. *Cell.* **2004**, 116, 281–297.

8. Basu, S.; Roychoudhury, A.; Saha, P.P.; Sengupta, D.N. Comparative analysis of some biochemical responses of three indica rice varieties during polyethylene glycol-mediated water stress exhibits distinct varietal differences. *Acta Physiologiae Plantarum.* **2010**, 32, 551–563.

9. Bian, H.; Xie, Y.; Guo, F.; Han, N.; Ma, S.; Zeng, Z.; Wang, J.; Yang, Y.; Zhu, M. Distinctive expression patterns and roles of the miRNA393/TIR1 homolog module in regulating flag leaf inclination and primary and crown root growth in rice (*Oryza sativa*). *New Phytol.* **2012**, 196, 149e161.

10. Bohnert, H.J.; Jensen, R.G. Strategies for engineering water-stress tolerance in plants. *Trends Biotechnol.* **1996**, 14, 89–97.

11. Bohnert, H.J.; Shen, B. Transformation and compatible solutes. *Sci Hort.* **1999**, 78, 237–260.

12. Bosl, B.; Grimminger, V.; Walter, S. The molecular chaperone Hsp104 – a molecular machine for protein disaggregation. *J. Struct. Biol.* **2006**, 156, 139–148.

13. Bouche, N.; Fromm, H. GABA in plants: just a metabolite? *Trends in Plant Science.* **2004**, 9, 110–115.

14. Bowler, C.; Montagu, M.V.; Inze, D. Superoxide Dismutase and Stress Tolerance. *AnnualReview of Plant Physiology and Plant Molecular Biology.* **1992**, 43, 83–116.

15. Chen, J.G. Sweet sensor, surprising partners. *Sci STKE.* **2007**, 373, 7.

16. Chen, T.H.; Murata, N. Glycinebetaine protects plants against abiotic stress: mechanisms and biotechnological applications. *Plant, Cell and Environment.* **2011**, 34, 1–20.

17. Chen, X. A microRNA as a translational repressor of APETALA2 in Arabidopsis flower development. *Science.* **2004**, 303, 2022–2025.

18. Chen, Z.H.; Bao, M.L.; Sun, Y.Z.; Yang, Y.J.; Xu, X.H.; Wang, J.H.; Han, N.; Bian, H.W.; Zhu, M.Y. Regulation of auxin response by miR393- targeted transport inhibitor response protein 1 is involved in normal development in Arabidopsis. *Plant Mol. Biol.* **2011**, 77, 619e629.

19. Cheng, Y.J.; Deng, X.P.; Kwak, S.S.; Chen, W.; Eneji, A.E. Enhanced tolerance of transgenic potato plants expressing choline oxidase in chloroplasts against water stress. *Bot Stud.* **2013**, 54, 30.

20. Cohen, S.S.A Guide to Polyamines, 1st edn. Oxford University Press, New York. **1998**.

21. Coruzzi, G.M.; Bush, D.R. Nitrogen and carbon nutrient and metabolite signaling in plants. *Plant Physiol.* **2001**, 125, 61-64.

22. Cromwell, B.T.; Rennie, S.D. The biosynthesis and metabolism of betaines in plants. 1. The estimation and distribution of glycinebetaine (betaine) in Beta vulgaris L. and other plants. *Biochemical Journal.* **1953**, 55, 189–192.

23. Cuevas, J.C.; Lopez-Cobollo, R.; Alcazar, R.; Zarza, X.; Koncz, C.; Altabella T.; et al. Putrescine is involved in Arabidopsis freezing tolerance and cold acclimation by regulating ABA levels in response to low temperature. *Plant Physiol.* **2008**, 148, 1094-105.

24. Cutler, S.R.; Rodriguez, P.L.; Finkelstein, R.R.; Abrams, S.R. Abscisic acid: Emergence of a core signaling network. *Annu Rev Plant Biol.* **2010**, 61, 651–679.

25. Dalton, T.P.; Shertzer, H.G.; Puga, A. Regulation of gene expression by reactive oxygen. *Annu. Rev. Pharmacol. Toxicol.* **1999**, 39, 67-101.

26. Ding, Y.; Tao, Y.; Zhu, C. Emerging roles of microRNAs in the mediation of drought stress response in plants. *J. Exp. Bot.* **2013**, 64, 3077–3086.

27. Duan, J.J.; Li, J.; Guo, S.R.; Kang, Y.Y. Exogenous Spermidine affects polyamine metabolism in salinity- stressed *Cucumissativus*roots and enhances short-term salinity tolerance. *J Plant Physiol.* **2008**, 165, 1620-35.

28. El-bassiouny, H.M.S.; Bekheta, M.A. Effect of salt stress on relative water content, lipid peroxidation,polyamines, amino acids and ethylene of two wheat cultivars.*International Journal of Agriculture and Biology.* **2005**, 7, 363–368.

29. Faize, M. Involvement of cytosolic ascorbate peroxidase and Cu/Zn-superoxide dismutase for improved tolerance against drought stress. *J Exp Bot.* **2011**, 62(8), 2599-613.

30. Feder, M.E.; Hofmann, G.E. Heat-shock proteins, molecular chaperones, and stress response: evolutionary and ecological physiology. *Annu. Rev. Physiol.* **1999**, 61, 243–282.

31. Feng, H.; Zhang, Q.; Wang, Q.; Wang, X.; Liu, J.; Li, M.; Huang, L.; Kang, Z. Target of tae-miR408, a chemocyanin-like protein gene (TaCLP1), plays positive roles in wheat response to high-salinity, heavy cupric stress and stripe rust. *Plant Mol. Biol.* **2013**, 83, 433-443.

32. Floris, M.; Mahgoub, H.; Lanet, E.; Robaglia, C.; Menand, B. Post transcriptional regulation of gene expression in plants during abiotic stress. *Int. J. Mol. Sci.* **2009**, 10, 3168–3185.

33. Foyer, C.H.; Noctor, G. Redox sensing and signalling associated with reactive oxygen in chloroplasts, peroxisomes and mitochondria. *Physiologia Plantarum.* **2003**, 119(3), 355-364.

34. Fujihara, S; Abe, H.; Yoneyama, T. A new polyamine 4-aminobutylcadaverine: occurrence and its biosynthesis in root nodules of adzuki bean plant *Vignaangularis. Journal of Biological Chemistry.* **1995**, 270, 9932–9938.

35. Fujii, H.; Chinnusamy, V.; Rodrigues, A.; Rubio, S.; Antoni, R.; Park, S.Y.; Cutler,S.R.;Sheen, J.; Rodriguez, P.L.; Zhu,J.In vitro reconstitution of an abscisic acid signalling pathway. Nature. **2009**, 462, 660–4.

36. Gao, P.; Bai, X.; Yang, L.; Lv, D.; Pan, X.; Li, Y.; Cai, H.; Ji, W.; Chen, Q.; Zhu, Y. osa-MIR393: A salinity- and alkaline stress related microRNA gene. *MolBiol Rep.* **2011**, 38, 237-242.

37. Goel, D.; Singh, A.K.; Yadav, V.; Babbar, S.B.; Murata, N.; Bansal, K.C. Transformation of tomato with a bacterial codA gene enhances tolerance to salt and water stresses. J *Plant Physiol.* **2011**, 168, 1286–1294.

38. Groppa, M.D.; Benavides, M.P.; Tomaro, M.L. Polyamine metabolism in sunflower and wheat leaf discs under cadmium or copper stress. *Plant Sci.* **2003**, 161, 481-8.

39. Groppa, M.D.; Benavides, M.P. Polyamines and abiotic stress: recent advances. *Amino Acids.* **2008**, 34, 35–45.

40. Gubis, J.; Vanková, R.; Cervená, V.; Dragunová, M.; Hudcovicová, M.; Lichtnerová, H. Transformed tobacco plants with increased tolerance to drought. *S Afr J Bot.* **2007**, 73, 505–11.

41. Gupta, S.C.; Sharma, A.; Mishra, M.; Mishra, R.; Chowdhuri, D.K. Heat shock proteins in toxicology: how close and how far? *Life Sci.* **2010**, 86, 377–384.

42. Gurley, W.B. HSP101: a key component for the acquisition of thermotolerance in plants. *Plant Cell.* **2000**, 12, 457–460.

43. Gutteridge, J.M.C.; Halliwell, B. The measurement and mechanism of lipid peroxidation in biological systems. *Trends in Biochemical Sciences.* **1990**, 15, 129–135.

44. Hajiboland, R.; Joudmand, A. The K/Na replacement and function of antioxidant defence system in sugar beet (*Beta vulgaris* L.) cultivars. *Acta Agriculturae Scandinavica Section B – Soil and Plant Science.* **2009**, 59, 246–259.

45. Han, K.H.; Hwang, C.H. Salt tolerance enhanced by transformation of a P5CS gene in carrot. *J Plant Biotechnol.* **2003**, 5, 149–53.

46. Hayashi, H.; Alia.; Mustardy, L.; Deshnium, P.; Ida, M.; Murata, N. Transformation of Arabidopsis thaliana with the codA gene for choline oxidase; accumulation of glycine betaine and enhanced tolerance to salt and cold stress. *The Plant Journal*. **1997**, 12, 133–142.

47. He, C.; Yang, A.; Zhang, W.; Gao, Q.; Zhang, J. Improved salt tolerance of transgenic wheat by introducing betA gene for glycine betaine synthesis. *Plant Cell Tiss Organ Cult*. **2010**, 101, 65–78.

48. Hmida-Sayari, A.; Gargouri-Bouzid, R.; Bidani, A.; Jaoua, L.; Savouré, A.; Jaoua, S. Overexpression of Ä1-pyrroline-5-carboxylate synthetase increases proline production and confers salt tolerance in transgenic potato plants. *Plant Sci*. **2005**, 169, 746–52.

49. Hong, Z.; Lakkineni, K.; Zhang, Z.; Verma, D.P. Removal of feedback inhibition of delta(1)-pyrroline-5-carboxylate synthetase results in increased proline accumulation and protection of plants from osmotic stress. *Plant Physiology*. **2000**, 122, 1129–1136.

50. Janmohammadi, M.; Enayati, V.; Sabaghnia, N. Impact of cold acclimation, de-acclimation and re-acclimation on carbohydrate content and antioxidant enzyme activities in spring and winter wheat. *Icelandic Agricultural Sciences*. **2012**, 25, 3–11.

51. Jannatul, Ferdous.; Syed S, Hussain.; Bu-Jun, Shi. Role of microRNAs in plant drought tolerance. *Plant Biotechnology Journal*. **2015**, 13, 293–305.

52. Jing, J.; Li, H.; He, G.; Yin, Y.; Liu, M.; Liu, B.; et al. Over-expression of the codA gene by Rd29A promoter improves salt tolerance in Nicotianatabacum. *Pak J Bot*. **2013**, 45, 821–7.

53. Jonak, C.; Ligterink, W.; Hirt, H. MAP kinases in plant signal transduction. *Cell Mol Life Sci*. **1999**, 55, 204-213.

54. Joshi, R.K.; Kar, B.; Nayak, S. Characterization of mitogen activated protein kinases (MAPKs) in the Curcuma longa expressed sequence tag database *Bioinformation*. **2011**, 7,180-183.

55. Kaplan, F.; Guy, C.L. beta-Amylase induction and the protective role of maltose during temperature shock. *Plant Physiology*. **2004**, 135, 1674–1684.

56. Kasukabe, Y.; He, L.; Nada, K.; Misawa, S.; Ihara, I.; Tachibana, S. Overexpression of spermidine synthase enhances tolerance to multiple environmental stresses and upregulates the expression of various stress regulated genes in transgenic *Arabidopsis thaliana*. *Plant Cell Physiol*. **2007**, 45, 712-22.

57. Kasukabe, Y.; He, L.; Watakabe, Y.; Otani, M.; Shimada, T.; Tachibana, S. Improvement of environmental stress tolerance of sweet potato by introduction of genes for spermidine synthase. *Plant Biotechnol*. **2006**, 23, 75-83.

58. Kim, T.H.; Bohmer, M.; Hu, H.; Nishimura, N.; Schroeder, J.I. Guard cell signal transduction network: Advances in understanding abscisic acid, CO^2 and Ca^{2+} signaling. *Annu Rev Plant Biol.* **2010**, 61, 561–591.

59. Kim, H.J.; Hwang, N.R.; Lee, K.J. Heat shock responses for understanding diseases of protein denaturation. *Mol. Cells.* **2007**, 23, 123–131.

60. Kinnersley, A.M.; Turano, F.J. Gamma aminobutyric acid (GABA) and plant responses to stress. *Critical Reviews in Plant Sciences.* **2000**, 19, 479–509.

61. Kishor, K.P.B.; Hong, Z.; Miao, G.H.; Hu, C.A.A.; Verma, D.P.S. Overexpression of [delta]-pyrroline-5-carboxylate synthetase increase proline production and confers osmotolerance in transgenic plants. *Plant Physiol.* **1995**, 108, 1387–94.

62. Kishor, P.; Hong, Z.; Miao, G.H.; Hu, C.; Verma, D. Overexpression of [delta]-pyrroline-5-carboxylate synthetase increases proline production and confers osmotolerance in transgenic plants. *Plant Physiology.* **1995**, 108, 1387–1394.

63. Kuehn, G.O., Bagga, S., Rodríguez-Garay, B., Philipps, A.C. Biosynthesis of uncommon polyamines in higher plants and their relationship to abiotic stress responses. In: Flores, H.E., Arteca, R.N., Shannon, J.C. (Eds.), Polyamines and Ethylene: Biochemistry, Physiology and Interactions. American Society of Plant Physiology, Rockville, Maryland. **1990**, pp. 190–202.

64. Kumar, S.V.; Sharma, M.L.; Rajam, M.V. Polyamine biosynthetic pathway as a novel target for potential applications in plant biotechnology. *PhysiolMolBiol Plant.* **2006**, 12, 13–28.

65. Kusano, T.; Yamaguchi, K.; Berberich, T.; Takahashi, Y. Advances in polyamine research in 2007. *J Plant Res.* **2007**, 120, 345-50.

66. Kuznetsov, V.; Radukina, N.L.; Shevyakoval, N.I. Metabolism of polyamines and prospects for producing stress-tolerant plants: an overview. In: Thangadurai, D., Tang, W., Song, S.Q. (Eds.), Plant Stress and Biotechnology. Oxford Book Company, Jaipur, India. **2007**, pp. 256.

67. Lawlor, D.W. Genetic engineering to improve plant performance under drought: physiological evaluation of achievements, limitations, and possibilities. *J. Exp. Bot.* **2013**, 64, 83–108.

68. Li, H.W.; Zang, B.S.; Deng, X.W.; Wang, X.P. Overexpression of the trehalose-6-phosphate synthase gene OsTPS1 enhances abiotic stress tolerance in rice. *Planta.* **2011**, 234, 1007–1018.

69. Lindquist, S.; Crig, E.A. The heat-shock proteins. *Annu. Rev. Genet.* **1988**, 22, 631–677.

70. Liu, J.H.; Kitashiba, H.; Wang, J.; Ban, Y.; Moriguchi, T. Polyamines and their ability to provide environmental stress tolerance to plants. *Plant Biotechnol.* **2007**, 24, 117-26.

71. Liu, K.; Fu, H.H.; Bei, Q.X.; Luan, S. Inward potassium channel in guard cells as a target for polyamine regulation of stomatal movements. *Plant Physiol.* **2000**, *124*, 1315-25.

72. Liu, H.H.; Tian, X.; Li, Y.J.; Wu, C.A.; Zheng, C.C. Microarray-based analysis of stress-regulated microRNAs in *Arabidopsisthaliana*. *RNA.* **2008,** 14, 836–843.

73. Liu, Z.; Kumari, S.; Zhang, L.; Zheng, Y.; Ware, D. Characterization of miRNAs in response to short-term waterlogging in three inbred lines of Zea mays. *PLoS ONE.* **2012**, 7, e.0039786.

74. Lubben, T.H.; Donaldson, G.K.; Viitanen, P.V.; Gatenby, A.A. Several proteins imported into chloroplasts form stable complexes with the GroEL-related chloroplast molecular chaperone. *Plant Cell.* **1989**, 1, 1223–1230.

75. Lv, S.; Yang, A.; Zhang, K.; Wang, L.; Zhang, J. Increase of glycinebetaine synthesis improves drought tolerance in cotton. *Mol Breed.* **2007,** 20, 233–248.

76. Ma, Y.; Szostkiewicz, I.; Korte, A.; Moes, D.; Yang, Y.; Christmann, A.; Grill, E.Regulators of PP2C phosphatase activity function as abscisic acid sensors. *Science.* **2009**, 324, 1064–1068.

77. Maizel, A.; Jouannet, V. Trans-acting small interfering RNAs: biogenesis, mode of action, and role in plant development. In: Sunkar, R., (Ed.), MicroRNAs in Plant Development and Stress Responses, Heidelberg, Berlin, Springer-Verlag. **2012**, pp. 83–108.

78. Mittler, R. Oxidative stress, antioxidants and stress tolerance. *Trends in Plant Science.* **2002**, 7(9),405-410.

79. Moller, I.M. Plant mitochondria and oxidative stress: Electron transport, NADPH turnover, and metabolism of reactive oxygen species. *Annual Review of Plant Physiology and Plant Molecular Biology.* **2001,** 52, 561-591.

80. Morimoto, R.I.; Tissieres, A.; Georgopoulos, C. Heat Shock Proteins: Structure, Function and Regulation. Cold Spring Harbor Lab. Press, Cold Spring Harbor, NY. **1994**.

81. Nanjo, T.; Fujita, M.; Seki, M.; Kato, T.; Tabata, S.; Shinozaki, K. Toxicity of free proline revealed in an Arabidopsis T-DNA-tagged mutant deficient in proline dehydrogenase. *Plant Cell Physiol.* **2003**, 44, 541–8.

82. Nishimura, N.; Sarkeshik, A.; Nito, K.; Park, S.Y.; Wang, A.; Carvalho, P.C.; Lee, S.; Caddell D.F.; Cutler, S.R.; Chory, J.; Yates, J.R.; Schroeder, J. I.

PYR/PYL/RCAR family members are major in-vivo ABI1 protein phosphatase 2C-interacting proteins in *Arabidopsis*. *Plant J.***2010**, 61, 290–299.

83. Ozhuner, E.; Eldem, V.; Ipek, A.; Okay, S.; Sakcali, S.; Zhang, B.; Boke, H.; Unver, T. Boron stress responsive microRNAs and their targets in barley. *PLoS ONE.* **2013**, 8, e59543.

84. Panaretou, B.; Zhai, C. The heat shock proteins: their roles as multi-component machines for protein folding. *Fungal biol. rev.* **2008**, 22, 110–119.

85. Park, E.J.; Jeknic, Z.; Sakamoto, A.; DeNoma, J.; Yuwansiri, R.; Murata, N.; Chen, T.H. Genetic engineering of glycinebetaine synthesis in tomato protects seeds, plants, and flowers from chilling damage. *The Plant Journal.* **2004**, 40, 474–487.

86. Park, S.Y.; Fung, P.; Nishimura, N.; Jensen, D.R.; Fujii, H.; Zhao, Y.; Lumba, S.; Santiago, J.; Rodrigues, A.; Chow, T.F.; Alfred, S.E.; Bonetta, D.; Finkelstein, R.; Provart, N.J.; Desveaux D.; Rodriguez, P.L.; McCourt, P.; Zhu, J.K.; Schroeder, J.I.; Volkman, B.F.; Cutler, S.R.Abscisic acid inhibits Type 2C protein phosphatases via the PYR/PYL family of START proteins. *Science.* **2009**, 324, 1068–71.

87. Parsell, P.A.; Lindquist, S. The function of heat-shock proteins in stress tolerance: degradation and reactivation of damaged proteins. *Annu. Rev. Genet.* **1993**, 27, 437–496.

88. Podazza, G.; Rosa, M.; González, J.A.; Hilal, M.; Prado, F.E. Cadmium induces changes in sucrose partitioning, invertase activities and membrane functionality in roots of Rangpur lime (*Citruslimonia L.* Osbeck). *Plant Biol.* **2006**, 8, 706-714.

89. Prabhavathi, V.R.; Rajam, M.V. Polyamine accumulation in transgenic eggplant enhances tolerance to multiple abiotic stresses and fungal resistance. *Plant Biotechnol.* **2007**, 24, 273-82

90. Pratt, W.B.; Toft, D.O. Regulation of signaling protein function and trafficking by the hsp90/hsp70-based chaperone machinery. *Exp. Biol. Med.* **2003**, 228, 111–133.

91. Rajam, M.V.; Dagar, S.; Waie, B.; Yadav, J.S.; Kumar, P.A.; Shoeb, F.; Kumria, R. Genetic engineering of polyamine and carbohydrate metabolism for osmotic stress tolerance in higher plants. *J Biosci.* **1998**, 23, 473–482.

92. Reinhart, B.J.; Weinstein, E.G.; Rhoades, M.W.; Bartel, B.; Bartel, D.P.MicroRNAs in plants. *Genes Dev.* **2002**, 16, 1616–1626.

93. Renault, H.; Roussel, V.; El Amrani, A.; Arzel, M.; Renault, D.; Bouchereau, A.; Deleu, C. The Arabidopsis pop2-1 mutant reveals the involvement of GABA transaminase in salt stress tolerance. *BMC Plant Biology.* **2010**, 1471-2229/10/20.

94. Rhodes, D.; Hanson, A.D. Quaternary ammonium and tertiary sulfonium compounds in higher-plants. *Annual Review of Plant Physiology and Plant Molecular Biology.* **1993,** 44, 357–384.

95. Rider, J.E.; Hacker, A.; Mackintosh, C.A.; Pegg, A.E.; Woster, P.M.; Casero, R.A. Jr. Spermine and spermidinemediate protection against oxidative damage caused by hydrogen peroxide. *Amino Acids.* **2007,** 33, 231- 40.

96. Ritossa, F. A. New puffing pattern induced by temperature shock and DNP in drosophila. *Experientia.* **1962,** 18(12), 571–573.

97. Rizhsky, L.; Liang, H.J.; Shuman, J.; Shulaev, V.; Davletova, S.; Mittler, R. When defense pathways collide. The response of Arabidopsis to a combination of drought and heat stress. *Plant Physiology.* **2004b,** 134, 1683–1696.

98. Rodriguez, M.C.; Petersen, M.; Mundy, J. Mitogen-activated protein kinase signaling in plants. *Annual Review of Plant Biology.* **2010,**61, 621–49.

99. Roitsch, T. Source-sink regulation by sugar and stress. *CurrOpin Plant Biol.* **1999,** 2, 198-206.

100. Rolland, F.; Baena-Gonzalez, E.; Sheen, J. Sugar sensing and signaling in plants: conserved and novel mechanisms. *Annu Rev Plant Biol.* **2006,** 57, 675-709.

101. Rosa, M.; Hilal, M.; González, J.A.; Prado, F.E.. Low-temperature effect on enzyme activities involved in sucrose-starch partitioning in salt-stressed and salt-acclimated cotyledons of quinoa (*Chenopodium quinoa* Willd.) seedlings. *Plant Physiol Biochem.* **2009,** 47, 300-307.

102. Rouch, J.M.; Bingham, S.E.; Sommerfeld, M.R. Protein expression during heat stress in thermo-intolerance and thermotolerance diatoms. *J. Exp. Mar. Biol. Ecol.* **2004,** 306, 231–243.

103. Roychoudhury, A.; Basu, S.; Sarkar, S.N.; Sengupta, D.N. Comparative physiological and molecular responses of a common aromatic indica rice cultivar to high salinity with non-aromatic indica rice cultivars. *Plant Cell Reports.* **2008,** 27, 1395–1410.

104. Santos, C.; Campos, A.; Azevedo, H. In situ and in vitro senescence induced by KCl stress: nutritional imbalance, lipid peroxidation and antioxidant metabolism. *Journal of Experimental Botany.* **2001,** 52, 351–360.

105. Sawahel, W.A.; Hassan, A.H. Generation of transgenic wheat plants producing high levels of the osmoprotectantproline. *Biotechnol Lett.* **2002,** 24, 721–5.

106. Scharf, K.D.; Berberich, T.; Ebersberger, I.; Nover, L. The plant heat stress transcription factor (Hsf) family: structure, function and evolution. *BiochimBiophys Acta.* **2012,** 1819(2), 104–119.

107. Schlesinger, M.J. Heat shock proteins. *J. Biol. Chem.* **1990,** 265, 12111–12114.

108. Schoffl, F.; Prandl, R.; Reindl, A. Regulation of the heat shock response. *Plant Physiol.* **1998,** 117, 1135–1141.

109. Schulze-Lefert, P. Plant immunity: the origami of receptor activation. *Curr. Biol.* **2004,** 14, R22–R24.

110. Seki, M.; Ishida, J.; Narusaka, M.; Fujita, M.; Nanjo, T.; Umezawa, T. Monitoring the expression pattern of around 7,000 Arabidopsis genes under ABA treatments using a full-length cDNA microarray. *FunctIntegr Genomics.* 2, 282–291.

111. Shelp, B.J.; Bown, A.W.; McLean, M.D. Metabolism and functions of gamma-aminobutyric acid. *Trends in Plant Science.* **1999,** 4, 446–452.

112. Shinozaki, K.; Yamaguchi-Shinozaki, K. Molecular responses to dehydration and low temperature: differences and cross-talk between two stress signaling pathways. *Curr Opin Plant Biol.* **2000,** 3, 217–223.

113. Shinozaki, K.; Yamaguchi-Shinozaki, K. Gene networks involved in drought stress response and tolerance. *J. Exp. Bot.* **2007,** 58, 221–227.

114. Shirasawa, K.; Takabe, T.; Kishitani, S. Accumulation of glycine betaine in rice plants that overexpress choline monooxygenase from spinach and evaluation of their tolerance to abiotic stress. *Ann Bot.* **2006,** 98, 565–71.

115. Sinha, A.K.; Jaggi, M.; Raghuram, B.; Tuteja, N. Mitogen-activated protein kinase signaling in plants under abiotic stress". *Plant Signaling and Behavior.* **2011,**6 (2), 196–203.

116. Song, H.M.; Xu, X.B.; Wang, H.; Wang, H.Z.; Tao, Y.Z. Exogenous gamma-amino butyric acid alleviates oxidative damage caused by aluminium and proton stresses on barley seedlings. *Journal of the Science of Food and Agriculture.* **2010,** 90, 1410–1416.

117. Stitt, M.; Krappe, A. The interaction between elevated carbon dioxide and nitrogen nutrition. The physiological and molecular background. *Plant Cell Environ.* **1999,** 22, 583-621.

118. Su, J.; Hirji, R.; Zhang, L.; He, C.; Selvaraj, G.; Wu, R. Evaluation of the stress-inducible production of choline oxidase in transgenic rice as a strategy for producing the stress protectant glycine betaine. *J Exp Bot.* **2006,** 57, 1129–35.

119. Su, P.H.; Li, H.M. Arabidopsis stromal 70-kD heat shock proteins are essential for plant development and important for thermotolerance of germinating seeds. *Plant Physiol.* **2008,** 146, 1231– 1241.

120. Sulpice, R.; Tsukaya, H.; Nonaka, H.; Mustardy, L.; Chen, T.H.; Murata, N. Enhanced formation of flowers in salt-stressed Arabidopsis after genetic engineering of the synthesis of glycine betaine. *Plant J.* **2003,** 36, 165–176.

121. Sung, D.Y.; Kaplan, F.; Guy, C.L. Plant Hsp70 molecular chaperones: protein structure, gene family, expression and function. *Physiol. Plantarum.* **2001,** 113, 443–451.

122. Sunkar, R.; Zhu, J.K. Novel and stress-regulated microRNAs and other small RNAs from *Arabidopsis. Plant Cell.* **2004,** 16, 2001-2019.

123. Sunkar, R.; Kapoor, A.; Zhu, J.K. Posttranscriptional induction of two Cu/Zn superoxide dismutase genes in *Arabidopsis* is mediated by down regulation of miR398 and important for oxidative stress tolerance. *Plant Cell.* **2006,** 18, 2051–2065.

124. Surekha, C.H.; Nirmala Kumari, K.; Aruna, L.V.; Suneetha, G.; Arundhati, A.; KaviKishor, P.B. Expression of the *Vignaaconitifolia* P5CSF129A gene in transgenic pigeon pea enhances proline accumulation and salt tolerance. *Plant Cell Tissue Organ Cult.* **2013,** 116, 27–36.

125. Szabados, L.; Savoure, A. Proline: a multifunctional amino acid. *Trends in Plant Science.* **2010,** 15, 89–97.

126. Trindade, I.; Capitao, C.; Dalmay, T.; Fevereiro, M.P.; Santos, D.M. miR398 and miR408 are up-regulated in response to water deficit in *Medicagotruncatula. Planta.***2010,** 231,705-716.

127. Tripp, J.; Mishra, S.K.; Scharf, K.D. Functional dissection of the cytosolic chaperone network in tomato mesophyll protoplasts. *Plant Cell Environ.* **2009,** 32, 123–133.

128. Tuteja N. Mechanisms of high salinity tolerance in plants. *Meth. Enzymol.:Osmosens. Osmosignal.* **2007,** 428, 419-438.

129. Verslues, P.E.; Agarwal, M.; Katiyar-Agarwal, S.; Zhu, J.; Zhu, J.K. Methods and concepts in quantifying resistance to drought, salt and freezing, abiotic stresses that affect plant water status. *Plant.* **2006,** 45, 523–39.

130. vonKoskull-Doring, P.; Scharf, K.D.; Nover, L. The diversity of plant heat stress transcription factors. *Trends in Plant Science.* **2007,** 12, 452–457.

131. Vranova, E.; Inze, D.; Van Breusegem, F. Signal transduction during oxidative stress. *Journal of Experimental Botany.* **2002,** 53(372), 1227-1236.

132. Waditee, R.; Bhuiyan, M.N.; Rai, V.; Aoki, K.; Tanaka, Y.; Hibino, T.; Suzuki, S.; Takano, J.; Jagendorf, A.T.; Takabe, T. Genes for direct methylation of glycine provide high levels of glycine betaine and abiotic-stress tolerance in *Synechococcus* and *Arabidopsis. Proceedings of the National Academy of Sciences, USA.* **2005,** 102, 1318–1323.

133. Wahid, A.; Gelani, S.; Ashraf, M.; Foolad, M.R. Heat tolerance in plants: an overview. *Environ. Exp. Bot.* **2007,** 61, 199–223.

134. Wang, W.X.; Vinocur, B.; Altman, A. Plant responses to drought, salinity and extreme temperatures: towards genetic engineering for stress tolerance. *Planta.* **2003,** 218, 1–14.

135. Wang, B.; Sun, Y.F.; Song, N.; Wei, J.P.; Wang, X.J.; Feng, H.; Yin, Z.Y.; Kang, Z.S. Micro RNAS involving in cold, wounding and salt stresses in *TriticumaestivumL.Plant Physiol. Biochem.* **2014,** 80, 90–96.

136. Wang, W.; Vinocur, B.; Shoseyov, O.; Altman, A. Role of plant heat-shock proteins and molecular chaperones in the abiotic stress response. *Trends Plant Sci.* **2004,** 9, 244–252.

137. Wen, X.P.; Pang, X.M.; Matsuda, N.; Kita, M.; Inoue, H.; Hao, Y.J. Over expression of the apple spermidine synthase gene in pear confers multiple abiotic stress tolerance by altering polyamine titers. *Transgenic Res.* **2008,** 17, 251-63.

138. Wi, S.J.; Kim, W.T.; Park, K.Y. Overexpression of carnation S adenosylmethionine decarboxylase gene generates a broad-spectrum tolerance to abiotic stresses in transgenic tobacco plants. *Plant Cell Rep.* **2006,** 25, 1111-21.

139. Xie, F.; Stewart, C.N. Jr.; Taki, F.A.; He, Q.; Liu, H.; Zhang, B. High throughput deep sequencing shows that microRNAs play important roles in switchgrass responses to drought and salinity. *Plant Biotechnol. J.* **2014,** 12, 354–366.

140. Yamada, M.; Morishita, H.; Urano, K.; Shiozaki, N.; Yamaguchi- Shinozaki, K.; Shinozaki, K.; Yoshiba, Y. Effects of free proline accumulation in petunias under drought stress. *Journal of Experimental Botany.* **2005,** 56, 1975–1981.

141. Yamaguchi, K.; Takahashi, Y.; Berberich, T.; Imai, A.; Takahashi, T.; Michael, A.J.; Kusano, T. A protective role for the polyamine spermine against drought stress in Arabidopsis. *BiochemBiophys Res Commun.* **2007,** 352, 486-90.

142. Yamaguchi-Shinozaki, K.; Shinozaki, K. Transcriptional regulatory networks in cellular responses and tolerance to dehydration and cold stresses. *Annu Rev Plant Biol.* 57, 781–803.

143. Yang, G.D.; Yan, K.; Wu, B.J.; Wang, Y.H.; Gao, Y.X.; C.C., Zheng. Genomewide analysis of intronic microRNAs in rice and *Arabidopsis. J. Genet.* **2012,** 91, 313–324.

144. Yang, X.; Liang, Z.; Wen, X.; Lu, C. Genetic engineering of the biosynthesis of glycine betaine leads to increased tolerance of photosynthesis to salt stress in transgenic tobacco plants. *Plant Mol Biol.* **2008,** 66, 73–86.

145. Yu, X.; Kikuchi, A.; Matsunaga, E.; Morishita, Y.; Nanto, K.; Sakurai, N. Establishment of the evaluation system of salt tolerance on transgenic woody plants in the special netted-house. *Plant Biotechnol.* **2009,** 26, 135–41.

146. Zhang, X.; Tang, W.; Liu, J.; Liu, Y. Co-expression of rice OsP5CS1 and OsP5CS2 genes in transgenic tobacco resulted in elevated proline biosynthesis and enhanced abiotic stress tolerance. *Chin J Appl Environ Biol.* **2014,** 717–22.

147. Zhang, L.; Chia, J.M.; Kumari, S.; Stein, J.C.; Liu, Z.; Narechania, A.; Maher, C.A.; Guill, K.; McMullen, M.D.; Ware, D. A genome-wide characterization of microRNA genes in maize. *PLoS Genet.* **2009,** 5(11), e1000716.

148. Zhang, X.; Zou, Z.; Gong, P.; Zhang, J.; Ziaf, K.; Li, H.; Xiao, F.; Ye, Z. Over-expression of microRNA169 confers enhanced drought tolerance to tomato. *Biotechnol. Lett.* **2011,** 33, 403–409

149. Zhao, B.T.; Liang, R.Q.; Ge, L.F.; Li, W.; Xiao, H.S.; Lin, H.X..; Ruan, K.C.; Jin, Y.X. Identification of drought-induced microRNAs in rice. *BiochemBiophys Res Commun.* **2007,** 354, 585-590.

150. Zhou, S.; Chen, X.; Zhang, X.; Li, Y. Improved salt tolerance in tobacco plants by co transformation of a betaine synthesis gene BADH and a vacuolar Na+/H+ antiporter gene SeNHX1. *BiotechnolLett.* 30, 369–76.

151. Zhu, B.; Su, J.; Chang, M.; Verma, D.P.S.; Fan, Y.L.; Wu, R. Over expression of a Ä1-pyrroline- 5-carboxylate synthase gene and analysis of tolerance to water and salt stress in transgenic rice. *Plant Sci.* **1998,** 139, 41–8.

Abiotic Stress Tolerance Mechanisms in Plants, Pages 291–320
Edited by: Gyanendra K. Rai, Ranjeet Ranjan Kumar and Sreshti Bagati
Copyright © 2018, Narendra Publishing House, Delhi, India

8

BREEDING APPROACHES TO OVERCOME ABIOTIC STRESS

**Rakhi Bhardwaj, Gyanendra K. Rai, Sreshti Bagati
and Abida Parveen**

*School of Biotechnology,
Sher-e- Kashmir University of Agricultural Sciences and Technology of Jammu-180009 (J&K)
E-mail: rakhibhardwaj0908@gmail.com*

Abstract

Agriculture throughout the world has been providing food, feed, fiber and more recently bio-fuel to meet the global demands. Introduction of new higher yielding and improved cultivars form the bases of the modern agriculture. Recent scientific literature enlists theimmense contribution and role of plant breeding in agriculture and finally to food production worldwide. Presently, agriculture is facing new and huge challenges, due to population growth, the pressure on agriculture liability on the environmental conservation, and climate change. To cope with these new challenges, many plant breeding programs have rejuvenated their breeding scope towards stress tolerance during the last few years. So, book chapter deals withseveral advancements and discoveries applied to abiotic stresses, explaining various new physiological concepts, breeding methods, and modern molecular biological approaches towards the development of improved cultivars bearing tolerancetowards most sorts of abiotic stresses.

Keywords: Agriculture, Breeding, Abiotic stress, Cultivars

1.   Introduction

Plant breeders consider plant breeding as an art as well as a science for changing the characters of plant so as to produce the plants with desirable characters. Mendel who is considered as the father of genetics laid the foundation stone of plant breeding in 1900's with the discovery of laws of genetics. The selection carried by earlier farmers are easy as they select the best looking plants and harvested their seeds for the next growth season. But when the science of genetics became better understood by the breeders they took the advantage of the gene of a plant to select the desirable traits with the purpose to develop improved verities. As all the characters of the plant are controlled by the genes which are ultimately positioned on the chromosomes. Thus, the plant breeding is considered as the manipulation of combination of the chromosomes. Plant breeding is therefore a considerable option for improving crop productivity, reducing farmers' risks and cultivating marginal land. Plant breeding has been practiced for thousands of years ago and still continues so as to produce the new verities which will be proved much more superior to their native ancestors when compared. Plant breeding often results into domestication. Plant breeding is practiced throughout the world right from a common man viz. farmers and gardeners to professional plant breeders employed by the universities, government institutions etc. plant breeding is carried out by various researchers or so as to fulfill these four objectives: i) the yield ii) the quality of crop products iii) the agronomic suitability iv) the resistant to the biotic and abiotic stresses. Out of these four, the last one has proved to be the most devastating and is of the great concern in this study. Abiotic stresses have largely been described as the negative effect of the non living factors on the living organisms viz. plants and animals in a particular environment. These abiotic factors include the following: Water stress (eg. Drought or Flooding), extreme temperature (eg. Cold, Chilling, Freezing or High), light (eg. Low or High), the soil (eg. Salinity, Mineral deficiency or toxicity) and lastly the pollutants (eg. Water and air pollutants viz. SO_2, NO_2 and Ozone). All these abiotic factors ultimately effect or we can say reduce the plant growth and yield. Pinstrup-Andersen et al., during the year 1999 predicted that the world population is expected to reach around 8 billion by the year 2025 which represents an addition of nearly 80 million people to the present population (6 billion) every year. It is estimated that this increases in world population will occur almost exclusively in developing countries, where the problem related to food and feed already exists and this increasing population ultimately results in creation of pressure on the agricultural soils. To feed the increasing world population and sustain well being of humankind, food production must be increased by up to 100% over the next 25 years (Borlaug and Dowswell,1993; Dyson,1999). As the land needed to carry out agricultural practices are limited. However, recent trends indicate that productivity and fertility of soils are globally declining due to degradation and intensive use of soils without consideration of proper soil-management practices (Gruhn et al., 2000; Cakmak, 2002). It is estimated that around 60% of cultivated soils have growth-limiting problems associated with mineral-nutrient deficiencies and toxicities (Cakmak, 2002). According to Byrnes and Bumb (1998), in the next 20 years fertilizer consumption has to increase by around 2-fold to achieve the needed increases in food production.

With the increase in the world population and exclusive use of natural resources, the problems related to environmental or in broader terms we can say that the stresses caused by the abiotic factors (e.g. water deficiency, extreme temperatures, salinity, flooding, soil acidity, and pathogenic infections) are increasing which ultimately results in the reduced crop yields . According to one researcher team i.e. Bray *et al.* (2000), the relative decreases in potential maximum crop yields (i.e., yields under ideal conditions) associated with abiotic stress factors vary between 54% and 82%. Most of yield decreases caused by abiotic stresses result from drought, salinity, high or low temperatures, excess light, inadequate mineral nutrient supply, and soil acidity.Therefore, for providing our society a food security, it is very much important to minimize the adverse effects of these abiotic stresses on crop production by adopting the two approaches: one is applying the breeding techniques and the biotechnological tools and the other is increasing the fertility along with maintaining the appropriate and balanced supply of mineral nutrients. But there are some constraints in breeding for abiotic stress and is considered to be more difficult because of: (a) complexity of conditions causing abiotic stresses, (b) complex nature of abiotic resistance in a variety, (c) occurrence of one stress more often in conjunction with the other, (d) low heritability of abiotic resistance, and (e) variable intensity of such stress under field condition (Choudhary and Vijayakumar, 2012). Plant breeders therefore face the endless task of continually developing new crop varieties (Evans, 1997). Thus the breeders need to develop such varities which are high yielding along with the resistance to the biotic as well as abiotic stresses. In the present study we will focus majorly on the breeding approaches to overcome abiotic stresses.

2. History of Plant Breeding in India

Before discussing the breeding approaches let's have a brief overview of the history of plant breeding. In the year 1871, the Government of India has established the Department of Agriculture which is considered as the first step for starting the organized agricultural practices in India. After the establishment of the department first scientist appointed there was agricultural chemist. Approximately after about ten years i.e. in the year 1905, the Imperial Agricultural Research Institute was established in Pusa, now in Bihar. Later this institute was shifted to New Delhi and renamed as Indian Agricultural Research Institute (IARI) in the year 1946.

With the aim of providing the agricultural knowledge and training various agricultural colleges were established in the various parts of the country like at Kanpur, Pune, Llyalpur, Sabour and Coimbatore between the year 1901 and 1950 and now it has occupied almost every state of our country in the form of agricultural universities with the first agricultural university in 1960 at Pantnagar, U.P. With a concept of increasing the yield and development of new varities, Indian Central Cotton Committee was constituted in 1921. This committee carried out many researches on cotton and developed approximately 70 improved varities of cotton. Satisfied by their results many other central community committees like on sugarcane, tobacco, coconut, spices, maize etc.were constituted.

In order to fasten the agricultural practices Imperial Council of Agricultural Research was established in 1929 which was renamed in 1946 as Indian Council of Agricultural Research (ICAR). So as to fasten the research a Project for Intensification of Regional Research on cotton, oilseeds and millets was started in 1956 and now it has also focused on castor, groundnut, brassica spp. til, toria, jowar and bajra. Inspired by the result of this project another project named All India Coordinated Maize Improvement Project was started in 1957 with the aim to exploit heterosis. As a result of this project first hybrid maize variety was released in 1961. Now a day's many projects are running under various scientists in various agricultural institutions with the aim to develop improved varities and to provide agricultural education.

3. Breeding Approaches

During the earlier times farmers used to collect the seeds and plants which are of the better morphology i.e. better shape, size etc for the next season. But this procedure of selection suffers from one major advantage that they are not permanent and get influenced by the environment. Moreover they may or may not be heritable. But as soon the farmers become familiar with the concept of genes and genetics, it has revolutionized the ways in which the plant breeding is carried out. Researchers started manipulating the plant chromosomes combination so as to get the desirable traits. The advantage of selecting the genes responsible for a desirable trait helps them in the development of new and improved varieties. With the passage of time the new and much more efficient approaches have developed which fasten and sharpens the progress necessary to overcome the problems of food and feed faced by the farmers, researchers and scientists. Depending upon their time of origin breeding approaches has been categorized into two main types (Figure 1):

3.1. Conventional breeding approaches

3.2. Nonconventional breeding approaches

3.1. Conventional Breeding Approaches

The ultimate aim of any breeding programme is to develop varieties superior to the existing ones in yielding ability, resistance to biotic and abiotic stress and other characteristics. Conventional plant breeding has been practiced from thousands of years and is still in use. The conventional methods of plant breeding are further classified into following:

i)   Domestication

ii) Mass Selection

iii) Pureline Selection

iv) Backcross Method

v) Hybridization

    a) Inter varietal hybridization

    b) Interspecific hybridization

vi) Polyploidy

vii) Mutational breeding

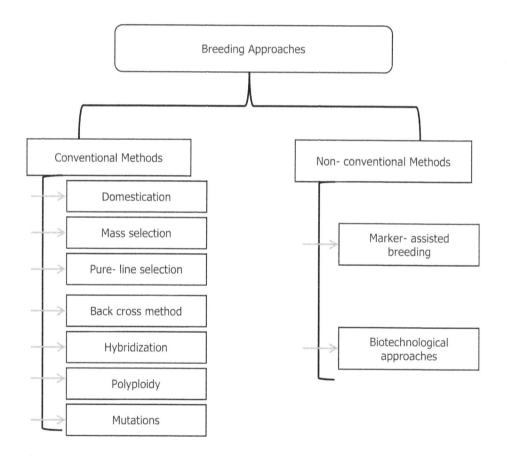

Fig. 1: Classification of various Breeding Approaches

3.1.1. *Domestication*

The process of domestication began somewhere 11,000 years ago during the prehistoric era when the wild species are brought under the human management. It is more or less like the first step taken by the humans while moving towards agricultural economy which lead them to spent a sedentary life thus changing the hunting and gathering mode of life to a permanent herding, selective hunting and settled agriculture. The process of domestication was carried out by the humans either knowingly or unknowingly. In terms of plant breeding domestication is defined as the process by which the plants are genetically modified over a time so as to achieve the desirable traits and characters which may be large size of the seed, fruit or resistance to various biotic and abiotic factors. Among the plants which we grow today cereals, legumes were the first that were domesticated by the humans. Domestication was carried out from the ancient time and is likely to continue in the future as the needs of the humans are continuously changing. So, the scope for the domestication will always be there as long as human civilization will show its presence on the earth. When coming to the examples of domesticated plants we came to know that it includes a very long list which categorizes not only the fruits but the vegetables, cereals, legumes, berries, the oil producing plants and even the plants producing the herbs and spices.

3.1.2. *Selection under Domestication*

One rule of selection has been working right from the ancient time and is still in use that is the survival of the fittest and the one who is the fittest of the all survives in this world. Depending upon the agency responsible for the selection, selection is divided into:

i)   **Natural Selection**: Natural selection is operative when natural forces like abiotic factors e.g. soil and weather, competition and biological factors are the main operative force to eliminate the plants which are least resistant to abiotic stress. The biological factors include the insect pests. This type of selection always occurs in natural populations. With time a particular population becomes much more adapted to a particular environment thus reproducing in that vary atmosphere. On the other hand completion arises when one plant grows at the expense of the other that is takes up more nutrient, minerals, water, light etc more than the other thus limiting the growth of the other. Thus the better adapted plant will grow and reproduce at a much faster rate giving competition to the others.

ii) **Artificial Selection**: This type of selection is carried out by the humans with the purpose to domesticate the plant species that are resistant to abiotic stress for his own use. It allows only the selected plants to grow and reproduce under the properly managed conditions thus decreasing the genetic variability. Most of the crops which we grow today are the outcome of continuous and well managed artificial selection.

Changes in Crop Plants under Domestication

Along with the development of resistance to abiotic stresses crop plants also experienced some other changes under domestication. These include:

i)   Elimination of dormancy

ii)   Decrease in plant height e.g., cereals, millets etc.

iii) Decrease in toxins

iv)   Reduction in shattering of pods

v)   Modification in plant type

vi)   Increase in plant height e.g. jute, sugarcane etc.

vii) Increase in yield

viii) Preference for polyploidy

ix)   Promotion of the asexual reproduction

x)   Shortened life cycle of crop plants e.g., cotton, pigeonpea etc.

3.1.3. *Mass Selection*

Selection is the most ancient and basic procedure in plant breeding. It is completed in three distinct steps. First, a large number of plants are selected from a group of genetically different population followed by the growth of individual plant in rows for several years in the second step. This will lead to the elimination of the undesirable plant. After that the field traits at the different locations at different environment was carried out to further eliminate the undesired plants. In the final and the third step the selected lines are compared with the already existing varities for their performance, yield and other agronomic traits. The process of selection is operative both in the self- pollinated as well as in the cross- pollinated crops.

Mass selection method of plant breeding is one of the earliest practiced method of crop improvement in which the large number of similar plant are selected on the basis of better phenotype and their seeds are harvested and mixed so as to constitute the new variety which must be proved superior to other plants. But here we are mostly concerned with the abiotic stress. This method holds the equal importance in the self and the cross- pollinated crop plants. The variety which is developed by passing through the procedure of mass selection holds considerable genetic variation and is amenable to further mass selection or pure line selection. The plants selected by this mode selection do not need to undergo the progeny test but some researchers believe that the selected plants should be undergone the progeny test so that the poor and the defective progenies should be discarded in the very first stage of selection. There are two methods of mass selection:

a) **HalletsMethod:** This method of mass selection is devised in 1869 in which the crop plants are grown in the best environmental conditions that is no abiotic stress is given to the plant and maximum amount of water and fertilizer is provided and then the mass selection is practiced.

b) **RimparMethod:** This method of mass selection is also useful in itself. During this process the crop has given the unfavorable conditions or we can say that the plants are exposed to the abiotic stresses with minimum amount of water and fertilizer and then the mass selection has been practiced. It has been observed that the plants grow better when further exposed to these conditions.

Procedure of Mass Selection

The general procedure of mass selection is divided into years and is outlined as:

1. **First year:** Large number of plants with the similar or we can say with desirable traits have been selected and their seeds have been harvested for the next season.

2. **Second year:** The collected seeds of the previous season have been planted with the standard check. Once grown the selected population is critically evaluated for that particular trait.

3. **Third to Fifth year:** Evaluation in coordinated yield trails at several locations.

4. **Sixth year:** Seed multiplication if the variety found to be superior.

5. **Seventh year:** Seed distribution.

Merits of Mass Selection

1. Less demanding approach
2. Prolonged yield trails are not required
3. Adaptation to original variety remains unchanged
4. Retention of considerable genetic variability

Demerits of Mass Selection

1. It utilizes the genetic variability already present and cannot generate variability.
2. Less amount of improvement through this approach

3. Varieties developed by this method show variation and are not uniform
4. Difficulty in the identification of improved varities.

3.1.4. *Pure Line Selection*

This method of selection is called as the individual plant selection and is operative in self- pollinated crops. In this method of selection, large number of plants are selected and harvested individually. The individually harvested plants are grown separately in the next season. These individually harvested plants are then evaluated separately and the best progeny is released as a pureline variety. Thus a pureline variety may be stated as a variety obtained from a single homozygous plant as a result of selfing and ismaintained by selfing. One of the most important features of pure line is that they are homozygous and have identical genotype. If any variation is present with in a pureline is solely because of the prevailing environment.

Important Features of Pureline

a) All the purelines of a particular plant are homozygous and have same genotype.
b) Variations present in the purelines are not heritable.
c) Purelines becomes genetically variable with time if care would not be taken.

Procedure of Pureline Selection

1) First year: Selection of large number of plants on the basis of better phenotype.
2) Second Year: Growing individual progenies and harvesting the seed of better progenies out of them.
3) Third year: Preliminary yield trials are conducted that is selected progenies are grown and undesirable progenies are eliminated or discarded.
4) Fourth- Sixth year: Replicated multilocation yield trials with suitable checks followed by the quality checks like resistance to biotic and abiotic stress.
5) Seventh year: Multiplication of the seeds of the superior progeny and is released as a new variety.

Advantages of Pureline Selection

1) Uniform
2) Easily identifiable
3) Best of the best

Disadvantages of Pure line

1) Time consuming
2) Least stability and adaptation
3) Dependency of improvement on the genetic variation present in the original population

Achievements

A large number of improved varieties have been developed in self pollinated crop like wheat, barley, rice, pulses, and oilseeds, cotton and many vegetables etc. Many wheat varieties developed include NP-4, NP-6, NP-12, NP-28, MungVar, T-1, B-1, tobacco chatham special-9, etc.

3.1.5. *Backcross Breeding*

Now a day's many molecular techniques of plant breeding has been introduced but a traditional backcrossing method holds equal importance. The typical reason behind this may be its rate of success. The backcross method is applied for transferring the qualitative traits that can be easily observed in the progeny. Backcross is defined as the method in which the F_1 hybrid and its subsequent generations are repeatedly backcrossed with one of the parent of F_1 hybrid so as to achieve the desirable result in the progeny. As a result of repeated backcrossing the genotype of backcrossed progeny becomes almost similar to that of parent to which it is backcrossed. This method is particularly applicable for the improvement of one or two traits in otherwise desirable variety without changing the genotype of the desirable variety except for the one or two characters. The variety which serves as donor is called nonrecurring parent whereas which serves as recipient is called as the recurrent parent.

Factors Determining the Success of Backcross Method

a) Identification of potent donor parent with particular trait of interest.
b) Identification of receptive plants lacking in that particular trait of interest.
c) Trait transferred form donor to recipient parent must have high heritability.
d) Appropriate number of backcrosses should be made so that the genotype of the donor parent should be recovered.

Procedure of Backcross Breeding Method

The methodology of backcross breeding is defined as:

1) **First Year:** In the very first step the identification of the recurrent and the non- recurrent parent occurred followed by their hybridization so as to achieve F1 hybrid.

2) **Second Year**: First backcrossing of F1 hybrid with recurrent parent. Growing of plant and seed harvesting.

3) **Third Year**: During the third year, the seeds are sown and the selected plants are subjected to second backcrossing with the recurrent parent.

4) **Fourth- Seventh Year:** Repeated backcrossing has been carried out for the successive seventh year.

5) **Eighth Year:** During the eighth year the desired plants are self- pollinated and their seeds are harvested separately.

6) **Ninth Year:** Growing individual plant progenies followed by selection.

7) **Tenth Year:** Growing of individual plant progenies and selection has been carried out.

8) **Eleventh Year:** Replicated yield trails have been conducted with one of the recurrent parent as check.

9) **Twelfth Year:** Multiplication of seeds so that the better raised variety can be distributed to the farmers.

Genetic Consequences of Backcross Method

1) Increase in the homozygosity.

2) Similarity between the genotype of the progeny raised by backcrossing with that of the recurrent parent.

3) Chances for crossing over to occur between the gene under transfer and the genes which are tightly linked.

Advantages of Backcross Method

1. Rate of success is high.

2. If the genotype of the recurrent parents is agronomically much more important, it gets retained or recovered in progenies.

3. If recombination does not occur, the genotype of the combination is not lost.

Disadvantages of Backcross Method

1. Time consuming as it requires twelve years to raise an improved variety.

2. Backcross method works properly for the dominant traits but is less effective for the recessive traits.

3. It is our assumption that the genotype of the non- recurrent parent is lost but in actual practice the some part of the genome of the non- recurrent parent may be retained.

4. Moreover, this approach works poorly for qualitative traits.

3.1.6. *Hybridization*

The credit for the discovery of plant hybridization goes to Josef Kolrueter who in performed his experiment on tobacco in 1760. Stebbins (1958) defined hybridization as the "crossing between individuals belonging to separate populations which possess different adaptive norms. In simple words hybridization is a process by which the characters of two or more different parents are brought together in one individual that we called the hybrid. Sometimes the F1 hybrid shows superiority over the parents. The superiority of F_1 over the parents is called 'heterosis' or the 'hybrid vigour'. Hybridization is a phenomenon increasingly recognized as important in the evolution of plants, animals, and fungi (Gross and Rieseberg, 2005; Mallet, 2007; Schwenk *et al.*, 2008; Mavarez and Linares, 2008; Giraud *et al.*, 2008; Paun *et al.*, 2009; Soltis and Soltis, 2009). Among many other potential and demonstrated effects, hybridization can result in new species of the same ploidy level (e.g., Rieseberg *et al.*, 2003; Gompert *et al.*, 2006) or different ploidy levels (e.g., Cronn and Wendel, 2004), the transfer of adaptive traits between species (e.g., Whitney *et al.*, 2006, 2010; Campbell *et al.*, 2009), and, in general, the "release" of genetic constraints on phenotypic evolution (Kalisz and Kramer, 2008). Some have even argued that hybridization is more important than mutation in generating genetic novelty within populations (Stebbins, 1959; Knobloch, 1972). In plants, hybridization has long been considered widespread, but estimates of its prevalence can vary dramatically among regions and sources (Kenneth *et al.*, 2010). Mallet (2005) used data from Mayr (1992) and Stace (1975, 1997) to estimate that 3.2% and 25.0% of plant species are involved in hybridization in the floras of Concord, Massachusetts and Great Britain, respectively. Hybridization plays a significant role in producing new genetic combinations. Thus, the phenomenon of hybridization can be can be utilized for the creation of such varities which are high yielding and show potential resistance to both biotic as well as abiotic stresses.

Types of Hybridization

The hybridization is divided into following types:

a) Inter- varietal hybridization.

b) Inter- specific hybridization.

Inter- Varietal Hybridization: Inter- varietal hybridization is also known as intra- specific hybridization. This of hybridization involves crossing of the parents of the same species which may be the two strains, varities or races of the same species. Depending upon the number of parents involved it may be called as simple (involvement of two parents) or complex (involvement of more than two parents).

Inter- Specific Hybridization: Also known as the distant hybridization involves the crossing between the individuals that belong to the different species of the same genus or the different genera. When two species of the same genus are crossed, it is known as inter-specific hybridization; but when they belong to two different genera it is termed as intergeneric hybridization. The main objective of this type of crosses is to transfer one or few simply inherited character like resistance to disease or drought etc., a type of abiotic stress. Sometimes, it may be used for developing a new variety.

Procedure of Hybridization

The process of hybridization involves following steps:

a) **Selection of parents:** The plants which are used as parents should be perfect and posses all the characters that are needed in the progeny or we can say in the first flial generation.

b) **Emasculation:** The process of removal of anthers from a bisexual flower before the anthers get mature is called as emasculation. The main aim behind emasculation is the prevention of crop plant to undergo self- pollination.

c) **Bagging and Tagging:** After conducting the process of emasculation flowers are immediately covered with the plastic, paper or the polythene. The purpose of bagging is to prevent the unwanted cross- pollination. After performing bagging, the emasculated flower should be tagged by writing the date and time during which the emasculation has been done.

d) **Artificial/ Cross- Pollination:** Pollen grains from the desired resistant plants are collected and then dusted over the emasculated flower. The flowers are then again covered until the stigma remains receptive. Bags are then removed when the flowers matures into fruits. On maturity the seed of those crops have been collected so as to perform testing. These seeds are sown in the next season.

e) **Selection and Testing of Superior Recombinants:** The seed which has been collected in the previous season has been sown. The grown plants are then analysed for the traits for which it has been raised. The seeds from the plants superior to both the parents are harvested where as inferior ones are discarded. The process of selection for the self- pollinated and the cross-pollinated crops has also been described previously.

f) **Testing, Release and Commercialization of New Cultivars:** The newly selected lines are evaluated for their yield and other agronomic traits such as quality, resistance to biotic and abiotic factors etc. After that multilocation yield trials have been performed for the three growing seasons. Finally if the variety found to be superiorthan the two parents, the seeds of new variety are multiplied and made available to the farmers.

Examples of some improved varieties:

(1) Wheat— KalyanSona, Sonalika.

(2) Rice— Jaya and Ratna

(3) Sugarcane— Saccharumbarberi, Sachharumofficinarum

(4) Rapeseed mustard Brassica— Pusaswarnim

Advantages of Hybridization

1. High yielding varieties can be obtained.

2. Varities resistant to biotic and abiotic stresses can be obtained.

3. Varities raised by hybridization possess lone life.

Disadvantages of Hybridization

1. Reduction in the gene pool due genetic erosion.

2. Hybridization of the plants those have different flowering time, different pollinators, inhibition of pollen tube growth, somatoplastic sterility, cytoplasmic-genic male sterility and structural differences of the chromosomes possess at potential barrier (Hermsen and Ramanna, 1976).

Examples of the Some of the Hybrid Plants

- In Rice: IR8, Jyothi Jaya.
- In Chilies: Solar Hybrid 1, Solar Hybrid 2, Jawalamukhi.
- In Brinjal: Soorya, Swetha, haritha, Neelima.

- In Tomato: Shakti, Mukthi, Anakha.
- In Cucumber: Burpee hybrid, Gemini hybrid, Arunima.
- In Pea: Alaska, Sabre, Serge.
- In Coconut: PB121, KHINA- 1, Kerasanka.

3.1.7. *Polyploidy*

Living organisms are bestowed with a feature of having the ability to reproduce whereas the nature and the other selective forces like evolution and speciation are always operative so as to maintain a continuously check upon them. In this context Polyploidy, a prime facilitator of speciation and evolution in plants and to a lesser extent in animals, is associated with intra and inter-specific hybridization (Levin, 1983). It refers to a definite arithmetic relationship between the chromosome numbers of related organisms (Grant, 1981). It has been defined as the possession of three or more complete sets of chromosomes and has been an important feature of chromosome evolution in many eukaryotic taxa including plants, yeasts, insects, amphibians, reptiles, fishes and even the mammalian genome (Ramsey *et al.*, 1998). Polyploidy is present to at least to some extent in most members of the plant kingdom being more common in some and rather rare in others(Stebbins, 1950). The fact that it is widespread many plants is also kind of exemplified by the wide variations in chromosome numbers with chromosome numbers ranging from 2n = 4 to500 in angiosperms to 2n=6 to 226 in monocots (Grant, 1981). Thousands of angiosperm species have 14 to 15 pairs of chromosomes (Grant, 1981). One of the early examples of a natural polyploid was one of De Vries's original mutations of *Oenotheralamarckiana* (mutat. *gigas*) (Ramsey *et al.*, 1998). The first example of an artificial polyploid was by Winkler (1916) who in fact introduced the term polyploidy (Ramsey *et al.*, 1998; Grant, 1981). Winkler was working on vegetative grafts and chimeras of *Solanumnigrum* and found that callus regenerating from cut surfaces of stem explants were teratploid. Digby (1912) had discovered the occurrence of a fertile type *Primulakewensis*from a sterile inter-specific hybrid through chromosome doubling but failed to realize its significance in the context of polyploidy (Stebbins, 1971). Though unaware of the 'Primula type" fertile hybrid, Winge (1917), from his studies on the chromosomal counts of *Chenopodium* and *Chrysanthemum* found that chromosome numbers of related species were multiples of some common basic number; he subsequently proposed a hypothesis that chromosome doubling in sterile inter-specific hybrids is a means of converting them into fertile offsprings (Swanson *et al.*, 1981; Hieter *et al.*, 1999). This was subsequently verified by various workers in artificial inter-specific hybridizations of Nicotiana, Raphanobrassica and Gaeleopsis(Grant, 1981).Finally the colchicine method of chromosome doubling was developed by

Blakeslee and Avery (1937) and became an important tool for the experimental study of polyploidy (Stebbins, 1950).

Origin of Polyploidy

There are various modes by which polyploids have been evolved:. Some of them are discussed below:

1. **ChromosomalDoubling:** Chromosomal doubling can either in the zygotic or the apical meristem. Somatic polyploidy is seen in some non-meristematic plant tissues as well.(eg: tetraploid and octoploid cells in the cortex and pith Vicia*faba*)(Ramsey *et al.*, 1998).In somatic doubling the main cause is mitotic non-disjunction(Grant, 1981).This doubling may occur in purely vegetative tissues(as in root nodules of some leguminous plants) or at times in a branch that may produce flowers or in early embryos (and may therefore be carried further down) (Grant,1981). The phenomenon of chromosome doubling in the zygotes was best described from heat shock experiments in which young embryos were briefly exposed to high temperatures (Lewis, 1980). Zygotic chromosome doubling was first proposed by Winge and the spontaneous appearance of tetraploids in *Oenotheralamarckiana* and amplidiploid hybrids in *Nicotiana* were shown to be a result of zygotic chromosome doubling (Lewis, 1980; Grant, 1981).

2. **Non- reduction in meiosis:** It occurs during theprocess of microsporogenesis and megasporogenesis thus resulting in unreduced (2n) gametes. Non reduction could be due to meiotic non-disjunction (failure of the chromosome of separate and subsequent reduction in chromosome number), failure of cell wall formation or formation of gametes by mitosis instead of meiosisThe classic example, *Raphanobrassica*, originated by a one step process of fusion of two non- reduced gametes (Elloit,1958; Grant, 1981). The production of non- reduced gametes has been shown to be rather common in *Solanumsps.* (Lewis, 1980).

3. **Polyspermy:** Polyspermy has been also considered as the major factor in the formation of polyploids.

4. **Endoreduplication:**Endoreduplication is a form of nuclear polyploidization resulting in multiple uniform copies of chromosomes. It has been known to occur in the endosperm and the cotyledons of developing seeds, leaves and stems of bolting plants (Larkins *et al.*, 2001).

Classification of Polyploids

On the basis of their chromosomal composition, polyploids have been classified into two main types:

i) Euploids

ii) Aneuploids

i) **Euploids:**The organisms which posses multiples of the complete set of chromosomes that are specific to species are called as euploids or eupolyploids and the phenomenon of occurrence of eupolyploids is known as eupolyploidy. Based on the composition of their genomes euploids are largely classified as:

a) Autopolyploids

b) Allopolyploids

a) **Autopolyploids:** Also known as the autoploids, they possess the multiples copies of chromosomes of the same genome (Acquaah, 2007; Chen, 2010). Autoploidy can occur natural as well as can be induced artificially. Natural autoploids include tetraploid crops such as alfalfa, peanut, potato and coffee and triploid bananas. They occur through the process of chromosome doubling. For example, induced autotetraploids in the watermelon crop are used for the production of seedless triploid hybrids fruits (Wehner, 2008). Such polyploids are induced through the treatment of diploids with mitotic inhibitors such as dinitroaniles and colchicine (Compton *et al.*, 1996). Another example includes a triploid clone of tea named TV29 has been released by the Tea Research Association, India for the commercial cultivation in the Northern parts of the country. This variety of tea produces larger shoots and is more tolerant to drought, a type of abiotic stress.

b) **Allopolyploids:** Allopolyploids include the organisms which are formed as a result of combination of different species (Acquaah, 2007). They result from hybridization of two or more genomes followed by chromosome doubling or by the fusion of unreduced gametes between species (Acquaah, 2007; Chen, 2010; Jones *et al.*, 2008; Ramsey and Schemske, 1998). This process acts as a key in the process of speciation for angiosperms and ferns (Chen, 2010) and occurs often in nature. Economically important natural alloploid crops include strawberry, wheat, oat, upland cotton, oilseed rape, blueberry and mustard (Acquaah, 2007; Chen, 2010).

Depending upon the number of chromosomes autopolyploids are named as autotriploid (3n), autotertaploids (4n), autopentaploid (5n) etc. whereas allopolyploids are named as allotetraploid (2n1+2n2), allohexaploids

(2n1+2n2+2n3). Apart from these there is one allopolyploid which contains two copies of each genome and behaves as diploid during meiosis. Such allopolyploids are called as amphiploids whereas the segmental allopolyploids contain two or more genomes which differ from each other to a very small extent.

List of major crops with its ploidy level is depicted in Table 1:

Table 1: Ploidy level in some of the major crops

Crop	Ploidy level	Method of propagation
Wheat	6n= 42; Hexaploid	Out crossing
Rice	2n=24; Diploid	Selfing
Cassava	2n=36; Diploid	Out crossing; Vegetative
Maize	2n=20; Diploid	Out crossing
Sugar beet	2n=18; Diploid	Out crossing
Watermelon	2n=22; Diploid	Out crossing
Sugarcane	8n=80; Octoploid	Out crossing; Vegetative
Maize	2n=20; Diploid	Out crossing
Tomatoes	2n=24; Diploid	Selfing
Barley	2n=14; Diploid	Selfing

ii) **Anueploids:** Aneuploids are polyploids that contain either an addition or subtraction of one or more specific chromosome(s) to the total number of chromosomes that usually make up the ploidy of a species (Acquaah, 2007; Ramsey and Schemske, 1998). Aneuploids result from the formation of univalents and multivalents during meiosis of euploids (Acquaah, 2007). For example, several studies have found that 30-40% of progeny derived from autotetraploid maize are aneuploids (Comai, 2005). With no mechanism of dividing univalents equally among daughter cells during anaphase I, some cells inherit more genetic material than others (Ramsey and Schemske, 1998). Similarly, multivalents such as homologous chromosomes may fail to separate during meiosis leading to unequal migration of chromosomes to opposite poles. This mechanism is called non-disjunction (Acquaah, 2007). These meiotic aberrances result in plants with reduced vigor. Aneuploids are classified according to the number of chromosomes gained or lost as shown:

Classification of aneuploids is shown Figure 2.

The role of polyploidy in the origin of bread wheat is discussed in Figure 3. Figure 3 shows the origin of bread wheat.

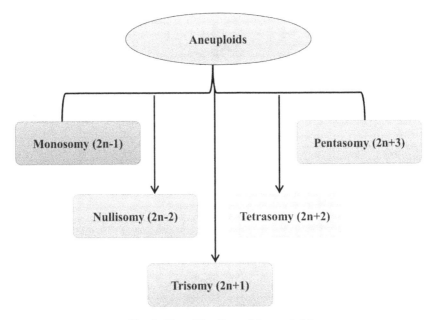

Fig. 2: Classification of Aneuploids

Advantages of Polyploidy

1. **Heterosis:** Polyploids exhibit the phenomenon of heterosis that is the superiority of F_1 hybrid over its both the parental forms.

2. **Gene redundancy**: Another advantage conferred by the polyploids is the gene redundancy that is the masking the effect of recessive lethal allele by the dominant wild type allele. Another advantage conferred by gene redundancy is the ability to diversify gene function by altering redundant copies of important or essential genes (Comai, 2005).

3. Polyploidy helps in the loss of self-incompatibility thus allowing the self fertilization to occur along with the gain of asexual reproduction.

Disadvantages of Polyploidy

1. Large size: The size of the autopoloids is larger with large water content and donot produce more dry matter.

2. Newly generated polyploids have many defects like high sterility or low fertility, or poor seed set.

3. The effects of polyploids cannot be predicted.

4. Polyploidy is a problem in the successful completion of meiosis and mitosis.

5. As a result of polyploidy there occurs the change in the regulation of gene expression.

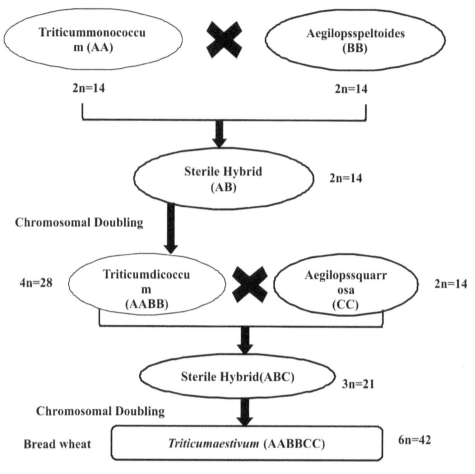

Fig. 3: Orign of bread wheat using polyploidy.

3.1.8. *Mutational Breeding*

Mutational breeding is also named as the variation breeding and is used for the improvement of crop plants. Mutational breeding is the process of exposing the seeds to the chemical or radiations so as to induce the variations in the crop plants. Harten gives his working definition of mutation in plants as "any heritable change in the idiotypic constitution of sporophytic or gametophytic plant tissue, not caused by normal genetic recombination or segregation" (Harten, 1998). A mutation is a sudden heritable change in the DNA in a living cell, not caused by genetic segregation or genetic recombination (Harten *et al.*, 1998).From 1930 to 2014 more than 3200 mutagenic plant varietals have been released(Schouten *et al.,* 2007; FOA/ IAEA, 2014) that have been derived either as direct mutants (70%) or from their progeny (30%) (Maluszynsk *et al.,* 2000). Crop plants account for

75% of released mutagenic species with the remaining 25% ornamentals or decorative plants (Ahloowali *et al.*, 2004). However, although the FAO/ IAEA reported in 2014 that over 1,000 mutant varietals of major staple crops were being grown worldwide (FOA/ IAEA, 2014). The variation created by these mutagens holds very much importance in the plant breeding as these variations are the source by which the new and important cultivars are generated. The agents which are used to generate mutations are called as mutagens and these agents which create mutations are only useful and are desirable in plant breeding if the changes created by these mutagens are heritable. The mutations can improve yield, quality, taste, size and resistance to disease and can help plants adapt to diverse climates and conditions.

Depending upon the mode of occurrence mutation is of following two types:

1. Natural mutations
2. Artificially Induced mutations

Natural Mutations: The mutations which occur naturally or spontaneously in nature are called as natural mutations but the frequency of mutations that occurs naturally have very low frequency and occur rarely in nature. The main reason behind natural mutation is the error made during cell replication or due to sudden or the unintentional exposure of plants to physical or chemical mutagens. It is estimated that the frequency of mutation to occur is every 10^{-8} base pair per generation in eukaryotic genomes (Drake *et al.*, 1998). In corn (*Zea mays*), mutations occur from 10^{-6} to 5×10^{-4} per base pair per generation (Stadler, 1930). Apart from the error made during replication, the mutations can be induced naturally with the transposable elements. Transposable elements are nothing but the self-replicating segments of DNA that excise and/or insert themselves within the genome. These transposable elements are also known as transposons. These strange sequences were first proposed by the pioneering Barbara McClintock working on maize (McClintock, 1948). Transposons have certain locations in the genome which are more likely to insert and can replicate these transposons. The mutational characteristics of transposable element are that they got inserted in the intron and cause gene disruption, alteration in the protein product rearrangements in the genome. If inserted into the intron of a gene, they can cause transcriptional inefficiency (Hartwell *et al.*, 2008).

Artificially Induced Mutations: These mutations are created by the humans or the scientists deliberately so as to create variations which will ultimately leads to the better genotype resulting in better phenotype. These mutations can be induced by using i) physical and ii) chemical mutagens.

i) **Physical Mutagens:** The physical type of mutagens includes ionizing radiations like ultra-violet (UV) light, X-ray, Gamma rays, and neutrons. HJ Muller first discovered and used the mutagenic properties of X-ray radiation to study the genetics of Drosophila flies and the mechanics of heredity (Muller, 1928). The mutagen activity of these radiations are that they cause single or the double stranded breaks in the DNA thus causing the frame shift mutations, inactive protein products or creates wrong transcripts.

ii) **Chemical Mutagens:** These include the chemical agents like EMS(Ethyl-methane Sulphonate), intercalating agents and nitrous acid etc. These chemicals differ from one another in their mode of action. While discussing the EMS, it reacts with guanine or thymine by adding an ethyl group which causes the DNA replication machinery to recognize the modified base as an adenine or cytosine, respectively thus resulting in the altered forms of a triplet sequence. The change in the triplet sequence will lead to either changes the amino acid transcribed, which can de-activate, reduce efficiency of, or produce a new protein or will lead to the formation of stop codon thereby suspending the transcription. Another chemical mutagen named intercalating agent intercalates themselves in the DNA strand between the adjacent base pair thus providing hindrance to replication and transcription machinery. While discussing the last one, Nitrous acid, a deaminating agent, removes the amine group from adenine or cytosine. These deaminated adenine and cytosine matches the cytosine and adenine respectively during replication resulting in disruption of central dogma.

Application of Mutational Breeding

1. Mutational breeding helps in the crop domestication and adaptability to new environment.

2. Mutational breeding helps in providing resistance to biotic and abiotic stress. The examples of mutational breeding to overcome biotic stress include resistance to powdery mildew in barley by utilizing chemical mutagens. The example of abiotic stress include the use of gamma radiations in 'Zhefu 802'cultivar of rice to make it tolerant to cold and rice blast disease along with reduction in the growing season and high yield potential.

3. Mutation helps in the improvement of crop quality and nutritional traits. One such example includes generation of high- quality durum wheat mutant in Italy and bread wheat in India (Ahloowalia *et al.*, 2004; Kharkwal *et al.*, 2009).

3.2. Unconventional Methods

The unconventional method of plant breeding works on the molecular level,that's why it is also called as molecular approach of plant breeding or simplified molecular breeding. It involves following approaches:

1. Marker Assisted Breeding
2. Biotechnological Approaches

3.2.1. *Marker Assisted Breeding*

As the name indicated marker assisted selection, is the selection that is done by using the markers. Markers are nothing but a short DNA sequences that can be used so as to identify the particular characteristic sequence within the genome of an organisms. They are just like the milestones standing at the side of the road that can be used to locate a particular position or place. Marker assisted selection is the most widely used application of marker system in the plant breeding. Marker assisted selection is also called as an indirect selection process where a marker linked to that very trait is used for selection rather than using the trait itself. While discussing about the markers, markers have been classified into i) morphological based on the morphology of plant like plant height, grain colour, shape etc. but they suffer from some limitation like they got effected by the environment ii) biochemical which are based on the isozymes which also suffers with some drawbacks of difficulty in their separation and their degradation over a course of time and iii) finally the DNA based molecular markers which are much more reliable and trust worthy. The commonly used molecular markers include RFLP (Restriction Fragment Length Polymorphism), RAPD (Random Amplified Polymorphic DNA), SSR (Simple Sequence Repeat), SNP (Single Nucleotide Polymorphism) etc. Markers can be used for the determination of quantitative trait loci (QTLs). QTLs are nothing but the single traits which are controlled by the multiple genes with each gene having the minor effect whereas the traits which are controlled by the single gene are called as qualitative traits. These single genes have major effect on the qualitative traits.

There are five main considerations for the use of DNA markers in MAS: (i) reliability that is the marker should be tightly linked with the trait under study; (ii) quantity and quality of DNA required means different markers use different amount of DNA; (iii) technical procedure for marker assay that is the markers used should be high throughput, simple and quick; (iv) level of polymorphism that is should be highly polymorphic; and (v) finally it should be cost effective (Mackill and Ni 2000; Mohler and Singrun 2004).

QTL Mapping using MAS

The detection of genes or QTLs controlling traits is possible due to genetic linkage analysis, which is based on the principle of genetic recombination during meiosis (Tanksley, 1993). This permits the construction of linkage maps composed of genetic markers for a specific population (Collard *et al.*, 2008). For the conduction of QTL analysis, the population under investigation should be segregating that is should be in F_2, F_3 state or the population should be backcross.

Using statistical methods such as single-marker analysis or interval mapping to detect associations between DNA markers and phenotypic data, genes or QTLs can be detected in relation to a linkage map (Kearsey 1998). The identification of QTLs using DNA markers was a major breakthrough in the characterization of quantitative traits (Paterson *et al.*, 1988). Reports have been numerous of DNA markers linked to genes or QTLs (Mohan *et al.*, 1997; Francia *et al.*, 2005).

The Advantages of using MAS over Conventional Breeding Methods are outlined as:

1. **Selection at the earlier stage:** Markers help us to identify the desired plants even at the seedling stage along with the identification of the genes which are expressed at the later stages of development.

2. **Simplicity:** The simplicity of the process attracts most of the breeders as this method of selection is simple and does not need a expert so as to analyze the phenotype.

3. **Time and labour effective:** This method of plant breeding is quite effective as it saves the time and labour of the breeders as they can discard the undesirable plant at the early stage of development.

4. **Non requirement of large population:** By utilizing the MAS, a single plant can be selected on the basis of better genotype which can be further multiplied so as to raise the generation.

5. **Selection of recessive genes and mutants:** By using the markers it is even possible to select recessive genes as well as mutants.

These advantages can be exploited by breeders to accelerate the breeding process (Ribaut and Hoisington, 1998; Morris *et al.*, 2003).

Application of MAS in Plant Breeding

1. **Marker assisted backcrossing:** Usually backcrossing is done to transfer one or few traits from undesirable variety to a variety which lacks in one or few character and is otherwise economically or agronomically important. The method was first described in 1922 and was widely used between the 1930s and 1960s (Stoskopf *et al.*, 1993). The use of DNA based markers in backcrossing is of great importance and increase the frequency of the selection of appropriate genotype.

2. **Marker assisted pyramiding:** pyramiding is the process of transfer of more than two genes in a single parent. Pyramiding can be done by using the classical method of plant breeding but it is not possible to identify such plant with all the introgression. But with the use of molecular markers successful screening of introgressed plants can be done with full efficiency. A notable example of the combination of quantitative resistance was the pyramiding of a single stripe rust gene and two QTLs (Castro *et al.*, 2003).

3. **Selection at the earlier stage:** By using markers it is possible for breeders to select the plant with the desired character at the earlier stage thereby saving their time and effort. One strategy proposed by Ribaut and Betran (1999) involving MAS at an early generation was called single large-scale MAS (SLS–MAS). They proposed that a single MAS step could be performed on F_2 or F_3 populations derived from elite parents. They used flanking markers (less than 5 cM, on both sides of a target locus) for three QTLs in a single MAS step.

4. **Marker assisted recurrent selection:** Marker-assisted recurrent selection (MARS) involves crossing and selecting the desired individual at each cycle thereby increasing the frequency of desired alleles in a population. These desire alleles can be brought from various different sources in selected individuals. This technique has been applied to sunflower, soybean, and maize to bring desirable alleles at several target loci into single elite lines (Eathington *et al.*, 2007).

3.2.2. *Biotechnological Approaches*

Biotechnology involves the use of microorganisms to get the desired product. But in true sense biotechnology is a multidisciplinary field which involves the use of living systems so as to get the desired product which may be the modified animals or a plants in terms of yield, biomass, and resistance to various biotic and abiotic stresses. The use of biotechnological approaches for crop improvement is beyond the scope of this chapter and will be discussed in the later chapters of this book.

4. Conclusion and Future Perspectives

Plant breeding has proved to be the continuously emerging field in the improvement of crops viz. increase in the yield, change in the morphological traits, and biochemical functions in addition to resistance to the biotic and abiotic stresses like water stress (eg. Drought or Flooding), extreme temperature (eg. Cold, Chilling, Freezing or High), light (eg. Low or High), the soil (eg. Salinity, Mineral deficiency or toxicity) and lastly the pollutants (e.g. Water and Air pollutants viz. SO_2, NO_2 and ozone). All these abiotic factors ultimately effect or we can say reduce the plant growth and yield. While analyzing the history of plant breeding we came to know that the plant breeding starts from domestication done by the common man followed by conventional breeding approaches which include mass selection, pureline selection, hybridization, polyploidy and mutations done by the farmers and breeders lead by the use of molecular markers and other biotechnological approaches by agricultural scientists as a part of the unconventional plant breeding. By using these approaches agricultural scientists have made considerable efforts in the development of high yielding and resistant varities thus making the world to cope up with the increasing food and feed demands with increase in the population. Enormous progress has been made in the field of plant breeding and still there lies a scope of improvement.

5. References

1. Acquaah, G. Principles of plant genetics and breeding. Wiley-Blackwell, Malden. **2007**.

2. Ahloowali, B.S. Global impact of mutation-derived varieties.*Euphytica.* **2004**. 135, 187–204.

3. Ahloowalia, B.S.; Maluszynski, M.; Nichterlein, K. Global impact of mutation-derived varieties. *Euphytica.* **2004**, 135(2), 187–204.

4. Akademikerverlag GmbH and Co. KG Heinrich-Böcking-Str. 6-8, 66121, Saarbrücken, Germany, **2012**.

5. Bertrand, C.Y.; Collard.; Mackil, D.J. Marker-assisted selection: an approach for precision plant breeding in the twenty-first century. *Phil. Trans. R. Soc.B.* **2008**, 363, 557–572.

6. Borlaug, N.E.; Dowswell, C.R. Fertilizer: To nourish infertile soil that feeds a fertile population that crowds a fragile world. *Fert. News.* **1993**, 387, 11–20.

7. Bray, E.A.; Bailey-Serres, J.; Weretilnyk, E. Responses to abiotic stresses. In: Buchanan, B., Gruissem, W., Jones, R.(Ed.), Biochemistry and Molecular Biology of Plants. American Society of Plant Physiologists. **2000**, pp. 1158–1203.

8. Byrnes, B.H.; Bumb, B.L. Population growth, food production and nutrient requirements. In: Rengel, Z. (Ed.), Mineral Nutrition of Crops: Mechanisms and Implications. The Haworth Press, New York USA. **1998**, pp. 1–27.

9. Cakmak, I. Plant nutrition research: Priorities to meet human needs for food in sustainable ways. *Plant Soil.* **2002**, 247, 3–24.

10. Campbell, L.G.; Snow, A.A.; Sweeney, P.M. When divergent life histories hybridize: insights into adaptive life-history traits in an annual weed. *New Phytol.* **2009**, 173, 648–660.

11. Castro, A.J. Mapping and pyramiding of qualitative and quantitative resistance to stripe rust in barley. *Theor. Appl. Genet.* **2003**, 107, 922–930.

12. Chen, Z. Molecular mechanisms of polyploidy and hybrid vigor. *Trends in plant science.* **2010**, 15, 57-71.

13. Choudhary, A.K.;Vijayakumar, A.G. Glossary of Plant Breeding, A Perspective. LAP LAMBERT, Academic Publishing, AV. **2012**.

14. Comai, L. The advantages and disadvantages of being polyploid. *Nature Reviews Genetics.* **2005**, 6, 836-846.

15. Compton, M.; Gray, D.; Elmstrom, G. Identification of tetraploidregenerants from cotyledons of diploid watermelon cultured in vitro. *Euphytica.* **1996**, 87, 165-172.

16. Cronn, R.; Wendel, J.F. Cryptic trysts, genomic mergers, and plant speciation. *New Phytol.* **2004**, 161, 133–142.

17. Drake, J.W.; Charlesworth, B.; Charlesworth, D.; Crow, J.F. () Rates of Spontaneous Mutation. *Genetics.* **1998**, 148, 1667-1686.

18. Dyson, T. World food trends and prospects to 2025. *Proc. Natl. Acad. Sci. USA* **1999**, 96, 5929–5936.

19. Elliot, F.C. Plant Breeding and Cytogenetics. McGraw Hill Book Company, Inc. **1958**, pp.136-178.

20. Evans, L.T. Adapting and improving crops: the endless task. *Phil. Trans. R. Soc. B.* **1997**, 352, 901–906.

21. Francia, E.; Tacconi, G.; Crosatti, C.; Barabaschi, D.; Bulgarelli, D.; Dall'Aglio, E.; Vale', G. Marker assisted selection in crop plants. *Plant Cell Tissue Org.***2005**, 82, 317–342.

22. Giraud, T.; Refregier, G.; Le Gac, M.; de Vienne, D.M.; Hood, M.E. Speciation in fungi. *Fungal Genet. Biol.* **2008**, 45, 791–802.

23. Gompert, Z.; Fordyce, J.A.; Forister, M.L.; Shapiro, A.M.; Nice, C.C. Homoploid hybrid speciation in an extreme habitat. *Science.* **2006**, 314, 1923–1925.

24. Grant, V.P. Polyploidy. Columbia University Press, New York. **1981**, pp.283-352.

25. Gross, B.L.; Rieseberg, L.H. The ecological genetics of homoploid hybrid speciation. *J. Hered.* **2005**, 96, 241–252.

26. Gruhn, P.; Goletti, F.;Yudelman, M. Integrated nutrient management, soil fertility, and sustainable agriculture: current issues and future challenges. Food, Agriculture, and the Environment Discussion Paper 32, International Food Policy Research Institute, Washington, D.C. **2007**.

27. Harten, A.M. Mutation breeding: theory and practical applications. Cambridge University Press. **1998**.

28. Hartwell, L.; Hood, L.; Goldberg, M.L.; Reynolds, A.E.; Silver, L.; Veres, R.C. Genetics : from genes to genomes (3rd Ed.) McGraw-Hill, New York. **2008**.

29. Hermsen, J.G.T.; Ramanna, M.S. Barriers to hybridization of SolanumbulbocastanumDun. and S. VerrucosumSchlechtd. and structural hybridity in their F1 plants. *Euphytica.* **1976**, 25(1), 1-10.

30. Hieter, P.; Griffith, T. Polyploidy- more is more or less. *Science*, **1999**, 285, 210-211.

31. Jones, J.R.; Ranney, T.G.; Eaker, T.A. A novel method for inducing polyploidy in Rhododendron seedlings. *J. Amer. Rhododendron Soc.* **2008**, 62, 130-135.

32. Kalisz, S.; Kramer, E.M. Variation and constraint in plant evolution and development. *Heredity.* **2008**, 100, 171–177.

33. Kearsey, M. J.; Farquhar, A.G.L. QTL analysis in plants; where are we now? *Heredity.* **1998**, 80, 137–142.

34. Kearsey, M. J. The principles of QTL analysis (a minimal mathematics approach). *J. Exp. Bot.***1998**, 49, 1619–1623.

35. Kenneth, D.; Whitney, N.; Jeffrey, R.; Ahern.; Lesley, G. Campbell, Loren P. Albert, Matthew S. King. Patterns of hybridization in plants. *Perspectives in Plant Ecology, Evolution and Systematics.* **2010**, 12, 175–182.

36. Kharkwal, M.C.; Shu, Q.Y. The role of induced mutations inworld food security. In: Induced Plant Mutations in the Genomics Era. Proceedings of an International JointFAO/IAEA Symposium. International Atomic Energy Agency, Vienna, Austria. **2009**, pp. 33–8

37. Knobloch, I.W. Intergeneric hybridization in flowering plants. *Taxon.* **1972**, 21, 97–103.

38. Larkins, B.A.; Dilkes, B.P.; Dante, R.A.; Coelho, C.M.; Woo,Y.M.; Liu,Y. Investigating the hows and whys of DNA endoreduplication. *J. Exp. Bot.* **2001**, 52, 183-192.

39. Levin, D.A. Polyploidy and novelty in flowering plants, *Amer. Nat.* **1983**, 122(1), 1-24.

40. Lewis, W.H. (Ed.) Polyploidy: biological relevance. Plenum Press, New York. **1980.**

41. Mackill, D. J.; Ni, J. Molecular mapping and marker assisted selection for major-gene traits in rice. In: Khush, G.S., Brar, D.S., Hardy, B. (Eds.), *Proc. Fourth Int. Rice Genetics Symp.* International Rice Research Institute. Los Baños, The Philippines. **2000**, pp. 137–151.

42. Mallet, J. Hybridization as an invasion of the genome. *Trends Ecol. Evol.* **2005**, 20, 229–237.

43. Mallet, J. Hybrid speciation. *Nature.* **2007**, 446, 279–283.

44. Maluszynsk, M.K.; Nichterlein, K.L.; Zanten, V.; Ahloowalia, B.S. Ofûcially released mutant varieties – the FAO/IAEA Database. *Mutation Breeding Review.* **2000**, 12, 1–84.

45. Mavarez, J.; Linares, M. Homoploid hybrid speciation in animals. *Mol. Ecol.* **2008**, 17, 4181–4185.

46. Mayr, E. A local flora and the biological species concept. *Am. J. Bot.* **1992**, 79, 222–238.

47. McClintock, B. Mutable loci in maize. Carnegie Inst. Washington Year Book, **1948**, 47,155-169.

48. Mohan, M.; Nair, S.; Bhagwat, A.; Krishna, T.G.; Yano, M.; Bhatia, C.R.; Sasaki, T. Genome mapping, molecular markers and marker-assisted selection in crop plants. *Mol. Breed.* **1997**, 3, 87–103.

49. Mohler, V.; Singrun, C. General considerations: marker-assisted selection. In: Lorz, H., Wenzel,G. (Eds.), Biotechnology in agriculture and forestry, vol. 55: Molecular marker systems. Springer, Berlin, Germany. 2004, pp. 305–317.

50. Morris, M.; Dreher, K.; Ribaut, J.M.; Khairallah, M. Money matters (II): costs of maize inbred line conversion schemes at CIMMYT using conventional and marker assisted selection. *Mol. Breed.* **2003**, 11, 235–247.

51. Muller, H. The production of mutations by X-rays. Proceedings of the National Academy of Sciences of the United States of America. **1928**, 14, 714.

52. Paterson, A.H.; Lander, E.S.; Hewitt, J.D.; Peterson, S.; Lincoln, S.E.; Tanksley, S.D. Resolution of quantitative traits into Mendelian factors by using a complete linkage map of restriction fragment length polymorphisms. *Nature.* **1988**, 335, 721–726.

53. Paun, O.; Forest, F.; Fay, M.F.; Chase, M.W. Hybrid speciation in angiosperms: parental divergence drives ploidy. *New Phytol.* **2009**, 182, 507–518.

54. Pinstrup-Andersen, P.; Pandya-Lorch, R.; Rosegrant, M.W. World food propects: Critical issues for the early twenty-first century. 2020 Vision Food Policy Report, International Food Policy Research Institute, Washington, D.C. **1999**.

55. Plant Breeding and Genetics . Joint FAO/IAEA Division of Nuclear Techniques in Food and Agriculture. 2014.

56. Ramsey, J.; Schemske, D.W. Pathways, mechanisms, and rates of polyploid formation in flowering plants. *Annual Review of Ecology and Systematics.* **1998**, 29, 467-501.

57. Ribaut, J.M.; Betran, J. Single large-scale marker assisted selection (SLS–MAS). *Mol. Breed.* **1999**, 5, 531–541.

58. Ribaut, J.M.; Hoisington, D. Marker-assisted selection: new tools and strategies. Trends Plant *Sci.* **1998**, 3, 236–239.

59. Rieseberg, L.H.; Raymond, O.; Rosenthal, D.M.; Lai, Z.; Livingstone, K.; Nakazato, T.; Durphy, J.L.; Schwarzbach, A.E.; Donovan, L.A.; Lexer, C. Major ecological transitions in wild sunflowers facilitated by hybridization. *Science.* **2003**, 301, 1211–1216.

60. Eathington, S.R.; Crosbie, T.M.; Edwards, M.D.; Reiter, R.S.; Bull, J.K. *Crop Sci.* **2007**, 47 (3), S154.

61. Schouten, H.J.; Jacobsen, E. Are Mutations in Genetically Modified Plants Dangerous? *Journal of Biomedicine and Biotechnology.* **2007**, 1.

62. Schwenk, K.; Brede, N.; Streit, B. Introduction. Extent, processes and evolutionary impact of interspecific hybridization in animals. *Philos. Trans. R. Soc. Lond., Ser. B: Biol. Sci.* **2008**, 363, 2805–2811.

63. Stace, C.A. Hybridization and the Flora of the British Isles. Academic Press, London & New York. **1975**.

64. Stace, C.A. New Flora of the British Isles, second ed. Cambridge University Press, Cambridge, UK. **1997**.

65. Stadler, L. The frequency of mutation of specific genes in maize. A*natomical Record.* **1930**, 47, 381.

66. Stebbins, G.L. The role of hybridization in evolution. *Proc. Am. Philos. Soc.* **1959**, 103, 231–251.

67. Stebbins, G.L. Jr. The inviability, sterility and weakness of inter-specific hybrids. *Adv. Genet.* 1958, 9, 147-215.

68. Stebbins, G.L.Jr. Variation and Evolution in plants, Columbia University Press, New York. **1950**.

69. Stebbins, G.L. Jr. Processes of Organic Evolution. Englewood Cliffs, N, Prentice-Hall. **1971**.

70. Stoskopf, N. C.; Tomes, D. T.; Christie, B.R. Plant breeding: theory and practice. CA; Westview Press Inc., Oxford, San Francisco. **1993**.

71. Swanson, C.P.; Merz, T.; Young,W.J. Cytogenetics: The Chromosome in Division, Inheritance and Evolution, Englewood Cliffs, Prentice-Hall. **1981**.

72. Tanksley, S. Mapping polygenes. *Ann. Rev. Genet.***1993**, 27, 205–233.

73. VanHarten, A.M. Mutation Breeding: Theory and Practical Applications Cambridge University Press, Cambridge, UK. **1998**.

74. Wehner, T.C. Watermelon. Springer Science, New York. **2008**.

Abiotic Stress Tolerance Mechanisms in Plants, Pages 321–339
Edited by: Gyanendra K. Rai, Ranjeet Ranjan Kumar and Sreshti Bagati
Copyright © 2018, Narendra Publishing House, Delhi, India

9

MORPHOLOGICAL, PHYSIOLOGICAL, BIOCHEMICAL AND MOLECULAR RESPONSES OF PLANTS TO DROUGHT STRESS

Abida Parveen[1], Gyanendra K. Rai[1], Sreshti Bagati[1], Pradeep Kumar Rai[2] and Praveen Singh[3]

[1]*School of Biotechnology,*
Sher-e- Kashmir University of Agricultural Sciences and Technology of Jammu-180009 (J&K)
[2]*ACHR,*
Sher-e- Kashmir University of Agricultural Sciences and Technology of Jammu-180009 (J&K)
[3]*Division of PBG,*
Sher-e- Kashmir University of Agricultural Sciences and Technology of Jammu-180009 (J&K)
E-mail: abichoudhary21@gmail.com

Abstract

Plants in nature are continuously exposed to several biotic and abiotic stresses. Among these stresses, drought stress is one of the most adverse factors of plant growth and productivity and considered a severe threat for sustainable crop production in the conditions on changing climate. Drought triggers a wide variety of plant responses, ranging from cellular metabolism to changes in growth rates and crop yields. Understanding the biochemical and molecular responses to drought is essential for a holistic perception of plant resistance mechanisms to water-limited conditions. This chapter describes some aspects of drought induced changes in morphological, physiological and biochemical changes in plants. Drought stress progressively decreases CO_2 assimilation rates due to reduced stomatal conductance. It reduces leaf size, stems extension and root proliferation, disturbs plant water relations and reduces water-use efficiency. It disrupts photosynthetic pigments and reduces the gas exchange leading to a reduction in plant growth and productivity. The critical roles of osmolyte accumulation under drought stress conditions have been actively researched to understand the tolerance of plants to dehydration. In addition, drought stress-induced generation of active oxygen species is well recognized at the cellular level and is tightly controlled at both the production and consumption levels, through increased antioxidative systems. This chapter focuses on the ability and strategies of higher plants to respond and adapt to drought stress.

Keywords: Drought stress, growth, yield, gas exchange, photosynthetic pigments, antioxidative system.

1. Introduction

When a region receives below-average precipitation, resulting in prolonged shortages in its water supply, whether atmospheric, surface or ground water is termed as drought. A drought can last for months or years, or may be declared after as few as 15 days. It can have a substantial impact on the ecosystem and agriculture of the affected region. Although droughts can persist for several years, even a short, intense drought can cause significant damage and harm to the local economy. Annual dry seasons in the tropics significantly increase the chances of a drought developing and subsequent bush fires. Periods of heat can significantly worsen drought conditions by hastening evaporation of water vapor.

Many plant species, such as those in the family Cactaceae (or cacti), have adaptations like reduced leaf area and waxy cuticles to enhance their ability to tolerate drought. Some others survive dry periods as buried seeds. Semi-permanent drought produces arid biomes such as deserts and grasslands. Prolonged droughts have caused mass migrations and humanitarian crises. Most arid ecosystems have inherently low productivity.

Types

As a drought persists, the conditions surrounding it gradually worsen and its impact on the local population gradually increases. People tend to define droughts in three main ways:

1. Meteorological drought is brought about when there is a prolonged time with less than average precipitation. Meteorological drought usually precedes the other kinds of drought.
2. Agricultural droughts are droughts that affect crop production or the ecology of the range. This condition can also arise independently from any change in precipitation levels when soil conditions and erosion triggered by poorly planned agricultural endeavors cause a shortfall in water available to the crops. However, in a traditional drought, it is caused by an extended period of below average precipitation.
3. Hydrological drought is brought about when the water reserves available in sources such as aquifers, lakes and reservoirs fall below the statisticalaverage. Hydrological drought tends to show up more slowly because it involves stored

water that is used but not replenished. Like an agricultural drought, this can be triggered by more than just a loss of rainfall. For instance, Kazakhstan was recentlyawarded a large amount of money by the World Bank to restore water that had been diverted to other nations from the Aral Sea under Soviet rule. Similar circumstances also place their largest lake, Balkhash, at risk of completely drying out.

Drought is a co-occurrence of several environmental stresses

- Low soil moisture ability.
- Limiting the supply of water to roots.
- High evaporate rate due to low humidity.
- High temperature; high insulation causes high respiration and damage to metabolic processes and cell structure.
- High solar radiation leads to photo-inhibition, photo-oxidation and eventually death of plants.
- Soil hardness increases as soil dried thus adversely affecting the root growth which reduces growth of leaves and photosynthesis.
- Accumulation of salt in upper region of soil around the roots leads to osmotic and toxic stress.

2. Effects of Drought Stress on Crop Plants

Plants in nature are continuously exposed to several biotic and abiotic stresses. Among these stresses, drought stress is one of the most adverse factors of plant growth and productivity and considered a severe threat for sustainable crop production in the conditions on changing climate (Figure 1). Drought triggers a wide variety of plant responses, ranging from cellular metabolism to changes in growth rates and crop yields. Understanding the biochemical and molecular responses to drought is essential for a holistic perception of plant resistance mechanisms to water-limited conditions. Research on impact of drought stress on plants is taking a new aspect due to climate change scenario, which results, increase in aridity in many parts of globe. Drought occurs due to deficiency of water in soil and atmosphere as well as due to radiations and high temperature which effect the growth of plants and plant productivity. Drought, a period of abnormally dry weather, results in soil-water deficit and subsequently plant-water deficit. The lack of water in the environment constitutes a stress when it induces an injury in the plant. Water deficit in the plant disrupts many cellular and whole plant functions, having a negative impact on plant growth and reproduction. Crop yields are reduced by 69% on average when plants are exposed to unfavourable conditions in the field (Boyer, 1982). Availability of water is the most important factor in the environment that reduces the production of our crops. As water is increasingly needed for human populations and prime agricultural lands are used for housing, the availability of water will have a greater impact on our ability to produce crops. In nature, certain species are adapted to dry environments. The genotype determines the ability of the plant to survive and thrive in environments with low water availability. In addition, the duration of the water-deficit stress, the rate of stress imposition, and the developmental stage of the plant at the time of stress imposition also effect plant growth. Whether the amount of water present in the environment is a stress is different for each species. A sensing mechanism initiates the responses to water deficit, which occur at the molecular, metabolic, physiological and developmental levels. Many of these responses are driven by changes in gene expression.

Plants, when subjected to environmental stress, undergo alterations in their growth, metabolism, and production. Among these, drought is the most adverse environmental factor regarding growth and productivity of cultivars. Losses in agricultural yield due to water stress probably exceed the losses inflicted by all other causes combined (Kramer et al., 1980). It is known that drought has a profound impact on agricultural and ecological systems, and thus the capacity of plants to withstand this stress is of great economic importance (Shao et al., 2008). Therefore, at present, with the aim of improving agricultural yield within the earth's limited resources, it is necessary to develop crops able to give a high yield when growing in stressed environments.

Water stress influences plant growth in several ways. For example, shoot biomass significantly decreased in wheat under drought conditions (Loggini et al.,1999). In potato plants, stem length and dry weight diminished under water stress (Ravindra et al., 1991) and (Ierna et al., 2006), and in tomato plant, shoot weight and total leaf area were lower than well-watered (Tahi et. al., 2008). Also, water deficit diminishes leaf size, longevity, and number of leaves per plant (Shao et al., 2008)The damage caused by water stress has two primary causes: first, the formation of reactive oxygen species (ROS) and, second, the alteration of water relationships within the plant (Figure 2). The extent to which plants can avoid or buffer these physiological processes determines the degree of resistance to water stress. Therefore the study of the metabolic and biochemical responses to water deficit is vital to present-day agriculture in order to select plants with high yield and stability under this type of stress (Yordanov et al., 2000).

Fig. 1: Drought stress and associated effects

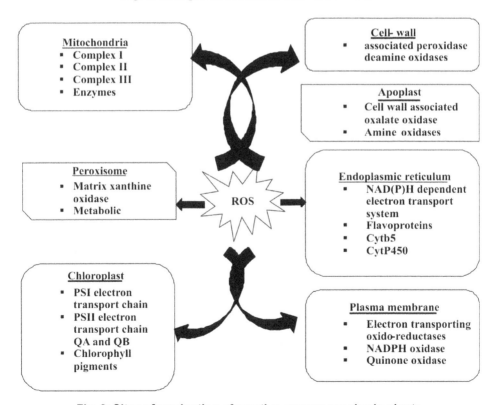

Fig. 2: Sites of production of reactive oxygen species in plants

3. Successful Strategies under Drought Stress - Resistance, Acclimatization and Plasticity

Plant resistance to drought has been divided into escape, avoidance and tolerance strategies (Levitt, 1972 and Turner, 1986). Resistance to water deficit may arise from the ability to tolerate water deficit or from mechanisms that allow avoidance of the water deficit. Some species, such as desert ephemerals, are able to escape drought by completing their life cycle when water is plentiful. Others avoid water deficit with the development of a large root system that permits improved extraction of water from the soil. Avoidance of water deficit may also be achieved by using mechanisms that save water as in succulents. Some plants may also have improved water use eficiency,such as found in crassulacean acid metabolism plants, in which stomata are open at night and an alternative form of carbon assimilation promotes the use of less water. However, these plants do not tolerate water deficit. Other plants have biochemical and morphological mechanisms such as found in mosses and resurrection plants that permit the plants to withstand dehydration. Morphological characteristics of some species permit continued survival in arid environments, while other plants must acclimate to the stress to permit survival. The respiration pathway, which breaks complex molecules into simple compounds to provide the energy required for plant development, is accelerated under drought stress (Haupt-Herting *et al.*, 2001). Protection systems such as the antioxidation pathway, which can reinforce plant cells to form reactive oxygen species scavengers, are also affected by drought stress (Apel and Hirt, 2004). Despite significant progress during the past decade in our understanding of pathways affected by drought stress, limited information is available regarding pathway dynamics in tomato under drought stress.

4. Response of Crop Plants to Drought Stress

- Morphological Response
- Physiological Response.
- Biochemical Response.
- Morphological Response.

4.1. Physiological Implication of Drought

- Water stress leads to reduced ability of water stress for vital cellular function and maintanance of turgor pressure.
- Dehydration or osmotic stress induces stomatal closure consequently reduction of biochemical capacity for carbon assimilation and use.

- One characteristic cellular feature activated by abiotic stress is the high production of ROS (Reactive oxygen species) in chloroplasts, mitochondria and in peroxisomescausing irreversible cellular and tissue damage however most plants have developed various adaptation and detoxification mechanisms to deal with stress condition. some of the most common responses for abiotic stress tolerance in plants is overproduction of several compatible organic solutes such as sucrose, betaine proline etc.

- Activation of genes involved in signal transduction pathway may lead to complex changes in gene expression resulting in plant adaptation to abiotic stress.

Younger leaves tend to be more resistant to drought than older leaves, and this increased tolerance may be particularly relevant in plants where a severe reduction in the size of the leaf canopy occurs as a result of shedding of older leaves, because it allows a fast recovery following rehydration (Pereira and Chaves, 1993). In addition to a plants' ability to avoid and/or endure water stress, photosynthetic recovery following rehydration is pivotal to dictate a plants' resistance to drought and to prevent dramatic declines in crop yield (Chaves *et al.*, 2009). It was shown that recovery from a severe stress was a two stage process: the first stage occurs during the first hours or days upon re-watering, corresponding to the improvement of leaf water status and stomatal reopening (Pinheiro *et al.*, 2005; Antonio *et al.*, 2008; Hayano-Kanashiro *et al.*, 2009); and the second stage lasts several days and requires de novo synthesis of photosynthetic proteins (Kirschbaum, 1988). Previous stress intensity and/or duration are crucial factors affecting both the velocity and the extent of recovery of photosynthesis (Miyashita *et al.*, 2005; Flexas *et al.*, 2006). Long-term down-regulation of gs after re-watering may be derived from limited recovery of leaf-specific hydraulic conductivity (Galme's *et al.*, 2007). Decrease in number of leaves and leaf area during drought is the first seen morphological character. Root modification during drought is important parameter .Increased root growth during drought stress depends on the location of osmolytes to the growing root tips. Roots penetrate deep into the soil and some plants also produce secondary roots to cope up with the drought.

4.1.1. *Stomatal Behaviour during Drought Stress*

Plants respond to water stress by producing abscisic acid (ABA), which stimulates the closure of the guard cells of the stomata to reduce water loss his process decreases CO_2 availability for photosynthesis, resulting in an imbalance between the generation and the use of electrons, provoking the overproduction of ROS. Stomatal closure can be considered as the third line of defence against drought. Guard cells of the stomata are exposed to atmosphere so can loss water directly by evaporation causing stomata to close by hydro passive closing mechanism/ Hydro active closure. Free flow of oxygen and carbon dioxide is dependent on stomatal opening which in turn control by turgor of both stomatal guard cells and epidermal cells. These two metabolic processes are affected by the dehydration within zone of cell turgor. Abscisic acid is synthesized at low rate in mesophyll cells and chloroplast accumulation. When the mesophyll cells are dehydrated these two things happen:

1. Some of the Abscisic acid stored in the chloroplast is released in Apoplast.
2. The net abscisic acid synthesis increases after closure begins and appears to enhance the initial effect of stored abscisic acid.

4.1.2. *Effect on Pigment Composition and Photosynthesis*

Environmental stresses have a direct impact on the photosynthetic apparatus, essentially by disrupting all major components of photosynthesis including the thylakoid electron transport, the carbon reduction cycle and the stomatal control of the CO_2 supply, together with an increased accumulation of carbohydrates, peroxidative destruction of lipids and disturbance of water balance (Allen and Ort, 2001).The ability of crop plants to acclimatize to different environments is directly or indirectly associated with their ability to acclimatize at the level of photosynthesis, which in turn affects biochemical and physiological processes and, consequently, the growth and yield of the whole plant (Chandra, 2003). Drought stress severely hampered the gas exchange parameters of crop plants and this could be due to decrease in leaf expansion, impaired photosynthetic machinery, premature leaf senescence, oxidation of chloroplast lipids and changes in structure of pigments and proteins (Menconi *et al.,* 1995). Anjum *et al.* (2011a) indicated that drought stress in maize led to considerable decline in net photosynthesis (33.22%), transpiration rate (37.84%), stomatal conductance (25.54%), water use efficiency (50.87%), intrinsic water use efficiency (11.58%) and intercellular CO_2 (5.86%) as compared to well water control. Many studies have shown the decreased photosynthetic activity under drought stress due to stomatal or non-stomatal mechanisms (Ahmadi, 1998; Del Blanco *et al.,* 2000; Samarah *et al.,* 2009). Stomata are the entrance of water loss and CO_2 absorbability and stomatal closure is one of the first responses to drought stress which result in declined rate of photosynthesis. Stomatal closure deprives the leaves of CO_2 and photosynthetic carbon assimilation is decreased in favor of photorespiration. Considering the past literature as well as the current information on drought-induced photosynthetic responses, it is evident that stomata close progressively with increased drought stress. It is well known that leaf water status always interacts with stomatal conductance and a good correlation between leaf water potential and stomatal conductance always exists, even under drought stress. It is now clear that there is a drought-induced root-to-leaf signaling, which is promoted by soil drying through the transpiration stream, resulting in stomatal closure. The "non-stomatal" mechanisms include changes in chlorophyll synthesis, functional and structural changes in chloroplasts, and disturbances in processes of accumulation, transport, and distribution of assimilates. Chlorophyll is one of the major chloroplast components for photosynthesis, and relative chlorophyll content has a positive relationship with photosynthetic rate. The decrease in chlorophyll content under drought stress has been considered a typical symptom of oxidative stress and may be the result of pigment photo-oxidation and chlorophyll degradation. Photosynthetic pigments are important to plants mainly for harvesting light and production of reducing powers. Both the chlorophyll a and b are prone to soil dehydration (Farooq *et al.*, 2009). Decreased or unchanged chlorophyll level during drought stress has been reported in many species, depending on the duration and severity of drought (Kpyoarissis *et al.*, 1995; Zhang and Kirkham, 1996).

4.1.3. *Effect on Osmotic Adjustment*

One of the most common strategies of plants for avoiding water stress is the accumulation of the so-called compatible solutes, also called osmoprotectors or osmolytes (Alonoso *et al.*, 2001) . During osmotic stress, plant cells accumulate solutes to prevent water loss and re-establish cell turgor. The solutes that accumulate during osmotic adjustment include ions such as K^+, Na^+, and Cl^-, or organic solutes that include compounds that contain N, such as proline and other amino acids, polyamines, and QAC (Tamura *et al.*, 2003). Proline accumulates in a great variety of plant species in response to stress such as drought, salinity, and extreme temperatures. Although its osmotolerant role in plants is not clear, under stress conditions, proline can act as a mediator of osmotic adjustment, stabilizer of subcellular structures, eliminator of free radicals, and as a buffer of redox potential (Molnari *et al.*, 2004), (Schobert *et al.*, 1978), (Matysik *et al.*, 2002) and (Hare *et al.*, 1997). There is currently great controversy on the protective properties of proline accumulation. Hanson (Hanson *et al.*, 1980) concluded that proline accumulation is not an adaptive feature, but rather only a symptom of stress. In agreement with this contention, findings show that drought-tolerant wheat plants have a higher LRWC related to a lower proline concentration(Rampino *et al.*, 2006).

4.2. The Molecular Responses Regulated During Drought

The molecular response of plants to water deficit defines a very interesting puzzle: how does a physical phenomenon, the loss of water from the cell, cause a biochemical response, the induction of specific genes? Or, how does the cell recognize the loss of water and respond to it? The answer to these questions is not known. Currently, it is thought that loss of turgor or change in cell volume resulting from different environmental stresses permits the detection of loss of water at the cellular level. One or both of these changes may activate stretch-activated channels, alter conformation or juxtaposition of critical proteins, or cause alterations in the cell wall-plasma membrane continuum (Ding and Pickard, 1993), thereby triggering a signal transduction pathway(s) that induces gene expression. Therefore, several different stresses may trigger the same or similar signal transduction pathways. The induction pathway(s) is also poorly understood, although there is evidence that there is more than one pathway. There may be a direct pathway, or additional signals may be generated that, in turn, alter the pattern of gene expression. The plant hormone ABA also accumulates in response to the physical phenomena of loss of water caused by different stresses, and elevation in endogenous ABA content is known to induce certain water-deficit-induced genes. Therefore, ABA accumulation is a step in one of the signal transduction pathways that induces genes during water deficit. At this time, a convenient way to categorize drought-induced genes is by response to ABA. Sets of genes have been identified that require ABA for expression, are ABA-responsive but may no require ABA for expression, and are not responsive to ABA.

4.2.1. *ABA induces Gene Expression during Water Deficit*

ABA concentration is altered when there are changes in the environment that result in cellular dehydration; especially well studied are changes that occur during periods of water deficit. ABA is synthesized through the carotenoid biosynthetic pathway. The cleavage of 9'-cis-neoxanthin results in the post cleavage intermediate to ABA, xanthoxin. Xanthoxin is oxidized to ABA-aldehyde, which is converted to ABA by ABA-aldehyde oxidase (Parry, 1993). The step in the ABA biosynthetic pathway that is likely to be regulated during drought stress is the cleavage step; the rate of ABA biosynthesis is limited by the production of xanthoxin, not the conversion of xanthoxin to ABA (Parry, 1993).

Transcription and translation are required for ABA biosynthesis during stress (Guerrero and Mullet, 1986), indicating that the ABA biosynthetic enzymes or other proteins in the pathway must be synthesized for elevated levels of ABA to accumulate and before ABA-requiring genes can be induced . Studies using ABA-deficient mutants have been used to conclusively demonstrate that elevated levels of ABA are required for the expression of specific drought-induced genes. ABA application studies can be used to determine the range of responses that a plant is capable of attaining in response to ABA. However, these types of studies do not provide evidence that the plant responds to changes in the concentration of endogenous ABA. Inhibitors of carotenoid biosynthesis result in a decreased level of ABA (Parry, 1993), and responses of the plant that are reduced by inhibitor application have been used to analyze the role of ABA. But, as with the use of other inhibitors, effects that are not directly caused by the reduction in ABA concentration may also occur, because carotenoid concentration is also reduced after application of these inhibitors. Mutants blocked in the last steps of the ABA biosynthetic pathway, ABA-deficient mutants, are the best tools available with which to examine the role of ABA during plant stress. With single-gene mutants, only a reduction in ABA occurs and pleiotropic effects can be reduced or eliminated. Experiments using mutants blocked in the ABA biosynthesis pathway demonstrated that specific genes require elevated levels of ABA for expression during water and low-temperature 'stress (Cohen and Bray, 3 990; Pla *et al.,* 1991; Ling and Palva, 1992). Several drought- induced genes are not expressed in the ABA-deficient mutant as they are in the wild type. These genes are referred to as ABA-requiring genes. Although sets of genes can be classified by responses to experimental conditions, the response of each gene to cellular signals is defined by the DNA elements acting within each gene. An ABRE, 5'-C/TACGTGGC-3', involved in controlling transcription in response to ABA, has been shown to bind a cloned transcription factor, EmBP-1 (Guiltinan *et al.,* 1990). This ABRE has been demonstrated to control ABA- regulated expression in several genes, Em, rabl6, rabl7; and rab28, that are preferentially expressed in seeds (Pla *et al.,* 1993). This DNA sequence is found in many genes that are expressed during seed development and in response to ABA application. However, it is not found in all genes that are in the ABA requiring category; for example, lel6, a gene expressed in wild-type tomato but not in the ABA-deficient mutant, does not contain a consensus ABRE (Plant *et al.,* 1991). Recently, (Michel *et al.,* 1993) have sit own that DNA regions required for ABA-induced expression in C. plantagineum do not contain the ABRE. Therefore, it is expected that there are multiple ABRES with different consensus DNA sequences. In addition, genes that are expressed during drought are also expressed during specific developmental stages and in specific cell types. Therefore, additional elements are required to control tissue- and organ-specific expression and other specific aspects of the expression pattern during water deficit (Figure 2). ABA-deficient mutants have been used to define an additional set of genes, those that are responsive to ABA but do not require ABA for expression (Gilmour and Thomas how, 1991; Nordin *et al.,* 1991; Yamaguchi-Shinozaki and Shinozaki, 1993). These genes are induced by ABA application, but are also induced by water deficit in the ABA-deficient mutant of Arabidopsis. Therefore, it has been concluded that these genes do not require elevated levels of endogenous ABA for expression but are ABA-responsive genes. These results indicate that there are two pathways that can be followed to induce these genes, but it is unknown if the pathways converge or if there are two entirely separate pathways. As promoter analyses are completed on these genes, it must be recognized that there may be multiple mechanisms of gene induction. ABA applications have also been used to show that there are a number of water-deficit-induced genes that do not respond to ABA application (Guerrero et al., 1990; Yamagu- chi-Shinozaki et al., 1992). These genes may be induced directly by the drought stress, or they may be controlled by other signaling mechanisms that may operate during water deficit. Although the responsiveness of a gene is controlled by its DNA elements, the pathways that lead to the formation of stable transcription complexes are equally important for gene induction. For ABA-requiring and ABA-responsive genes, ABA must be recognized at the cellular level triggering a pathway(s) leading to the formation of transcription complexes and gene induction. Little is known about the mechanism of recognition of the ABA molecule. ABA analogs are being used to identify molecular structures required for gene induction by ABA (Van der Meulen et al., 1993). After ABA recognition there are many possible pathways that could lead to active transcription complex formation. The pathway taken may be altered by the physiological state of the cell. The sensitivity to ABA is altered by the osmotic potential of the cells; there is increased sensitivity to ABA with increased osmotic stress. In some cases, osmoticum can completely replace exogenous ABA. For mRNA accumulation in rice cell cultures, increasing concentrations of NaCl increased the accumulation of mRNA in response to suboptimal concentrations of ABA (Bostock and Quatrano, 1992). Therefore, the cells' response to ABA may be altered by the water potential or water content of the cell. However, another possibility should be considered. Because the cells are induced to accumulate ABA in response to the osmotic stress, it becomes difficult to determine if the newly synthesized ABA is contributing to the induction of genes. ABA that is synthesized in the cell may not be located in the same compartment within the cell as ABA that is applied to the cell (Bray and Zeevaart, 1986). Therefore, the plant may be more sensitive to endogenous ABA than to applied ABA. It is known that high levels of ABA must be applied to elicit a response similar to that stimulated by endogenous ABA concentrations. Further studies with ABA-deficient mutants could be used to resolve this possibility.

4.2.2. *Acid Metabolism*

C_4 or CAM plantshave developed a strategy to fix carbon dioxide for sugar production with a minimum loss of water. This metabolism involves nocturnal assimilation of CO_2. C_4 plants have developed a mechanism to efficiently channel CO_2 to RuBisCO. They have also developed a particular leaf anatomy where bundle sheath cells have chloroplasts, besides mesophyll as in C_3 plants. Instead of direct fixation in the calvin cycle, CO_2 is converted to a 4 carbon organic acid with the ability to regenerate CO_2 in the chloroplasts of bundle sheath cells. These cells utilize CO_2 to synthesize carbohydrate by the conventional C_3 pathway. For this process stomata are open in night, allowing CAM plants to colonize hot environment, as their expenses of water is very low in comparison with other plants. Anatomical adaptations are observed in tolerant plants, consisting of spongy tissues, acting as water reservoirs; growth is also impaired and plants reduce their foliar area to limit evaporation. Similar strategies are the rolling of the leaves, floral abscission and alterations in cuticle permeability. Floral induction, associated to long distance movement of FT proteins, is also modulated by water limitation during plant development.

Drought Induced Proteins

Based on their functions, we divided the genes that are associated with abiotic stress tolerance into 4 groups such as osmolyte biosynthesis, antioxidant protectants, protection of cell integrity and ion homeostasis (Sangam *et al.*, 2005; Valliyodan and Nguyen, 2006). Research in genetic engineering for stress tolerance was limited in the pre-genomics era by the inadequate availability of genes and specific promoters (Zhu *et al.*, 1997). It is now possible to study many genes simultaneously on a genome-wide scale with respect to their structure and function. Thus, the current trend in stress-biology is to use large-scale genomics data to scrutinize and revalidate the protective mechanisms which were described based on transgenic approaches (sugar alcohol biosynthesis, osmolyte biosynthesis, antioxidants, LEA proteins, molecular chaperones, cell membrane proteins, aquaporins, ion homeostasis and transcription factors).

Thus, there is a good correlation among the gene sets identified by traditional and genomic studies. Moreover, many important components identified for abiotic stress tolerance were tested by transgenic technologies and confirmed the roles of these genes associated with the biosynthesis of osmolytes, stress proteins, antioxidants, aquaporins and ion homeostasis (Kishor *et al.*, 2005; Sangam *et al.*, 2005).

Protection of Cellular Structures

A number of water-deficit-induced gene products are predicted to protect cellular structures from the effects of water loss. These predictions are derived from the deduced amino sequence and expression characteristics. These genes, frequently called leu, were first identified as genes that are expressed during the maturation and desiccation phases of seed development. It has since been recognized that these genes are also expressed in vegetative tissues during periods of water loss resulting from water, osmotic, and low-temperature stress. At least six groups of leu genes have been identified, based on amino acid sequence similarities among several species (Dure, 1993b). The majority of the leu gene products are predominantly hydrophilic, biased in amino acid composition, and lacking in Cys and Trp and are proposed to be located in the cytoplasm.

The individual amino acid sequences and predicted protein structures have been used to propose specific functions for each group of LEA proteins (Dure, 1993b). These predicted functions include sequestration of ions, protection of other proteins or membranes, and renaturation of unfolded proteins (Figure 3).

One of the groups of LEA proteins (D-7 family, or group 3) is predicted to play a role in the sequestration of ions that are concentrated during cellular dehydration. These proteins have an 11-mer amino acid motif with the consensus sequence TAQAAI'KAGE, repeated as many as 13 times (Dure, 1993a). This motif is predicted to form an amphiphilic α-helix. The hydrophobic face may be important in forming a homodimer and the outside charged face may be involved in sequestering ions whose concentration is increased during water deficit (Dure, 1993a). Another group of LEA proteins (D-29 family, or group 5) are also predicted to sequester ions during water stress. This group also has an 11-mer repeat in which each amino acid in the motif has similar chemical properties to group 3, but unlike group 3 a high degree of residue specificity is lacking at each position. Another group of LEA proteins (13-19 family, or group 1) is predicted to have enhanced water-binding capacity. These proteins have a high percentage of charged amino acids and Gly. One member of this group, Em, is approximately 70% random coil, leading to the prediction that it has a high capacity for binding water. Group 4 LEA proteins (D-113 family) may replace water to preserve membrane structure. The amino acid sequence in the N tenninus, which is thought to form an α-helix, is conserved. These proteins have little conservation in the C- terminal amino acid sequence, although the C-terminal random coil structure is conserved.

5. Conclusion and Future Perspectives

Recently progress has been achieved in the isolation of water-deficit-induced genes. Amino acid sequences have been used to predict functions and the genes have been used to follow gene expression. In the coming years, more attention must be paid to the roles of these genes under relevant stress conditions to determine if the expression of these genes is accentuated by the experimental conditions employed. Like- wise, the same view point must be used to determine the signal transduction pathways. Pathways that are induced in the laboratory may not be successfully induced in the field.

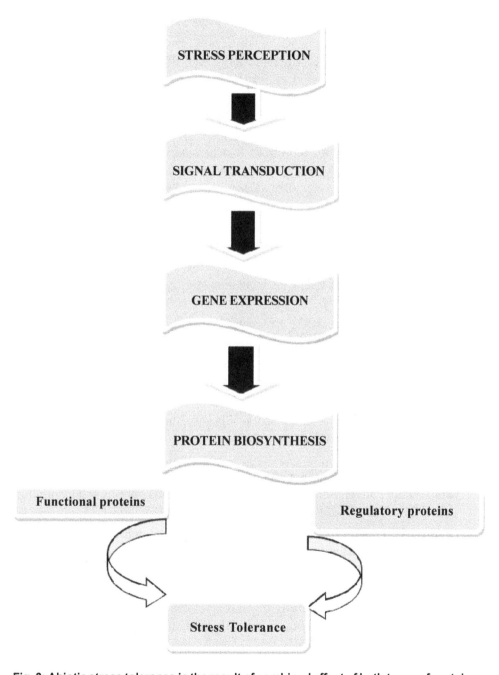

Fig. 3: Abiotic stress tolerance is the result of combined effect of both types of proteins.

6. References

1. Parry, A.D. Abscisic acid metabolism. *Methods Plant Biochem.* **1993**, 9, 381-402.

2. Yamaguchi-Shinozaki, K.; Koizumi, M.; Urao, S.; Shinozaki, K. Molecular cloning and characterization of 9 cDNAs for genes that are responsive to desiccation in Arabidopsis thaliana: sequence analysis of one cDNA clone that encodes a putative transmembrane channel protein. *Plant Cell Physiol.* **1992**, 33, 217-224.

3. Yamaguchi-Shinozaki K.; Shinozaki, K. Characterization of the expression of a desiccation-responsive rd 29 gene of Arabidopsis thaliana and analysis of its promoter in transgenic plants. *Mol. Gen. Genet.* **1993**, 236, 331-340.

4. Kishor, P.B.K. Regulation of proline biosynthesis, degradation, uptake and transport in higher plants: its implications in plant growth and abiotic stress tolerance. *Curr. Sci.* **2005**, 88, 424–438.

5. Sangam, S.; Jayasree, D.; Reddy, K.J.; Chari, P.V.B.; Sreenivasulu, N.; KaviKishor, P.B. Salt tolerance in plants—transgenic approaches. *J. Plant Biotechnol.* **2005**, 7, 1–15.

6. Zhu, J.K.; Hasegawa, P.M.; Bressan, R.A. Molecular aspects of osmotic stress in plants. *Crit. Rev. Plant Sci.* **1997**, 16, 253–277.

7. Valliyodan, B.; Nguyen, H.T. Understanding regulatory networks and engineering for enhanced drought tolerance in plants. *Curr. Opin. Plant Biotech.* **2006**, 9, 189–195.

8. Chaves, M.M.; Flexas, J.; Pinheiro, C. Photosynthesis under drought and salt stress: regulation mechanisms from whole plant to cell. *Annals of Botany.* **2009**, 103, 551–560.

9. Flexas, J.;Bota, J,;Galme´ s, J.; Medrano, H.;Ribas-Carbo, M. Positive carbon balance under adverse conditions :responses of photosynthesis and respiration to water stress. *Physiologia Plantarum.* **2006a**, 127, 343–352.

10. Galmes. J.; Flexas. J.; Save. R.; Medrano, H. Water relations and stomatal characteristics of Mediterranean plants with different growth forms and leaf habits: responses to water stress and recovery. **2007c**.

11. Hayano-Kanashiro. C.; Caldero´ n-Va´ zquez. C.; Ibarra-Laclette, E.; Herrera-Estrella, L.; Simpson, J. Analysis of gene expression and physiological responses in three Mexican maize landraces under drought stress and recovery irrigation. *PLoS ONE.* **2009**,4, e7531.

12. Kirschbaum, M.U.F. Recovery of photosynthesis from water stress in Eucalyptus pauciflora—a process in two stages. *Plant, Cell and Environment.* **1988**, 11, 685–694.

13. Miyashita, K.; Tanakamaru, S.; Maitani, T.; Kimura, K. Recovery responses of photosynthesis, transpiration, and stomatal conductance in kidney bean following drought stress. *Environmental and Experimental Botany,* **2005**, 53, 205–214.

14. Pereira, J.S.; Chaves, M.M. Plant water deficits in Mediterranean ecosystems. In: Smith J.A.C., Griffiths, H. (Eds.), Plant responses to peroxidase activity and hydrogen peroxide level in roots of rice seedlings. *Plant. Growth Regul.* **1993**, 37,177-183.

15. Pinheiro, C.; Kehr, J.; Ricardo, C.P. Effect of water stress on lupin stem protein analysed by two-dimensional gel electrophoresis. *Planta.* **2005**, 221, 716–728.

16. Antonio, C.; Pinheiro, C.; Chaves, M,M.; Ricardo, C.P.; Ortuno, M.F.; Thomas-Oates, J. Analysis of carbohydrates in Lupinusalbus stems on imposition of deficit, using porous graphitic carbon liquid chromatography- electrospray ionization mass spectrometry. *Journal of Chromatography.* **2008**, 1187, 111–118.

17. Boyer, J.S.Plant productivity and environment potential for increasing crop productivity, genotypic selection .*Science.* **1982**, 218, 443-448

18. Kramer, P.J. Drought, stress, and the origin of adaptation. In: Turner, N.C., Kramer, P.J. (Eds.), Adaptation of Plants to Water and High Temperature Stress, John Wiley and Sons, New York, NY, USA. **1980**, pp.7–20.

19. Shao, H.B.; Chu, L.; Jaleel, C.A.; Zhao, C.X.Water-deficit stress-induced anatomical changes in higher plants, *ComptesRendus Biologies.* **2008**, 331, 215–225.

20. Loggini.; Scartazza, A.; Brugnoli, E.; Navari-Izzo, F. Antioxidant defense system, pigment composition, and photosynthetic efficiency in two wheat cultivars subjected to drought, *Plant Physiol.* **1999**, 119, 1091–1099.

21. Ravindra, V.; Nautiyal, P.C.; Joshi, Y.C. Physiological analysis of drought resistance and yield in groundnut (*Arachishypogaea* L.), *Trop. Agric. Trinidad.* **1991**, 67, 290–296.

22. Ierna, A.; Mauromicale, G. Physiological and growth response to moderate water deficit of off-season potatoes in a Mediterranean environment, *Agric. Water Manage.* **2006**, 82, 193–209.

23. Tahi, H.; Wahbi, S.; El Modafar, C.; Aganchich, A.; Serraj, R. Changes in antioxidant activities and phenol content in tomato plants subjected to partial root drying and regulated deficit irrigation, *Plant Biosyst.* **2008**, 142, 550–562.

24. Yordanov, I.; Velikova, V.; Tsonev, T. Plant responses to drought, acclimation, and stress tolerance, *Photosynthetica.* **2000**, 38, 171–186.

25. Levitt, J. Responses of plants to environmental stress. Academic press, New York. **1972**.

26. Turner, N.C. Crop Water Deficits: A decade of progress. *Advances in Agronomy* **1986**, 39, 1-51.

27. Ajay, A.; Sairam, R.K.; Srivatava, G.C. Oxidative stress and antioxidative system in plants, *Curr. Sci.* **2002**, 82, 1227–1238.

28. Rampino, P.; Pataleo, S.; Gerardi,C.; Mita, G.; Perrotta, C. Drought stress response in wheat: physiological and molecular analysis of resistant and sensitive genotypes. *Plant Cell Environ.* **2006**, 29, 2143–2152.

29. Alonso, R.; Elvira, S.; Castillo, F.J.; Gimeno, B.S. Interactive effects of ozone and drought stress on pigments and activities of antioxidative enzymes in *Pinishalpensis. Plant Cell Environ.* **2001**, 24, 905–916.

30. Tamura, T.; Hara, K.; Yamaguchi, Y.; Koizumi, N.; Sano, H. Osmotic stress tolerance of transgenic tobacco expressing a gene encoding a membrane-located receptor-like protein from tobacco plants. *Plant Physiol.* **2003**, 131, 454–462.

31. Molinari, H.; Marur, C.J.; Bespalhok, J.C.; Kobayashi, A.K.; Pileggi, M.; Pereira, F.P.P.; Vieira, L.G.E. Osmotic adjustment in transgenic citrus rootstocks Carrizo citrange (*Citrus sinensis* Osb. × *Poncirustrifoliata* L. Raf.) overproducing proline. *Plant Sci.* **2004**, 167, 1375–1381.

32. Schobert, B.; Tschesche, H. Unusual solution properties of proline and its interactions with proteins. *Biochim. Biophys. Acta.* **1978**, 541, 240–277.

33. Matysik, J.; Alia, A.; Bhalu, B.; Mohanty, P. Molecular mechanisms of quenching of reactive oxygen species by proline under stress in plants. *Curr. Sci.* **2002**, 82, 525–532.

34. Hare, P.D.; Cress, W.A. Metabolics implications of stress-induced proline accumulation in plants. *Plant Growth Regul.* **1997**, 21, 79–102.

35. Hanson, A.D. Interpreting the metabolic responses of plant to water stress. *Hortic. Sci.* **1980**, 15, 623–629.

36. Parry, A.D. Abscisic acid metabolism. *Methods Plant Biochem.* **1993**, 9, 381-402.

37. Apel, K.; Hirt, H. Reactive oxygen species: metabolism, oxidativestress, and signal transduction. *Annual Review of Plant Biology.* **2004**, 55, 373–399.

38. Haupt-Herting, S.; Klug, K.; Fock, H.P. A new approach to measure gross CO_2 fluxes in leaves: gross CO_2 assimilation, photorespiration, and mitochondrial respiration in the light in tomato under drought stress. *Plant Physiology.* **2001**, 126, 388–396.

39. Dure, L. Structural motifs in Lea proteins. In: Close,T.J., Bray, E.A. (Eds.), Plant Responses to Cellular Dehydration during Environmental Stress. Current Topics in Plant Physiology, Vol 10. American Society of Plant Physiologists, Rockville, MD, **1993**, pp. 91-103.

Abiotic Stress Tolerance Mechanisms in Plants, Pages 341–356
Edited by: Gyanendra K. Rai, Ranjeet Ranjan Kumar and Sreshti Bagati
Copyright © 2018, Narendra Publishing House, Delhi, India

10

STOMATAL ACTIVITIES AGAINST ABIOTIC STRESS AND SELECTION OF POTENTIAL TOLERANT HORTICULTURAL CROPS

Deepak Kumar Sarolia, Mukesh K. Berwal and Ramkesh Meena

ICAR-Central Institute for Arid Horticulture, Bikaner-334006
E-mail: deephorti@gmail.com

Abstract

Abiotic stressful environments such as salinity, drought, and high temperature (heat) cause alterations in a wide range of physiological, biochemical, and molecular processes in plants. Abiotic stress, as a natural part of every ecosystem, will affect plants in a variety of ways. Plants also adapt very differently from one another, even from a plant living in the same area. A plant's first line of defense against abiotic stress is in its roots than leaf, on the leaf surface natural opening or stomata play important role in first level of adaptation towards survivality. All together plant against stress either escape or avoidance or tolerance mechanism maintain their turgor pressure in leaves that's depends on level on adaptation. Most practical approach in management of stress side is selection of crops and its varieties in arid environment.

Keywords: Stomatal activities, Abiotic stresses, Drought, Frost, Anti-transpirant

1. Introduction

Schimper explains drought resistance on the basis of water conservers of water actually loses water more rapidly than plants of moist habitat. Water saver lose as little as 1/4800 their height per day whereas, water spend lose as much as five times their height per hours. Thus water spender may lose water as much as 50,000 times as rapidly as water savers. The water savers are able to restrict transpiration long before wilting occurs.

Indian arid zone is one of the largest sub-tropical deserts of the world of which 20% is arid andrest is semi-arid with varied habitats. The arid region covers nearly 12% land surface of India and spread over 39.54 million ha area, out of which 31.71 million ha is under hot arid region. The hot arid region occupies major part of north-western India and occurs in small pockets in southern India. The present cropping pattern of arid region can be classified into two categories (i) where assured irrigation is available fruits like pomegranate, lime, kinnow mandarin, grapes, etc are grown, (ii) the areas which are under rainfed conditions, the crops like Kair (*Capparis decidua*), pilu (*Salvadora oleoide*s), leshua (*Cordia myx*a) and jharber (*Zizyphus nummularia*), bael (*Ageal marbalos*) and kaith. Similarly, among vegetables, watermelon, kachari are grown but okra, amaranthus, sweet potato etc.About 91% of Indian arid zone covers, North West Gujrat, Western Rajasthan, South West Punjab & Haryana. Among various states Rajasthan has the highest amount of arid region i.e. 62% or 1,96,150 Sq. km followed by Gujarat 20% (82,180 Sq.km)

Thus climatic and edaphic constraints are severe which limit preferable production of horticultural crops under arid conditions. Productivity is very low which can be increased by creating infrastructure and other facilities. In this region paucity of water stomatal transpiration play multifold activities.

2. Stomatal Stress and Effect on the Plants

(i) Stress occurs due to deficit water condition, when the plant is subjected to an artificially induced evaporative loss of water, commonly called desiccation stress. A Stress that is capable of inducing a loss of water in the liquid state called osmotic stress.

(ii) The water deficit or drought may be measured by vapour pressure deficit (vpd) $Sd = Po-P$ whereas, $Sd=$ drought stress $Po=$ vapour pressure of pure water at environmental temperature $p=$ vapour pressure of environment. In term of water potential terminology it is expressed as: $Sd=\Psi e$ (water potential of environmental bars/ atm.)Since water potential (Ψ) is always higher .water stress becomes positive.

(iii) Water stress to which a plant is exposed is not known unless both stress in the shoot environment $(Sd)s$ and that in the root environment $(Sd)r$ are known. Agriculture drought in fact defined on the role basis at of the root environment it is to occur when soil moisture in the root zone is at or below wilting point.

(iv) The drought stress where plants which show only survival have particular importance for environmental plant physiology (ecology). These plants survive under very high harsh condition. The adoptions of such plants are towards survival.

(v) Plant adaptation should towards higher yield (agronomical) eg at PWP (-Ψ 15 atm) stomata are close to minimize transpiration at the same time photosynthesis is decreased since CO_2 enhance is decreased because of stomatal closer. This productivity is decreased. Hence, drought has adverse effect on productivity.

(vi) Plants are poikilotherms and tend to follow closely the temperature changes of their environment. In contrast higher plants homoihydric and tends to maintain a steady. State water potential above environment. This state is maintained so long as water supply is available to plant.

(vii)If plant as a whole is exposed a stress it will gradually approach and finally attain equilibrium with external water potential. Evening equilibrium is finally reaches with environmental of 0% RH at room temperature. Development of internal water deficit with special reference to thermodynamic equilibrium (Ψsoil> Ψroot > Ψ leaf) general trend of water potential as the day increase. The Ψ gradient decrease between soil, root and leaf. Hence, flow of water from soil to leaf minimize. Finally the water deficit level reaches the PWP where the Ψ of soil, root and leaf are level and than a thermo equilibrium (TE) is usually established at -15bars of Ψ. In drought avoidance plant the TE was not attained because of certain mechanism of the plant (spender and saver). Water spender plants due to higher root to shoot ratio efficient absorption of water takes place from deep layers (also called phraetophyte plants). Water saver plants conserve water by minimum transpiration.

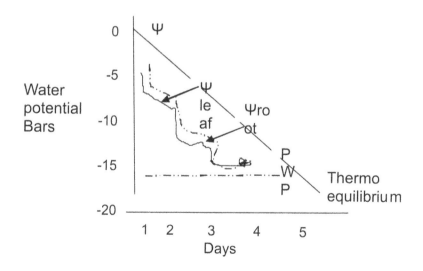

Fig. Water potential trends at day time advancement

Classification of adaptation of plant to water stress: According to environmental water supply required for normal completed of their life cycle.

(a) Hydrophytes: those plants adapted to partial or complete submerged in free water are called hydrophytes.

(b) Mesophytes: Land plants adapted to a moderate water supply.

(c) Xerophytes: those adapted to arid zones.

Hydrophytes not grow beyond -5 to -10bar, Mesophytes not beyond -20 bars and xerophytes not far beyond -40bar.

Stomata open during daylight for short time in early morning and they are able to close rapidly. More drought resistant species show a great degree of stomatal closure major control of water loss is through stomata. When soil dried changes in water vapour conductance of stomata is linearly related to the changes in soil water potential (from -1 to -20bar). The control mechanism appears depend on an inhibitor of stomatal opening which accumulates in leaves after a period of water stress which promote opening.

The two substances of appear to be abscisic acid and cytokine. The cytoknin decreases in quantity in stressed plants and is found to stimulate transpiration. The abscisic acid reduces stomatal closure. Once stomata are closured the cuticle on the leaf surface reduces the transpiration rate of the water savers to a much fraction of stomatal transpiration than in the case of Mesophytes. This difference in cuticular transpiration accounts for the superior resistance of pine over mulberry seedlings.

It has been reported that same drought resistant plant have very low rate of cuticular transpiration e.g. cactus holds fraction of water for long periods even if deprived of all contact with water. However, in surface lipid reduce cuticular transpiration. Even in relatively un-adapted plants exposure to moderate drought may result in pseudo-hardening due to deposit of lipids on leaves resulting in reduced cuticular transpiration.

The cuticle of pseudo-hardened plants contained greater proportion of wax if obtained from plants under going water stress and this wax was found to control permeably of cuticle to water. Similarly, rubbing away bloom from leaves of *Brassica oleracea* increased cuticular transpiration rate (Denna, 1998).

Various leaf modifications are responsible for cuticular transpiration eg high water content of succulent in responsible for high behind water and thus flows lower water loss. Similarly, high cell sap concentration has been suggested as method to reduce water lose by lowering vapour pressure.

Table 1: Rate of Stomatal and Cuticular Transpiration

Species	% reduction in initial transpiration rate	Time (Min.)	Cuticular transpiration (% of total)
Anona coriacea	60	30	2
Spondias tuberose	55	2	2-5
Caesalpima pyramidalis	50	10	15
Jatropha phylancantha	50	2	10-20
Ziziphus foazeiro	50	5	15

In spite of low cuticular transpiration of succulent their water loss/ unit area may sometimes be quite high. In these cases the principle adaptation responsible for reduced water loss is the reduction in better specific surface area. In case of non succulents, rolling, folding or shedding of leaves act as aid to drought resistance. Clemant (1937) found no adaptive response to drought in case of tomato and sunflower other than shedding of lower leaves and leaf folding in dry weather. Some reduction in relative surface may be achieved simply by compact form of foliage and crown. Eg. Leaves of apple trees with dense pyramidal crown had 20.2% higher water retaining capacity than trees with loose dispersed crown.

In case of many succulent the shallow spreading root system functions by quickly absorbing small amounts of water supplied by high rains "rain roots" develop within a few hours after shower and disappear as soon as soil dries. Thus root top ratio may be smaller in extreme water savers than in Mesophytes thus reducing root surface from which water can be lost.

Qankel et al. (1967) formation of metabolic water from intense respiration may be a mechanism for maintaining plant water content. In support of the concept drought hardened plants respire more rapidly than non hardened plants.

These adaptation that favour conservation of water by stomatal closure are unfavourable to photosynthesis, since stomata may be closed during day light hours preventing CO_2 from entering leaves. Most extreme water savers have reduced CO_2 absorption. Succulents partially counteract this deficiency by developing active dark CO_2 assimilation accompanished by opening of stomata at night. In some tropical grasses and some dicotyledonous tree is special C4-dicarboxylic acid pathway of photosynthetic CO_2 function. It is an adaptation for efficient rapid carbon fixation in environment where water stress frequently limits photosynthesis.

Transpiration Control

(i) Reducing transpiration is important:

- Enable plants to maintain adequate water balance until root system is reestablishment in transplanted plants.
- During drought reduction in transpiration will enable plant to survive with minimum injury increasing yield.
- Reduction of transpiration: this important for decreasing irrigation water requirement under arid-semiarid conditions and alleviate water stress.

(ii) Two important benefits from transpiration are expedition of mimeral uptake and cooling of leaves.

- Very low rate of transpiration needed for mineral uptake.
- Complete stoppage of transpiration results in 8-10oC rise in leaf temperature under extreme condition. Such increase in temperature even up to 3-5^{0}C may results in detrimental rise in respiration/ photosynthesis ratio which may be lethal.

(iii) Plant water stress occurs when transpiration exceeds water uptake

3. Rate of Transpiration

(i) Mesophytic plants growing under dry conditions transpire more than 99% of all the water they absorb.

(ii) Loss is unavoidable result of exposure of large leaf area and solar energy which necessary for photosynthesis.

(iii) Efficient absorption of radiant energy during day noon high potential rates of transpiration.

(iv) Kathju and Lahiri (1981) increased transpiration rate of younger and older leaves of *zizyphus nummularia* under dry condition of Jodhpur during last monsoon season. Results indicate younger leaves transpired more than older leaves.

3.1. Factors Controlling Rate of Transpiration

- Supply of water energy available to vaporized water.
- Vapour pressure gradient which constitute during force.
- Resistances in the vapour path way chief resistance are cuticle and stomata.
- Chief environmental factor are light mostly vapour pressure and temperature of air, wind and water supply to the root.

- Plant factors: extent and efficient of root system leaf area leaf arrangement and structure and stomatal behaviour.
- In most tree species stomata present usually on lower epidermis while in crop plant present on both surfaces.
- Size of stomata from 3-12u in width and 10-30 um length.

Stomatal mechanism: Stomatal open when guard cells are turgid and close when flaccid sequence of event leading to stomatal opening and closure.

a) With advent of light photosynthesis began in cell.

b) This reduces concentration of CO_2 and that of carbonic acid there by bringing about an increase in pH from 5-7.

c) Increase in pH promotes starch hydrolysis in combination with phosphate resulting in the production of glucose-1-phosphate which further converted to glucose-6-phosphate and then to glucose and phosphate.

d) Additional solutes (in form of glucose) reduces water potential of gourd cells causing water to enter them by osmosis from adjacent epidermal cells.

e) As a result of water intake the turgor potential of gourd cell increase causing stomata to open. In the darkness reverse of the above events results in stomatal closure.

Table 2: Distribution of stomata in the leaves of some species

Species	Average stomata number/cm^2	
	Upper epidermis	Lower epidermis
Apple	0	29400
Begonia	0	4000
Coleus	0	14100
Cabbage	14100	22600
Bean	4000	28000
Tomato	1200	13000
Wheat	3300	1400
Pea	10000	21600
Maize	5200	6580

4. Driving Force and Pathway of Transpiration

Loss of water through transpiration in the passive process in which water moves from leaves to the air by vapour differsion. Driving force in the vapour density gradient between air around the leaf or above the leaf canopy. Transpiration can be decreased by a diffusion equation:

$$T = eI\text{-}Ea/rI + ra$$

Where, T = Transpiration flux from leaf

rI = Resistance of leaf

ra = boundary layer resistance

eI = water vapour densities in leaf

ea = water vapour density in air

Way to reduce transpiration:

1. By reducing the net energy input. (eI)
2. By measuring humidity of air near leaf (ea)
3. By measuring resistance of the leaf to water lass. (rI)
4. By measuring resistance of air near leaf vapour transport out of their apparatus number 2 to 4 approaches (ea & ra) are primarily management practices involve use of wind break and shelter belts, respectively.
5. Energy input can be reduced by measuring plant reflectivity by the application of reflective materials either alone or mixed with anti-transparent.

The real physiological approaches involve rise of reflective material to reduce energy load on leaves and use of chemical causing stomatal closure to increase leaf resistance to water vapour loss.

Two approaches discussed below: Modification of leaf reflectance

1. Principle behind this approach is to regulate gas exchange of leaves by affecting their energy balance of reflective materials.
2. During day time energy to plant is from solar radiation which is often super optimal intensities (1.5 cal $cm^{-2}mm^{-1}$) as photosynthesis is usually saturated at 0.2 to 0.5 $calcm^{-2}$ mm^{-1}.

The reflective material are: zinc oxide, kaolinite, chalk etc. shese have v\been used to reduce trunk temperature of fruit trees often used material is kaolonite suspension in material increases reflectivity in variable height than in infra red region.

Increased reflectivity cools leaves decreased vapour pressure of leaves which is main cause of reducing transpiration rate. Abow Khaled *et al* (1970) coating of kaolinite on the upper surface of citrus leaves increased leaf albedo and reduce leaf temperature and the vapour pressure gradient between leaf and atmosphere therby reducing transpiration rate by about 25% with much less effect on photosynthesis.

Simultaneously with decrease in transpiration rate reflective materials present over heating of leaf. Any reflective material stick and spread evenly on plant surface and sufficiently permeable to gases so as not interfere with photosynthesis and respiration.

Increase in leaf resistance: coating of leaf with method to increase resistance of leaves to water vapour loss is the application of the leaf chemicals that indicate stomatal closure.

Table 3:   Change in photosynthesis and transpiration rates and leaf temperature induced by the application of kaolinite to *Citrus sinensis*.

Days after kaolinite application (22.5g m^{-2})	Transpiration %	Photosynthesis	Leaf temperature
1	68	70	98
2	74	82	95
3	75	91	93
4	78	96	91
5	76	104	90
6	74	107	90
7	71	117	88

Some film forming compounds have been used by many investigators. However a material with higher permeability for CO_2 than for water vapour is still lacking of H_2O has low molecular weight than CO_2.

Table 4: Ratio of permeability of plastic films to water vapour and CO_2

Plastic film material	Rating of permeability	
	H_2O	CO_2
Poly vinylidene chloride	48	3500
Rubber hydrochloride	145	1044
Poly vinyl chloride	175	255
Polystyrene	7.7	385
Poly ethylene	4.2	–
Poly propylene	6.8	7.6
Natural rubber	22.6	–
Silicon rubber	3.5	17.7

Other commercial products formed suitable anti-transparent are :wet proof, mobileaf, clear spray, vapour guard, folicote ect.

Certain chemicals e.g. Insecticides, fungicides, herbicides and metabolic inhibitors reduce transpiration.

(a) Phenyle Mercuric Acetate (PMA)

(b) Substituted ureas

(c) Decenyl Succinic Acid (DSA)

(d) Atrazine

(e) Simazine

(f) Alkenyl succinic Acid

(g) Alfa Hydroxysulphonates

(h) 8-hydroxy quinol

(i) Abscissic acid (ABA)

Chemical causing stomatal closure should follow following plants:

1. Low mobility within leaf
2. Should confine to the leaf epidermal to avoid toxic side effects.
3. persistent effects
4. photosynthetic not be affected
5. inexpensive

5. Anti-Transpirants

Antitranspirants are substances applied to decrease transpiration rates. They can be divided into film forming material and chemical inducing stomatal closure.

The method of decreasing water loss by coating leaf surfaces with materials impervious to water vapour. These materials act on the leaf surface as physical barriers to vapour diffusion. These are usually polymers such as polyvinyl uraces, polyethylene and vinyl acrylate as well as higher alcohols such as hexadecanol. In order to be effective, a film forming anti- transpirants must be non-toxic, easy to spread and sick on to surface, stable under solar radiation, elastic and durable and of course, impermeable to water vapour. The use of chemical antitranspirants appears to be more promising because they induce partial stomatal closure which increases the assimilation ratio. However, several precautions must be taken when considering a certain chemical as an anti- transpirants.

(i) The substance must not be toxic to the plant.

(ii) The closure of stomata must not be induced by increasing the CO_2 between air and mesophyll since the concentration gradient of CO_2 between air and mesophyll is also reduced. This results in a proportionally greater inhibition of photosynthesis in comparison to water vapour loss.

(iii) The effect of anti-transparent must be restricted to the guard cells without affecting the other cells of the leaf.

Various compounds have been used for the chemical control of stomatal opening. However, only a few of them fill the requirements mentioned above metabolic inhibitors such as sodium oxide and potassium cyanide induce over all toxic effects. Phenyl Mercuric Acetate (PMA) gave encouraging results by increasing the assimilation ration on many PMA is supplied to act solely on the guard cells by inhibiting photophosphorylation without entering the mesophyll. A further short coming of PMA is its human and animal toxicity, which restricts its use to non edible plants. More promising results are expected from the growth retardant abscisic acid (ABA), which seems to posers all the properties of an ideal anti- transparent. ABA was found to increase considerably the assimilation ratio by inducing partial stomatal closure. It act by reducing the solute content and hence the turgor of the guard cells without affecting the other epidermal cells. Other natural substances with chemical structure similar to that of ABA, for instance all –trans-farnesol are also being studied for their suitability as antitranspirants.

Enclosures

Covered enclosures reduce the transpiration of the growing plants by means of the high air vapour concentration which reduces the gradient for the movement of water vapour out of the leaf. In addition there is a further possibility to decreasing transpiration with a concurrent increase in the assimilation ratio by adding CO_2 in the closure. The increased CO_2 concentration induces partial stomatal closure and thus decreases water loss. At the same time, despite the increased resistance of the epidermis, photosynthesis is maintained at quite high levels because of the steeper gradient in CO_2 concentration between mesophyll and air. Although CO_2 itself is considered as an ideal anti-transparent, its application is limited only to restricted areas under covers.

5.1. Reflecting Materials

The use of reflective materials sprayed on to transpiring plant surfaces aims at modifying their energy balance. One approach is to use reflective materials to decrease the net radiation of the surface such a reduction also results in radiation in a decrease in the latent heat dissipation compound. It should also be possible to use a selective pigment transmitting the radiation between 400 and 700nm and reflecting all other parts of the spectrum. In this ways plant would avoid above 60% of the total incoming radiation which is mainly restricted to the long wave band.

5.2. Wind Break

A beneficial effect of wind breaks on the assimilation ratio is expected only when adventives heat increase considerably the energy balance of the leaf (oasis effect). In that case the lowering of wind speed decreases the amount of advective heat conveyed t leaves and increases the water vapour content of the air near the leaf.

5.3. Breeding Technique

It is possible to select for some morphological and physiological characteristics which enable plant to reduce water loss and to introduce them in breeding programme aimed at increasing crop.

6. Selection of Fruits Crops

Choice of fruit crops for arid areas important and should be done carefully for making their cultivation successfully:

(a) Fruit crops chosen for region must synchronous flowering fruit growth and development period with the season when maximum moisture is available in the soil and vapour pressure deficit in atmosphere in low.

(b) Fruit ripening and harvest period must complete before the onset of hot and arid summer.

(c) Fruit crops chosen must has deep and strong root system in ability to extract moisture from deeper layers of soil profile and must be capable to pierce hard pan commonly found in arid region.

(d) Fruit crops of arid region should have tolerance to salinity and saline irrigation water.

(e) Fruit crops should have ability of drought resistance.

7. Potential Fruit Crops for Dry land Arid Region

Ber

(i)   most hardy fruit crop, can withstand drought and salinity and saline water.

(ii) Its roots can penetrate through hard pan and can draw moisture from deeper layers of soil.

(iii) It is fruit crop sheds leaves with onset of dry and hot weather.

(iv) Its flowering and fruiting coincides with maximum moisture availability in soil and harvesting is over before onset of dry and hot.

(v) This crop requires little irrigation for establishment and its production will be enhanced if supplementary irrigation is provided.

(vi) Alternatively, certain amount of catchment area with slop for collection of runoff water needs to be provided for its successful cultivation.

(vii) Leaves xerophytic characteristics and buds are scaly

Date palm: withstand salinity and saline irrigation water, however, it must be grown where sufficient irrigation can be given and ripening of fruits must complete before onset of monsoon. Alternatively, the rainfall should be very scanty and there should be no humidity build up. Require sufficient heat unit by the end of June and high vapour pressure deficit (25 mb) during fruit development.

Pomegranate: (i) Tolerate salinity and saline irrigation and does well in shallow soil. **(ii)** Tolerates drought to some extent and fruiting can be tailored during period of maximum soil moisture availability. (iii) It is very sensitive to soil and atmosphere moisture fluctuation which result in cracking of fruits and will require supplementary irrigation even in area zones for its economic productivity.

Custard apple: (i) very hardy fruit crop tolerant to drought salinity and saline irrigation water to some extent. (ii) Requires shallow soils, fruiting coincides during maximum moisture availability and harvesting complete before onset of dry hot weather. (iii) It also sheds leaves during hot arid dry summer. (iv) Flowering in June July and harvesting by November –December at optimum environment conditions.

Guava: (i) Tolerate of salinity (ii) flowering fruiting can be regulated during maximum moisture availability (iii) Harvesting is over before onset of dry hot summer (iv) Fruit requires deeper soil and irrigation for establishment and for economic productivity.

Acid Lime: (i) relatively less drought tolerant and requires shallow soils. (ii) Can withstand salinity and saline irrigation water with salt tolerant rootstock. (iii) Requires irrigation for establishment and economic productivity. (iv) plants can be tailored to flower during monsoon major crop in late winter.

Lehsua: After the crop is over in early summer the tree sheds off leaves. Leaves thick and cuticular. Fruits are used for picking, harvested immature.

Aonla: Great tolerance to salt and alkalinity. Flowering in February –March in spring optimum moisture stage than after its embryo goes to dormancy under hot condition and than in July its start growth under good moisture condition.

Phalsa: Drought hardy. Very short reproduction phase. After harvesting loose folage. Flowering in December –January crop is ready by April.

Bael: Drought and salinity tolerant plant. Flowering coincides with the onset of monsoon and fruits mature before the onset of hot summer.

Wood Apple: Tolerance to drought and salinity. Extensive root system and synchronization of its reproductive phase with high moisture availability.

Jamun: Can stand drought which can be attributed to extensive root system. Flowering in spring find swell of fruits in monsoon.

Karonda: Drought tolerant and its reproductive phase synchronize to the period of moisture abundance.

Fig: Can tolerant salinity and drought. Fruiting can be monitored to synchronized with rainy season only so that crop is harvested in late winter and plants shed off leaves during summer.

Other fruit plants: Tamarind: deep root system

Kair: Scanty folage, musilagenous in plant parts will hold water

Pilu : deep root system.

8. Selection of Vegetable Crop and Importance of Vegetable Crops in Arid Region

Selection on the basis of water requirement, deep root system, short duration etc

(i) Cucurbits are eminently suitable for dry land because they long tap root system and some of them have got xerophytes habit.

(ii) Some of the cucurbits like watermelon have been found growing in dryland of Kalahari desert in South Africa and they have low moisture requirement.

(iii) Among cucurbits following are important: Kikoda (*Momordica dioica*), Gourd (Luffa acantangula), Wwater and musk melons, kachra ect.

Other vegetables: guar, sem, lobia, khejri pods

8.1. Ornamental Trees, Shrubs and Hedges Plants

Selection of plant species that require little management, less water, contrast flowering, drought hardy, salt tolerant and disease free etc.